Physics and Chemistry in Space Vol. 22
Planetology

Edited by M. C. E. Huber, Noordwijk, L. J. Lanzerotti, Murray Hill, and D. Stöffler, Münster

R. D. Hunsucker

Radio Techniques for Probing the Terrestrial Ionosphere

With 152 Figures

Springer-Verlag
Berlin Heidelberg New York
London Paris Tokyo Hong Kong
Barcelona Budapest

Professor ROBERT D. HUNSUCKER
Geophysical Institute, University of Alaska
Fairbanks, Alaska 99775-0800, USA

Series Editors:

Professor Dr. M. C. E. HUBER
European Space Research and Technology Centre
Keplerlaan 1, NL-2200 AG Noordwijk, The Netherlands

Dr. L. J. LANZEROTTI
AT&T Bell Laboratories, 600 Mountain Avenue
Murray Hill, NJ 07974-2070, USA

Professor D. STÖFFLER
Institut für Planetologie, Universität Münster
Wilhelm-Klemm-Str. 10, W-4400 Münster, FRG

ISBN-13: 978-3-642-76259-8 e-ISBN-13: 978-3-642-76257-4
DOI: 10.1007/978-3-642-76257-4

Library of Congress Cataloging-in-Publication Data. Hunsucker, R. D. Radio techniques for probing the terrestrial ionosphere / R. D. Hunsucker. p. cm. – (Physics and chemistry in space; v. 22) Includes bibliographical references and index. ISBN 3-540-52830-X (Berlin). – ISBN 0-387-53406-7 (New York) 1. Ionosphere – Technique. 2. Ionospheric radio wave propagation. 3. Radio meteorology. I. Title. II. Series. QC801.P46 vol. 22 [QC881.2.I6] 523.01 s–dc20 [5381.767] 90-23822

Softcover reprint of the hardcover 1st edition 1991

32/3145-543210 – Printed on acid-free paper

This monograph
is dedicated to
my wife Phyllis

Preface

In the years since the pioneering efforts of Sir Edward Appleton, M.A.F. Barnett, G. Breit, and M.A. Tuve, many radio techniques have been employed to investigate the terrestrial ionosphere. The purposes of this book are to examine the basic physical interaction process of radio waves with the ionosphere, scrutinize each of the radio techniques currently in use, and describe the elements of each technique, as well as assess their capabilities and limitations. I have included some of the history of each technique, since we often tend to forget the efforts of the "pioneers".

The interaction of radio waves with the terrestrial ionosphere has been described in considerable detail in several "classic" treatments, e.g., Ratcliffe (1959), Al'pert (1963), Budden (1961) and Davies (1965), Rishbeth and Garriott (1969), and in other more recent books, e.g., Flock (1979), Davies (1990), Hargreaves (1979), and Budden (1985). A few of the radio techniques have been described by Hargreaves (1979) and a book by Giraud and Petit (1978) has also included discussion of several of the techniques. The "WITS" handbook No. 2 (1989) also contains description of several radio techniques.

This book is primarily addressed to graduate students and beginning researchers in aeronomy, and hopefully, will also serve as a compendium for active workers in the field. It is hoped that the book will aid researchers in aeronomy and ionospheric propagation to gain an appreciation of "how radio techniques work", and describe the information one can expect to obtain from each particular technique. More specifically, the author has attempted to objectively appraise in some detail the capabilities and limitations of each particular radio technique. The "limitations" sections will probably result in "stepping on the toes" of the proponents of certain techniques, who are convinced that their particular method tells the "absolute truth" at the greatest efficiency and in a most elegant manner about the peculiar ionospheric pathogen which they are studying! One is sometimes reminded of the well-known Indian proverb about the "blind men and the elephant", wherein each of the men is certain that the particular portion of the beast he has grasped reveals the entire truth and the complete description of the pachyderm! Some of the material in this book is based on a course *Experimental Methods in Aeronomy and Space Physics* offered to graduate students at the University of Alaska. I am indebted to these students for valuable mutual exchange of ideas throughout the years. As many Professors discover, some of their best colleagues have been their students.

I believe that the timing of this volume is fortuitous for the following reasons: (1) radio techniques have, indeed, been utilized to study the terrestrial ionosphere for some 65 years now, and the passing decades have seen the elimination of a few techniques, the modification of some, the introduction of several new ones, and the persistence of some (such as the ionosonde). (2) The era of international research cooperations such as the International Geophysical Year (IGY) and following intervals was particularly dynamic for the development of new techniques and the refinement of others. (3) Likewise, the decade of the 1980's witnessed the worldwide deployment of several sophisticated systems (incoherent scatter radars, coherent VHF and HF Doppler radars, and satellite beacon systems); some of these systems and "first results" are described in special issues of the journals *Radio Science* (vol. 18, 1983), Greenwald and Hunsucker (1983), *Journal of Atmospheric and Terrestrial Physics* (vol. 46, 1984), and *Geophyiscal Research Letters* (vol. 11, 1984).

Before proceeding further, the reader should realize that the research area described by the title of this book is cursed with many acronyms, abbreviations, and nonstandardized symbols. Appendices A.1 and A.2 are provided in an attempt to alleviate this "alphabet-soup" problem. Similarly, when one is confronted with all the pathogenic problems of implementing a sophisticated radio technique at some of the hostile field sites of interest to ionospheric physicists, there is no one book which will provide all the solutions. This book is no exception. It is my hope, however, that the material in this monograph will at least provide a glimpse of the path to follow toward the partial solution of some problems, as well as describe the specific radio instrumentation techniques used by ionospheric physicists.

Before listing my acknowledgements for this book, I would like to give a salute to the scientists and engineers who actually *invented* these techniques! I would like to express my sincere appreciation to several people who helped to make this book possible. Two successive directors of the Geophysical Institute of the University of Alaska, Juan Roederer and Syun-Ichi Akasofu, offered considerable encouragement and financial support when most needed. Other colleagues at the Geophysical Institute, in particular Brenton Watkins, Paul Johnston, and Brett Delana reviewed parts of the book, as did graduate students Donald Rice and Ning Jing. Louis Lanzerotti of AT & T Bell Laboratories suggested that a book on *Techniques* should be written for this series, and invited me to attempt the project while I was working with him during sabbatical leave. He has been a constant source of encouragement. I am also indebted to a host of friends and colleagues I have had the pleasure of knowing and working with during the last three decades. They have reviewed each of my chapters as they were finished. In particular, Robert Benson of the NASA Goddard Space Flight Center, Richard Behnke of the National Science Foundation, Thomas Croft, then at SRI-I, Kenneth Davies of the NOAA Space Environmental Labs, Louis Duncan of Clemson University, Tony Ferraro of Penn State University, Warren Flock of the University of Colorado (who was also chairman of my Ph. D. committee), Ray Greenwald of the Johns Hopkins Applied Physics Lab, Dick Grubb of NOAA Space Environment Labs, John

Hargreaves of the University of Lancaster, Robert Helliwell of Stanford University, O. G. (Mike) Villard of Stanford University, and last (but certainly not least) Jim Wait of the University of Arizona. I would also be remiss if I did not acknowledge the support and encouragement from the Electrical Engineering Department of the University of Alaska Fairbanks. Specifically, thanks are due to Vince Haneman, Dean of the School of Engineering and to John Aspnes, Chairman of the Electrical Engineering Department, also to Robert Merritt and Thomas Roberts for stimulating discussions. Finally, my sincere thanks to Vickie Ivester and Debbie LaBarre for their painstaking and accurate preparation of the final version of the manuscript.

ROBERT D. HUNSUCKER

Contents

Part I Introduction 1

1 Terrestrial Atmospheric Structure and Properties of Electromagnetic Waves 3

1.1 The Radio Spectrum 3
1.2 Structure and Composition of the Terrestrial Ionosphere 3
1.3 Electrical Noise and Radio Frequency Interference 15
1.3.1 Thermal Noise in a Resistor 16
1.3.2 The Noise Factor, Figure, and Temperature of a Radio Receiver 17
1.4 Properties of Electromagnetic Waves and Their Interaction with Matter 20
1.4.1 Maxwell's Equations and the Wave Equation 20
1.4.2 Some Characteristics of Propagating Plane Waves 23
1.4.3 Properties of the Propagation Constant and Vector 25
1.4.4 Polarization of an Electromagnetic Wave in Free Space 31
1.4.5 The Radar Equation 36

2 Radio Propagation in a Plasma 40

2.1 Radio Wave Propagation in an Unmagnetized Plasma 40
2.1.1 Vertical and Oblique Propagation 41
2.1.2 The Virtual Height Concept 44
2.1.3 Vertical and Oblique Propagation 44
2.2 Radio Propagation in a Magnetized Plasma 46
2.2.1 Dispersion and Polarization 47
2.2.2 Absorption of Radio Waves in the Ionosphere 50
2.3 Ionospheric Scattering of Radio Waves 53
2.3.1 Coherent Scatter from Ionospheric Irregularities 54
2.3.2 Incoherent Scatter from the Ionosphere 58

Part II Earth-Based Radio Techniques 65

3 Vertical Sounders – the Ionosonde 67

3.1 History of Development and Basic Principles of Operation ... 67
3.2 Standard Ionosondes 68

3.3 Advanced Ionospheric Sounders 77
3.4 Capabilities of Ionosondes 90
3.5 Limitations 92

4 Oblique Backscatter Sounders 94

4.1 HF Pulsed Groundscatter Systems 95
4.1.1 The Development of Backscatter Sounding as a Remote Sensor of the Ionosphere 96
4.1.2 Basic Ionospheric Groundscatter Theory 98
4.1.3 Various Techniques Employed in Groundscatter Sounding 107
4.2 Characteristics of a Typical Scanned Groundscatter System with Variable Frequency 107
4.2.1 The ITS/ESSA OBS System (1960's Technology) 107
4.3 Direct Ionospheric Backscatter 107
4.3.1 History of Development of the Technique 109
4.3.2 Direct Backscatter from the E-Region 110
4.3.3 Backscatter from the F-Region 112
4.4 Capabilities and Limitations of OBS Systems 116
4.4.1 HF Groundscatter OBS Capabilities 116
4.4.2 Direct HF Backscatter OBS Systems 117
4.4.3 Auroral Radar Systems 118

5 Incoherent Scatter Radar (ISR) Systems 122

5.1 Introduction 122
5.2 Basic ISR System Considerations 122
5.3 Currently Operational ISR Systems 123
5.3.1 Descriptions of Individual Operational ISR's 124
5.3.2 How to Obtain Data and Information from Currently Operational ISR's 138
5.4 Capabilities and Limitations 139

6 Ionospheric Modification by High Power Radio Waves 142

6.1 Ionospheric Cross Modulation Techniques – Some Representative Results 144
6.1.1 Modulated Disturbing Wave, Unmodulated Wanted Wave at Oblique Incidence, and Gyro Interaction 144
6.1.2 Self-Interaction and Gyro Self-Interaction 145
6.1.3 Pulse Modulated Disturbing and Wanted Waves – Vertical Propagation 145
6.1.4 Pulse Modulated Disturbing Wave – Cosmic Noise Wanted Wave, Using Vertical Propagation 146
6.1.5 Outline of Cross Modulation Theory 146
6.2 Ionospheric Heating Using High Power HF/CW Transmitters . 152

6.2.1 Basic Theory of Ionospheric Heating 152
6.2.2 Descriptions of Facilities and Some Results from Ionospheric Heating Investigations at Mid-Latitudes 155
6.2.3 The Low-Latitude HF Heating Facility at the National Aeronomy and Ionosphere Center (NAIC), Arecibo, Puerto Rico .. 158
6.2.4 HF Heating Facilities at High Latitudes 159
6.3 Capabilities and Limitations of Radio-Wave Interaction and HF Heating Experiments 162
6.3.1 Capabilities 162
6.3.2 Limitations 163

7 Methods of Measuring Ionospheric Absorption and Other D-Region Parameters 165

7.1 A1 – The Pulse Reflection Method 166
7.1.1 Vertical Incidence 166
7.1.2 A1 Technique – Oblique Incidence 168
7.2 A2 – The Cosmic Noise Technique 169
7.3 Oblique Field Strength Measurements (the A3 Method) 177
7.3.1 The A3a Method – Oblique Incidence Field Strength Observation on Frequencies Above 2 MHz 179
7.3.2 Method A3b – Oblique Incidence Field Strength Observation on Frequencies in and Below the MF Broadcasting Band 179
7.4 Partial Reflection Experiments (PRE) 180
7.4.1 Theory 180
7.4.2 Experimental Technique 181
7.5 Other Techniques 182
7.6 Capabilities and Limitations 183

Part III Satellite, Rocket, and Other Techniques 185

8 Earth Satellite Radio Studies of the Ionosphere 187

8.1 Satellite Radio Experiments 187
8.1.1 A Short History of Radio Beacon Experiments (RBE) 188
8.1.2 Methods of Determining Electron Content 189
8.1.3 Global Variation of TEC 193
8.1.4 RBE Studies of Ionospheric Scintillation 196
8.2 Satellite Topside Sounders 200
8.2.1 Sweep Frequency Topside Sounders 201
8.3 Capabilities and Limitations of Satellite Radio Techniques ... 203

9 Other Radio Techniques 205

9.1 Ionospheric in Situ Measurements 205
9.2 HF Spaced-Receiver and Doppler Systems 207
9.3 Whistlers and Controlled Stimulation of VLF Emissions 213

9.3.1 Whistlers and Other Natural VLF Radiation 213
9.3.2 Controlled Stimulation of VLF Emissions 218
9.4 VLF and HF High Resolution Sounders 219
9.5 Ionospheric Imaging by Radio 225
9.6 Backscatter from Meteoric Ionization 227

10 Epilog 229

10.1 "Where We Have Been and Where We Are Now" 229
10.2 Specific Techniques and Their Applications 230
10.3 Complementarity 233
10.4 Concluding Remarks 235
10.4.1 The Present Status of Radio Instrumentation 235
10.4.2 Possible Future Trends 235
10.4.3 Excelsior! 235

A.1 List of Principal Symbols, Constants, and Conversion Factors 236

A.1.1 Principal Symbols 236
A.1.2 Selected Physical Constants 238
A.1.3 Some Selected Conversion Factors 238

A.2 Acronyms 241

A.3 A Radio Frequency Spectrum Chart Showing the ITU Nomenclature and Principal Uses 246

A.4 Annotated Bibliography 247

A.5 Selected Global Maps and Other Information on Electrical Noise 254

A.6 Some Representative Values of Ground Electrical Characteristics 257

A.7 Fresnel Zone Antenna Siting Fundamentals 258

A.7.1 Introduction 258
A.7.2 HF Antenna Siting Example 260
A.7.3 The Fresnel Zone in Three Dimensions 265

References 267

Subject Index 291

Part I Introduction

The use of radio waves to explore the terrestrial atmosphere began with the pioneering efforts of Appleton and Barnett (1926) and Breit and Tuve (1926), when they independently used different techniques to detect ionospheric layers. Their work was, of course, founded on Marconi's demonstration of transatlantic radio transmissions and on the hypotheses of Kennelly (1902) and Heaviside (1902), who independently concluded that there must be radio-reflecting layers in the upper atmosphere to explain certain experimental results. The foregoing discoveries rested upon the bedrock of the experimental and theoretical work of Heinrich Hertz (1893) and James Clerk-Maxwell (1873) respectively. The references at the end of this book include some of the salient articles describing the experimental and theoretical work mentioned above. Additional descriptions of historical development of the specific techniques and of the theory will be found in various sections of this book, and an excellent historical account of the origins of radio from Hertz into the 1920's may be found in two volumes by Aitken (1985a, b).

As mentioned in the Preface, it is assumed that the users of this book have studied electromagnetic theory at least at the undergraduate level and have a reasonable understanding of the principles of the generation of EM waves, their propagation in free space, transmission lines and waveguides, and basic antenna theory. Those who feel the need for a review of these topics will find ample material in the textbooks by Jordan and Balmain (1968), Ramo et al. (1965), Sander and Reed (1986), Kraus (1988), and Wait (1985). Researchers seeking specific information on radio engineering and antenna design should also find the annotated bibliography a useful starting point.

An essential lesson learned by the successful "journeyman ionospheric physicist of the experimental persuasion" in order to design the appropriate experiment is that he (or she) must know the basic ionospheric parameters and their expected variance which describe the particular ionospheric phenomenon to be studied. It is hoped that this book may be of some help in that task – particularly in the selection of the best technique or techniques.

The "road map" for this book starts in Chapter 1 with an outline of the structure and nomenclature of the earth's neutral atmosphere and ionosphere showing some "typical" and a few atypical features, including representative electron density profiles illustrating the high variance encountered. This is followed by a mathematical and physical description of the basic properties of electromagnetic waves propagating in free space and in other media and concluded by a discussion of the radar equation. Chapter 2 concerns the behavior of radio waves in the ionosphere and includes the elements of magnetoionic theory and incoherent scatter theory. The basic effects of the ionosphere upon a radio wave which "intrudes in its domain" – such as reflection, refraction, diffraction, scattering, attenuation, and changes in polarization – are discussed.

The remaining chapters examine each of the "generic" radio techniques used by ionospheric physicists to explore the terrestrial ionosphere remotely. Chapters 8 and 9 include discussions of in-situ ionospheric probing systems utilized in rockets and satellites as well as various radio techniques, and Chapter 10 is a summary of this monograph with concluding remarks.

Chapter 1 Terrestrial Atmospheric Structure and Properties of Electromagnetic Waves

Scanning the terrain of Chapter 1 reveals a somewhat "lumpy topography" wherein we discuss some diverse topics including the radio spectrum, atmospheric structure and composition, electrical noise, fundamental properties of electromagnetic waves, and the radar equation, in that order. A logical starting point, then, seems to be a short discussion of the electromagnetic spectrum in general and that particular region of the spectrum employed for ionospheric studies.

1.1 The Radio Spectrum

In general, the portion of the electromagnetic spectrum utilized to investigate the ionosphere of the earth includes approximately 10 kHz through 1300 MHz, or VLF-UHF (for acronyms see Appendix A.2). Figure 1.1 may be of use in visualizing the frequency spectrum used for the radio techniques described in this book. A more detailed spectrum chart is provided in Appendix A.3, including the International Telecommunications Union (ITU) band nomenclature, some of the international frequency band allocations, and other information useful to radio engineers.

1.2 Structure and Composition of the Terrestrial Ionosphere

Before proceeding with an abbreviated account of the details of EM wave interaction with the terrestrial ionosphere in Chapter 2, we will briefly describe the basic structure and properties of the terrestrial atmosphere. Figure 1.2 shows a schematic representation of the earth's atmosphere under rather idealized conditions, including the height variation of temperature, as well as standard atmospheric nomenclature and other properties. A more detailed description of the daytime electron density distribution versus height is shown in Fig. 1.3. The ionospheric layer nomenclature along with a qualitative picture of ionospheric radio propagation mode structure and penetration levels of solar radiation are also illustrated. Further details on the properties and pertinent physical and chemical parameters of the specific ionospheric layers are shown in Table 1.1. Physical and mathematical descriptions of the very complex processes and dynamics of the ionosphere are thoroughly covered in standard texts and monographs, e.g., Banks and Kockarts (1973), Bauer (1973), Kato (1980), Gurevitch (1978), Chamberlain (1978), and Rees (1989).

RADIO TECHNIQUES FOR PROBING THE IONOSPHERE

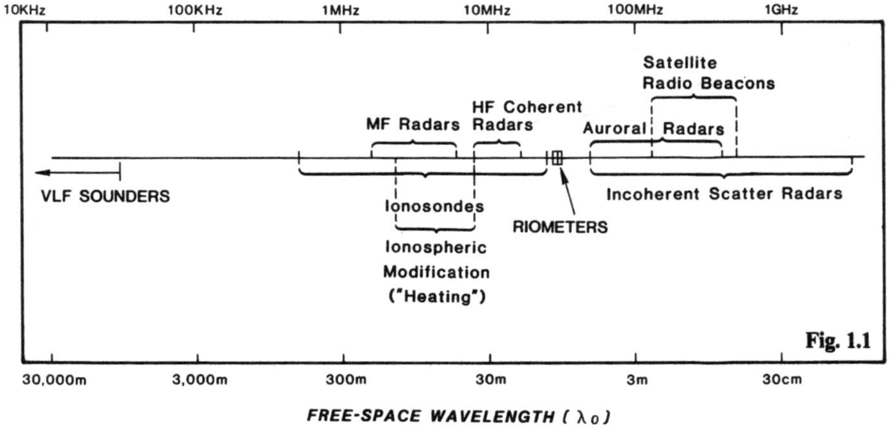

Fig. 1.1

FREE-SPACE WAVELENGTH (λ_0)

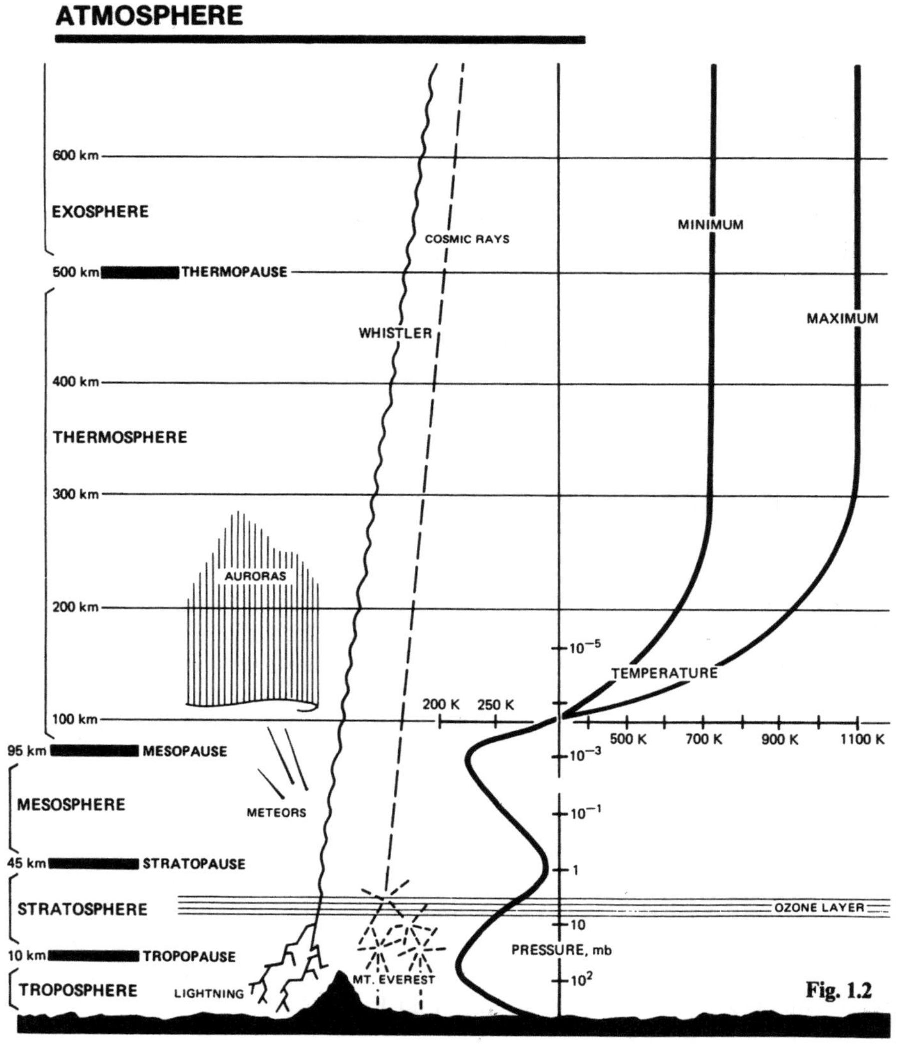

Fig. 1.2

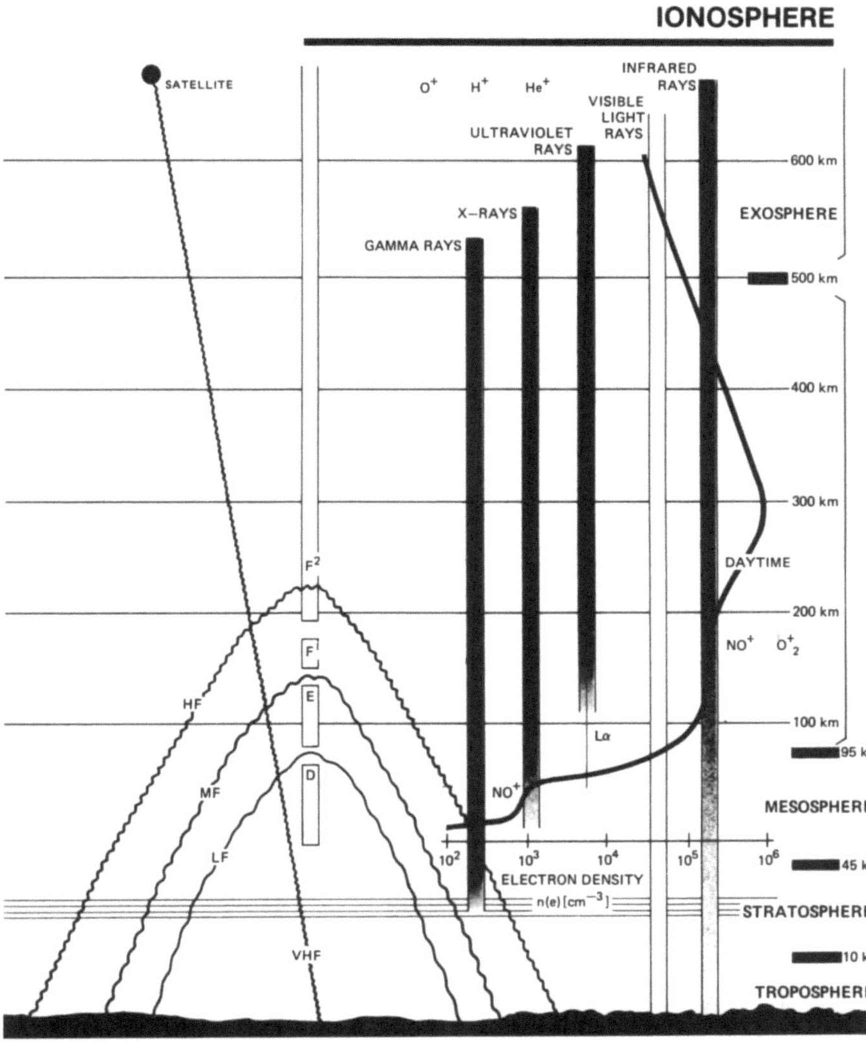

Fig. 1.3. Idealized representation of daytime ionospheric electron density distribution

Fig. 1.1. Portion of the EM spectrum useful for ionospheric probing

Fig. 1.2. Schematic representation of the earth's atmosphere under idealized conditions. (NRC Report Solar-Terrestrial Research for the 1980's. National Academy Press, 1985)

Table 1.1. Typical properties of various ionospheric regions

Parameter	D-Region		E-Region		F-Region					
	60 km		100 km		200 km		300 km		500 km	
	Day	Night	Day	Night	Day	Night	Day	Night	Day	Night
Neutral concentration (m^{-3})		7.2×10^{21}		1.2×10^{19}		7.6×10^{15}	9.8×10^{14}	6.6×10^{14}		6.0×10^{13}
Electron concentration (m^{-3})		10^{6}	1.7×10^{11}	10^{10}	3.5×10^{11}	3×10^{10}	1.1×10^{12}	3.5×10^{11}		$\sim 10^{10}$
Gas temperature (K)		250		210		1100	1360	980		1600
Ion-neutral collision freq. (s^{-1}) ν		6×10^{6}		7.6×10^{3}		5		0.75		6×10^{-2}
Electron-neutral collision freq. (s^{-1}) ν		4×10^{7}	48,000	42,000	150	100	18	7	0.9	0.16
Ion gyro frequency (rad s^{-1}) Ω		~ 140		100		100		210		190
Electron gyro frequency (rad s^{-1}) Ω		8.5×10^{6}		8.4×10^{6}		8.0×10^{4}		7.6×10^{6}		7.0×10^{6}
Scale height (km) H		7.5		6.3		37.0		54.7		73.5

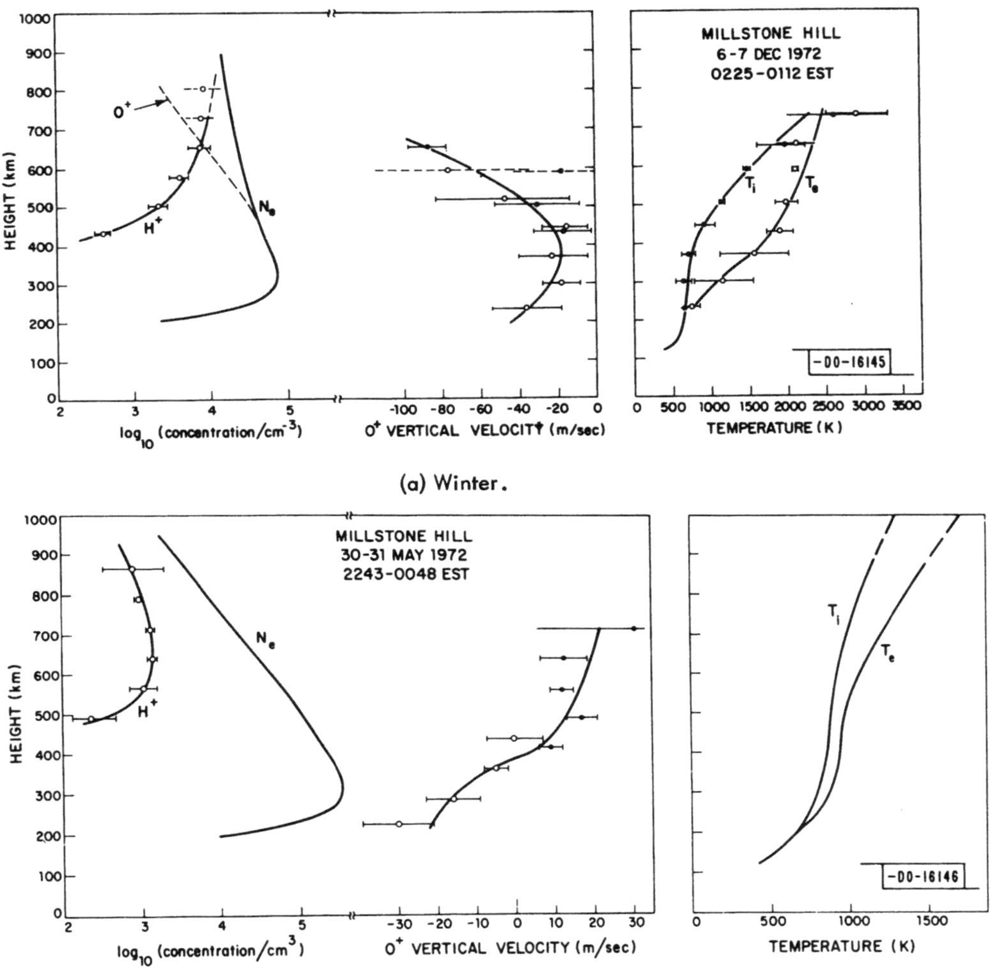

Fig. 1.4. Examples of electron density and temperature distribution – Millstone Hill winter from the Millstone incoherent scatter radar (ISR). (Evans 1979)

Ionospheric properties vary markedly with geographic and geomagnetic coordinates, local time, height, season sunspot cycle, and geomagnetic stormtime. Phenomena such as tides, thermospheric winds, and traveling ionospheric disturbances (TID's) also distribute some of the energy deposited at high latitudes equatorward, thus modifying the mid-latitude ionospheric structure. Therefore, the properties shown in Fig. 1.3 and Table 1.1 represent only "typical" values. Some idea of the extremes in variation of one ionospheric characteristic – the electron density profile – is shown in Figs. 1.4 through 1.13, and a summary of some temporal changes in the high latitude ionosphere is illustrated in Table 1.2. The *electron density profile* was chosen by the author to be representative of ionospheric structure because it was available from equatorial, mid-latitude auroral, and polar latitude incoherent scatter radars.

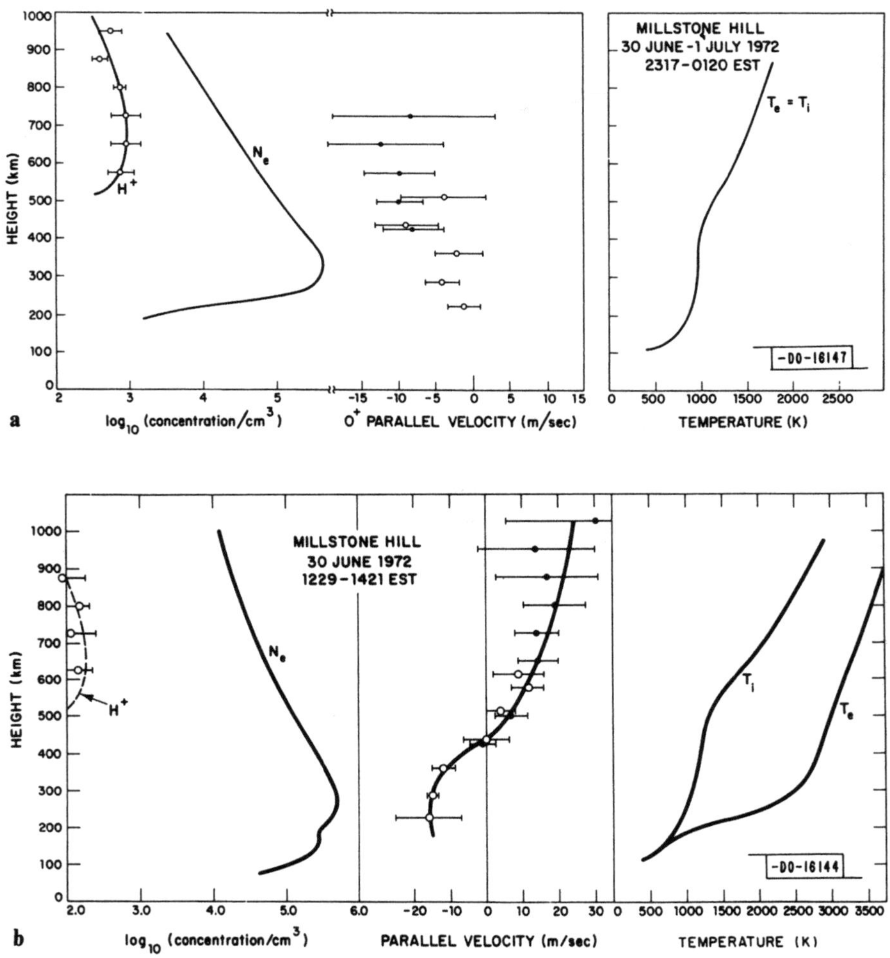

Fig. 1.5. **a** Example of electron density and temperature distribution – Millstone Hill, midsummer. (Courtesy of John Foster) **b** Example of electron density, parallel ion velocity, and temperature at Millstone Hill for midsummer afternoon conditions. (Courtesy to John Foster)

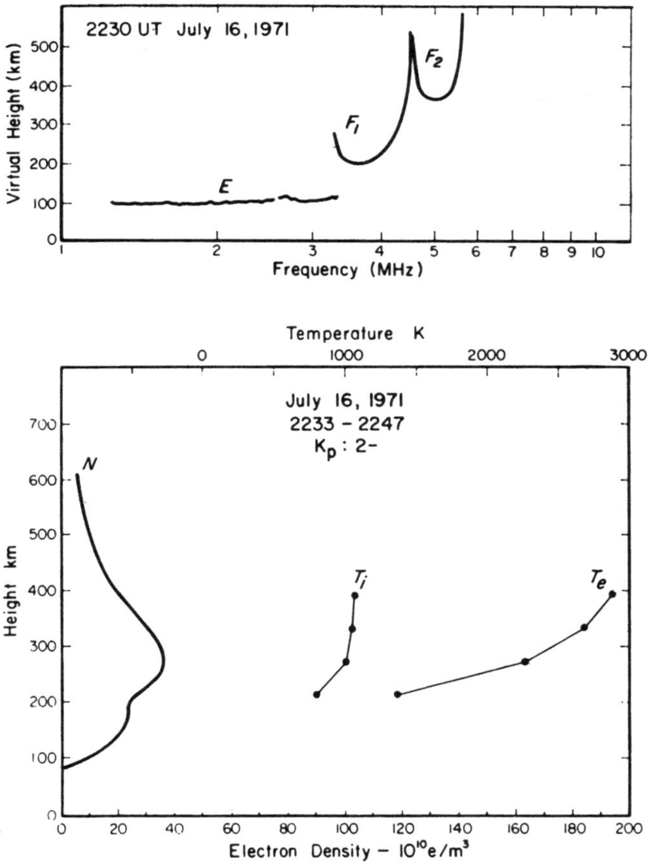

Fig. 1.6. Simultaneous profiles from the Chatanika ISR and College, Alaska, ionograms for midsummer magnetically quiet afternoon conditions (Bates and Hunsucker 1974). System height resolution was 50 km

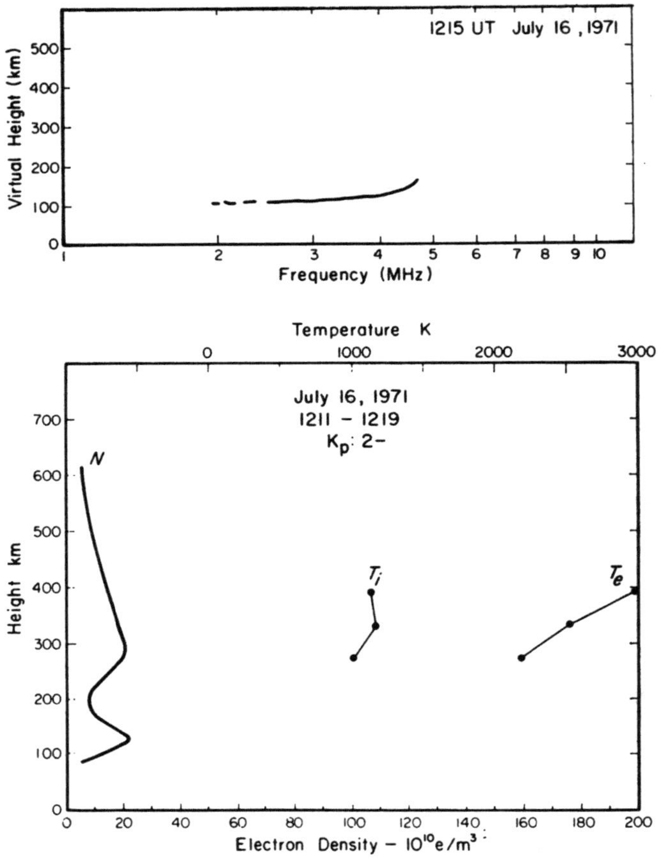
1215 UT July 16, 1971
Virtual Height (km)
Frequency (MHz)
Temperature K
July 16, 1971
1211 - 1219
K_p: 2-
N
T_i
T_e
Height km
Electron Density – 10^{10} e/m³

Fig. 1.7

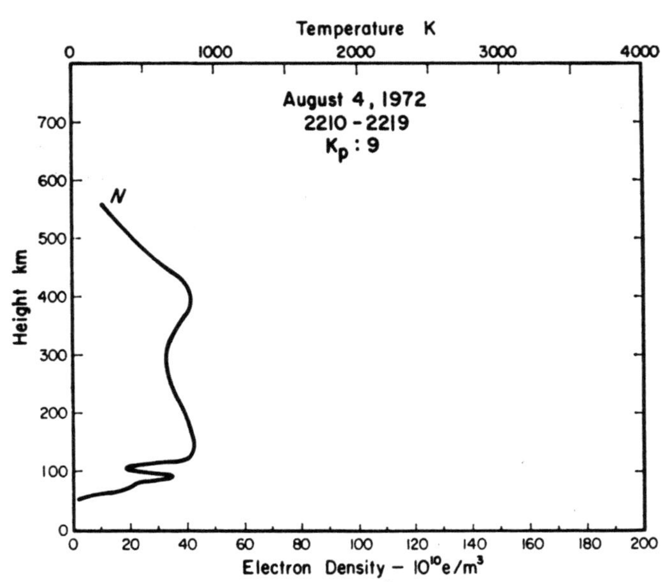
Temperature K
August 4, 1972
2210 - 2219
K_p : 9
N
Height km
Electron Density – 10^{10} e/m³

Fig. 1.8

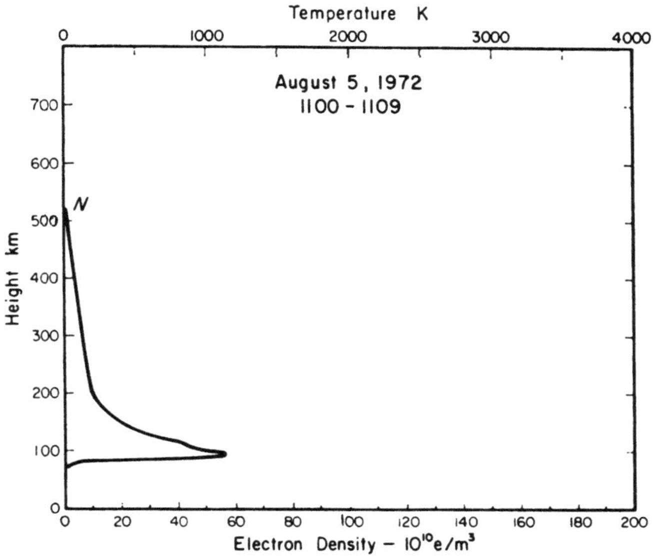

Fig. 1.9. Electron density profile showing strong auroral-E layer from the Chatanika ISR during the great geomagnetic storm of August 1972. (Bates and Hunsucker 1974)

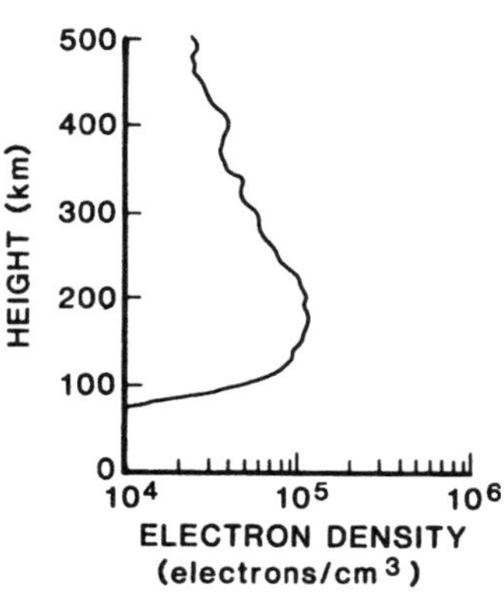

Fig. 1.10. Typical midsummer electron density profile – polar ionosphere – from the Sondrestrom ISR

Fig. 1.7. Simultaneous profiles from Chatanika ISR and College ionogram for midsummer, magnetically quiet, near-midnight conditions. (Bates and Hunsucker 1974)

Fig. 1.8. Electron density profile from the Chatanika ISR during the great geomagnetic storm of August 1972. (Bates and Hunsucker 1974)

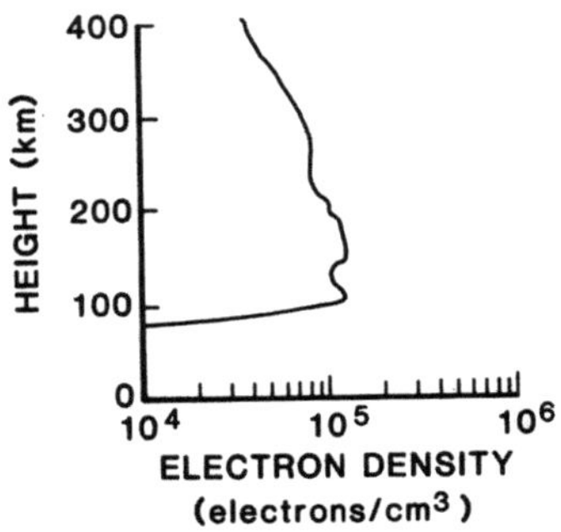

Fig. 1.11. Sondrestrom ISR electron density profile

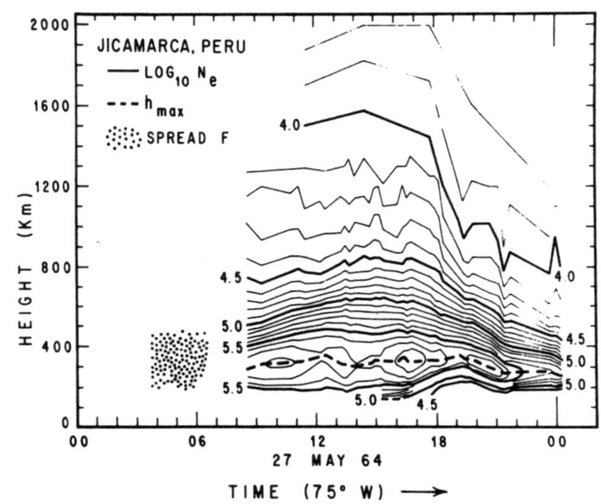

Fig. 1.12. Typical Jicamarca, Peru, ISR electron density profiles obtained from Faraday rotation of the scattered signal for 27 May 1964. (McClure et al. 1970)

The qualitative relationship between visual aurora and electron density distribution in the E-region is illustrated in Figs. 1.14 and 1.15. The solid line plot is the Chatanika radar electron density, the insert on the right is a tracing of the College, Alaska ionogram, and the insert at the lower left is the Chatanika all-sky camera (ASC) photograph. Figure 1.14, showing the simultaneous data for 0706 UT (2106 Alaska Standard Time), illustrates the "background level" of E-region ionization of 0.2×10^5 electrons/cm, the right insert an fE_s value of 7.5 MHz, and the ASC indicating a discrete auroral arc to the north. At 0800 UT in Fig. 1.15 the ASC shows a discrete arc in the radar beam, resulting in an E-region density of 1.2×10^6 electrons/cm and an fE_s value of 11.2 MHz. This illustrates the drastic change in E-region structure when discrete auroral forms are present. A more complete description of the behavior of the auroral E-layer is given by Hunsucker (1975).

There is also a wide variation in other ionospheric parameters such as scale height (H), electron and ion temperatures (T_e and T_i), mean molecular mass (m), plasma drift velocity (V_E), etc. with height, geographic location, and geomagnetic

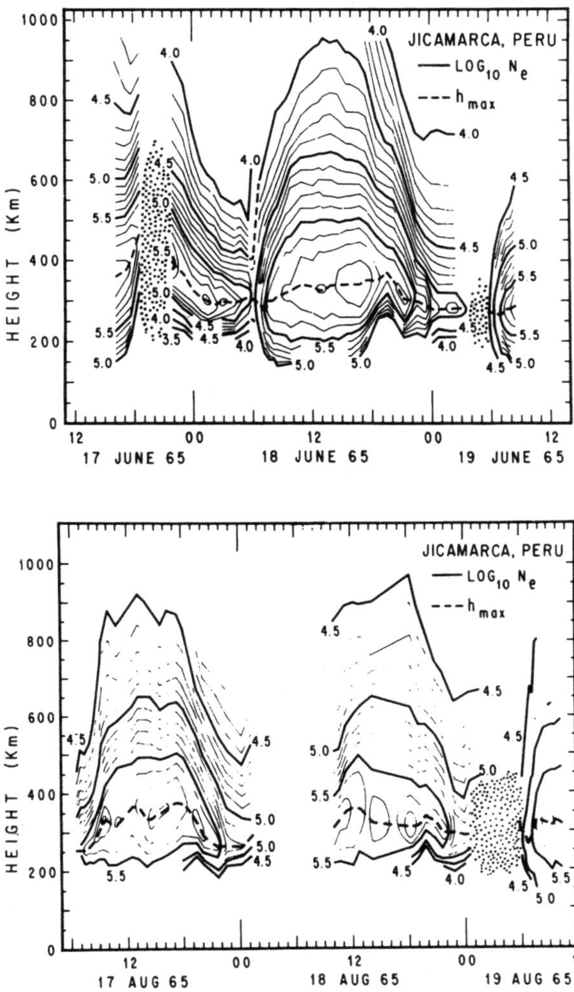

Fig. 1.13. Jicarmarca, Peru, ISR electron density contours for 18 June 1965 (*top*) and 18 August 1965 (*bottom*). (McClure et al. 1970)

Table 1.2. Summary of the effects of various disturbances on the auroral oval E-layer (Hunsucker 1975)

Date	Disturbance phenomenon	Representative values						Comments
		N_{max} (el/cm^3)	Height (km)	$\frac{N_{max}}{N_{min}}$	ΔN (el/cm^3)	Δt (s)	$\frac{\Delta N^a}{\Delta t}$ (cm^3/s)	
	Daily solar variation (EUV[b])	1.5×10^5	~ 120	5.0	1.2×10^5	3.24×10^4	3.7	May–Aug.
June 30, 1954	Solar eclipse	6.0×10^4	~ 100	1.7	2.4×10^4	3.0×10^3	8.0	Baker Lake, Canada
June 20, 1963	(EUV)	2.8×10^4	~ 110	4.0	7.3×10^4	3.9×10^3	18.7	Anchorage, Alaska
June 10, 1972		1.2×10^4	110	1.7	5.0×10^3	3.3×10^3	1.5	Chantanika, Alaska
				Avg = 2.5			Avg = 9.4	
Apr. 2, 1973	"Bow shock wave" associated with auroral infrasonic wave (AIW)	2.8×10^6	102	14.0	2.6×10^6	3.3×10^2	7.9×10^3	College, Alaska
Nov. 21, 1971	Auroral	~ 1.5×10^6	108	30.0	1.45×10^6	1.8×10^2	8.1×10^3	Chatanika ISR
Feb. 24, 1972	particle	7.5×10^5	112	9.3	6.8×10^5	1.2×10^2	5.7×10^3	
Mar. 16, 1972	precipitation	~ 1.5×10^6	109	10.0	1.35×10^6	2.4×10^2	5.6×10^3	
Mar. 2, 1973		1.2×10^6	100	24.0	1.14×10^6	1.8×10^2	6.3×10^3	
				Avg = 18.3			Avg = 6.4×10^3	
Aug. 6, 1972 (0809 UT)	Magnetic storm	4.48×10^6	101	3.9	3.29×10^6	4.8×10^1	6.8×10^4	Chatanika ISR (largest N_c observed by Chatanika radar)

[a] This is a measure of the "impulsive" character of N_e variations.
[b] EUV = Extreme ultraviolet radiation from the sun ($\lambda = 100$–400 A).

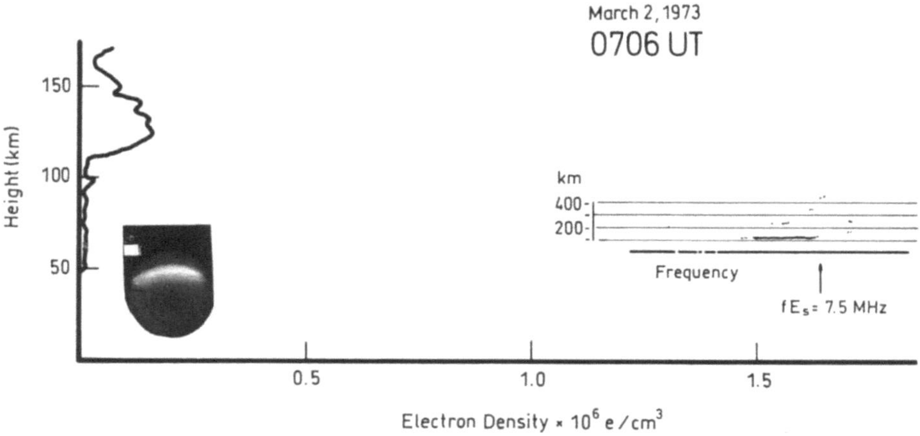

Fig. 1.14. Relationship between ISR, ionosonde, and all-sky camera (ASC) data at 070 UT on 2 March 1973 for an auroral arc north of the Chatanika ISR beam. (Hunsucker 1965)

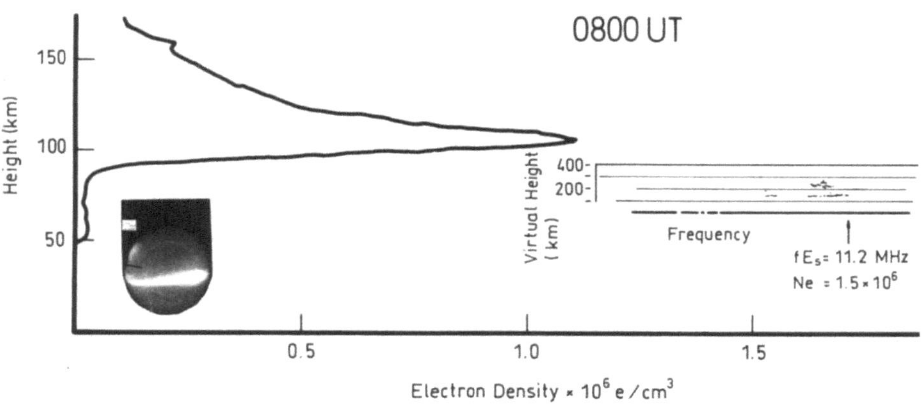

Fig. 1.15. Simultaneous Chatanika ISR, College ASC, and ionosonde data when a discrete auroral arc was in the field of view of the three instruments. (Hunsucker 1975)

activity. Therefore, the particular ionospheric phenomenon and the range of expected ionospheric parameters associated with it dictates which particular technique should be chosen to yield the most information about the phenomenon to be investigated.

1.3 Electrical Noise and Radio Frequency Interference

Although the topic of electrical noise is covered in the books listed in the Annotated Bibliography, a brief treatment will be presented in this section. To start, one can consider either noise *originating* in a particular system, or noise *coming into* the system from external sources. *Internally generated noise* causes contamination of the acquired data, reduces experiment efficiency, and/or interference to other services (TV, etc.). It can be alleviated by the use of high-gain, low-noise amplifiers in the input stages of a receiver; and by shielding components and filtering the signal until the output is of the optimum spectral shape (sometimes a nontrivial exercise!). The *external* component of noise is quite often of considerable importance, and requires one to at least estimate the expected noise levels at the experiment site. In addition, it is usually recommended that a careful electrical noise and radio-frequency interference (RFI) survey be made at the site, using a methodology similar to that suggested by Lauber and Pike (1979) or by Hagn (1978). Appendix A.6 contains selected maps showing representative global noise curves and contours, and CCIR Report 322 (1985), Volland (1982), and a report by Spaulding and Washburn (1985) can be consulted for further details concerning the relative levels of galactic and atmospheric radio noise. The characteristics of man-made radio noise may be found in CCIR report 258, Herman (1979), and Skomal (1978). Another component of external noise originates from the earth's surface due to black body radiation and can be quite noticeable when operating a UHF radar-receiving antenna at low elevation angles.

1.3.1 Thermal Noise in a Resistor

Consider a resistor of resistance R ohms at an absolute temperature T. The random motion of electrons in the resistor will produce a noise voltage V across its terminals, and the power spectral density of this noise voltage can be described by Planck's distribution law

$$P_n = \frac{4hfRB}{e^{hf/KT} - 1}, \tag{1.1}$$

where:

h = Planck's constant = 6.6254×10^{-34} joule/s
K = Boltzmann's constant = 1.38×10^{-23} joules/K
B = bandwidth in Hertz
f = frequency in Hertz

For normal terrestrial temperatures and frequencies in the radio portion of the spectrum, the quantity $(hf/KT) \ll 1$, and

$$e^{hf/KT} \approx 1 + hf/KT, \tag{1.2}$$

so Eq. (1.1) can be expressed as

$$P_n = 4KTRB \text{ (true noise power, in watts)} \tag{1.3}$$

and the RMS noise voltage across the terminals of the resistor is given by Nyquist's formula

$$V_n = 2\sqrt{KTRB}. \tag{1.4}$$

Since the noise-voltage spectrum [Eq. (1.4)] is independent of frequency, it describes a "white" noise spectrum, and the bandwidth (B) is the "noise bandwidth."

If this resistor is connected to a passive network through an ideal narrow-band lossless filter, which has an input impedance equal to R, it can be shown (see Kraus 1966 and Collin 1985) that the *available noise power* from this resistor, *under matched conditions*, is

$$P = KTB. \tag{1.5}$$

1.3.2 The Noise Factor, Figure, and Temperature of Radio Receiver

In order to quantitatively assess the characteristics of an RF amplifier in terms of signal and noise, it is helpful to utilize the quantities noise temperature, T, and noise factor, F. Let us consider a radio receiver which has several stages of amplification ahead of the detector, and examine the noise performance of the first RF amplifier stage. We will assume that this amplifier input resistance is matched to its source resistance, R, for maximum power transfer, that the amplifier has an effective bandwidth, B, power gain, G, and that the source resistance is at room temperature (T = 290 K). Under the stated conditions, we can define the amplifier noise factor, F, as

$$F = \frac{\text{signal-to-noise power ratio at the input}}{\text{signal-to-noise power ratio at the output}} \tag{1.6a}$$

$$= \frac{P_i/KT_oB}{G\,P_i/(GKT_oB + P_A)}, \tag{1.6b}$$

where:

P_i = input signal power

P_A = additional noise power contributed by the amplifier, measured at the amplifier output.

Equation (1.6b) can be simplified to

$$F = 1 + \frac{P_A}{GKT_oB}\,(\text{dimensionless}). \tag{1.7}$$

(The noise *figure* can be expressed in *decibles* as $f_{dB} = 10\log_{10}F$).

From Eq. (1.7), the *noise power* contributed by this amplifier at its output is

$$P_A = G(F - 1)KT_oB\,(\text{watts}). \tag{1.8}$$

This noise power, P_A, can be considered to be equivalent to the thermal noise in the input resistance, R, at a temperature $(F - 1)T_o$. We can then define the effective noise temperature as

$$T_F = (F - 1)T_o. \tag{1.9}$$

When a system operates under nonstandard conditions (not at the reference temperature, nor with matched impedances) the analysis above must be modified as outlined in Collin (1985), and the noise factor can be expressed as

$$F' = \frac{T}{T_o} + \frac{F-1}{M}, \tag{1.10}$$

where:

F' = the noise factor of a system under nonstandard conditions
T = the actual temperature
M = the impedance mismatch factor $= \dfrac{4 R_g R_{in}}{|Z_g + Z_{in}|^2}$
R_g = internal resistance of generator
R_{in} = system input resistance
Z_g = internal impedance of generator
Z_{in} = input impedance of system $= R_{in} + JX_{in}$
X_{in} = input reactance of system.

We can also employ the concepts of noise temperature to a resistive *attenuator* whose attenuation is $L = 1/G$, assuming that the input resistance of the attenuator is matched to its source. The equivalent *output noise temperature* of the attenuator is

$$T_{out} = (1 - G)T_o. \tag{1.11}$$

Collin (1985) illustrates the importance of considering attenuators in the system analysis by developing the example of the thermal noise due to a lossy transmission line.

When considering the entire receiver as a system as shown in Fig. 1.16, we can define the system noise temperature T_s as

$$T_s = T + \frac{F-1}{M} T_o = T + \frac{1}{M} T_F. \tag{1.12}$$

If we apply the system noise concept to a receiver consisting of N stages connected in series, and assume that the stages are all impedance-matched, the noise factor for the system is given by

$$F = F_1 + \frac{F_2 - 1}{G_1} + \frac{F_3 - 1}{G_1 G_2} + \cdots + \frac{F_N - 1}{G_s}, \tag{1.13}$$

where:

G_s = total system power gain
G_1 = gain of first stage
G_2 = gain of second stage
F_2 = noise factor of second stage.

Another essential component of the system, the antenna, also can be analyzed using the noise temperature concept. The antenna noise temperature, T_A, is due to reception of noise power through the main beams and sidelobes from a variety of sources, such as: the sun and other radio stars, terrestrial atmospheric effects,

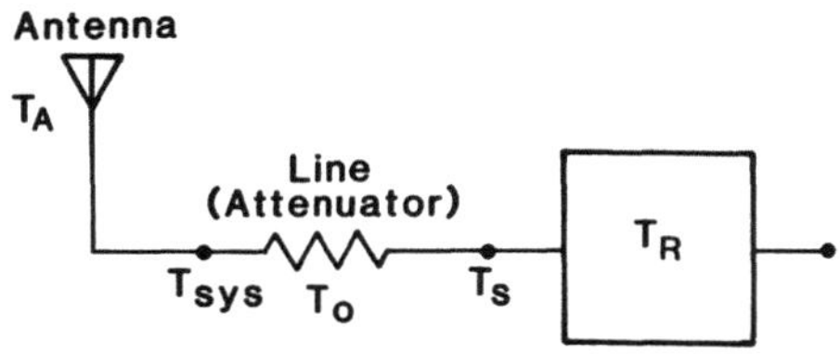

Fig. 1.16. Idealized antenna and receiver for noise analysis. (Flock et al. 1982)

thermal radiation from the earth's surface, man-made electrical noise, and even noise introduced by the antenna structure. These subjects are discussed in detail by Kraus (1966), Bell Telephone Laboratories (1982), Flock (1987), Flock et al. (1982), and Spaulding and Washburn (1985). Representative Plots of galactic, natural and manmade noise are given in Appendix A5 of this book.

Refering to Fig. 1.16, we define the system noise temperature (referred to the antennas terminals), as

$$T_{sys} = T_a + (L_a - 1)T_o + L_a T_R, \tag{1.14}$$

where:

T_{sys} = system noise temperature
T_a = noise temperature of attenuator
T_R = noise temperature of receiver
$L_a = 1/G_a$ = attenuator loss.

Some treatises on this subject define the system temperature, T_s, at the receiver terminals, as indicated in Fig. 1.16. The two approaches are related by the expression

$$T_s = T_{sys} G_a \tag{1.15}$$

then,

$$T_s = T_A G_a + (1 - G_a)T_o + T_R \tag{1.16}$$

the noise power at the *antenna terminals* is

$$N_A = KT_{sys} B \tag{1.17}$$

and at the *receiver* terminals by

$$N_A = K T_S B. \tag{1.18}$$

The effect of placing a high-gain, low-noise preamplifier (with noise temperature T_{PA}) as close as possible to the antenna terminals in order to reduce the total system noise figure is illustrated by including the preamplifier characteristics in Eq. (1.14),

$$T_{sys} = T_A + T_{PA} + \frac{(L_a - 1)T_o + L_a T_R}{G_{PA}}. \tag{1.19}$$

The term noise *factor* is also used in the treatment of electrical noise, and Spaulding and Washburn (1985) present an analysis of the system noise factor, while more detailed definitions and discussions of antenna temperature including spurious responses are presented in CCIR Report 413 (1966). The relationships between *noise power, the noise power spectral density*, and the *noise power bandwidth* developed by Hagn are given by Spaulding and Washburn (1985, Fig. 2, p.7).

1.4 Properties of Electromagnetic Waves and Their Interaction with Matter

Although this book is primarily concerned with the description of experimental techniques which depend upon radio wave interaction with the ionosphere, an orderly development of the subject requires proceeding from the general to the specific. This section, therefore, presents a review of the fundamental properties of electromagnetic waves and concludes with an example of the interaction of EM waves with matter as exemplified by the radar equation.

1.4.1 Maxwell's Equations and the Wave Equation

Since readers of this book are assumed to have taken a course in electromagnetic theory (at least at the undergraduate level), we will simply review Maxwell's equations and the constitutive relations, the wave equation and the Poynting vector. Derivations of equations will not be presented in detail, but references will be listed which contain complete derivations. The genius of James Clerk Maxwell was to synthesize and modify existing laws governing voltages induced by magnetic fields and continuity of charge and to introduce the concept of displacement current. He then assembled a self-consistent set of four equations which completely describe the time-dependent behavior of electromagnetic fields. The vector differential equation and the integral (large-scale) formulation are listed below using the MKS or SI system of units. This system will be used throughout this book since it employs volts, amperes, ohms, etc., which are the units actually measured by electronic instruments. For those somewhat rusty on vector calculus, the book by Schey (1973) and chapter 1 of Wait's (1985) book are highly recommended as lucid expositions of the subject. For a lossless medium we can then write Maxwell's equations,

Note: All quantities for Maxwell's equations and the constitutive relations are defined in this section. The superscript arrows define vector quantities.

Integral representation *Vector representation*

$$\oint_s \vec{D}\cdot d\vec{s} = \oint_v \rho\, dv \qquad \nabla\cdot\vec{D} = \rho, \tag{1.20}$$

which states that at each and every point, the electric flux diverging from a unit volume equals the charge contained in that volume.

$$\oint_s \vec{B}\cdot d\vec{s} = 0 \qquad \nabla\cdot\vec{B} = 0. \tag{1.21}$$

At every point, the net outward flux of B from a unit volume equals zero.

$$\oint \vec{E}\cdot d\vec{l} = -\frac{\partial}{\partial t}\oint_s \vec{B}\cdot d\vec{s} \qquad \nabla\times\vec{E} = -\frac{\partial\vec{B}}{\partial t}. \tag{1.22}$$

The electromotive force induced in any fixed closed loop equals minus the time rate of change of the magnetic flux through that loop (from Faraday's law).

$$\oint \vec{H}\cdot d\vec{l} = \oint_s \vec{i}\cdot d\vec{s} + \frac{\partial}{\partial t}\oint_s \vec{D}\cdot d\vec{s} \qquad \nabla \times \vec{H} = \vec{i} + \frac{\partial D}{\partial t}. \tag{1.23}$$

The work required to move a unit magnetic pole around any closed path equals the total (conduction and displacement) current linking that path (from Ampere's law).

The so-called constitutive or auxiliary relations for Maxwell's equations are:

1. The force law (definitions of $\vec{E}$ and $\vec{B}$)

   $\vec{f} = q[\vec{E} + \vec{v} \times \vec{B}]$ newtons. (1.24)

2. Ohm's law (definition of conduction current)

   $\vec{J} = \sigma\vec{E}$ amperes/m^2. (1.25)

   The more general case is $\vec{J} = (\sigma + j\varepsilon\omega)\vec{E}$, where $j\varepsilon\omega\vec{E}$ is the displacement current.

3. Convection current definition

   $\vec{j} = \rho\vec{v}_e$ amperes/m^2. (1.26)

4. Permitivity (dielectric constant) definition

   $\vec{D} = \varepsilon\vec{E} = \varepsilon_r\varepsilon_o\vec{E}$. (1.27)

5. Permeability definition

   $\vec{B} = \mu\vec{H} = \mu_r\mu_o\vec{H}$, (1.28)

where:

σ = conductivity (siemens/m)
q = charge (coulombs)
v = velocity (m/s)
$\vec{E}$ = electric field strength (vol/m) – a real vector
$\vec{B}$ = flux density of magnetic field (tesla)
ρ = charge density (coulombs/m)
$\vec{D}$ = electric flux density (coulombs/m)
ε_o = permitivity of free space = $1/36\pi \times 10^{-9}$ farads/m
ε_r = relative permitivity
μ_o = permeability of free space = $4\pi \times 10^{-7}$ henrys/m
μ_r = relative permeability
$\vec{J}$ = current density in amperes per square meter

$\oint_s d\vec{s}$ denotes the integral over a surface
$\oint d\vec{l}$ denotes the line integral around a closed path.

Capital letters denote peak values.

We may obtain Maxwell's equations for the very important time-periodic case, which describes the behavior of steady-state alternating fields varying sinusoidally with time, $[e(t) = \mathrm{Re}\{E(j\omega)e^{j\omega t}\}]$ simply by replacing $(\partial/\partial t)$ in the differential-equation form of Maxwell's equations with $(j\omega)$ to yield:

$$\nabla \cdot \vec{D} = \rho \tag{1.29}$$

$$\nabla \cdot \vec{B} = 0 \tag{1.30}$$

$$\nabla \times \vec{E} = -j\omega\vec{B} \tag{1.31}$$

$$\nabla \times \vec{H} = \vec{i} + j\omega\vec{D}, \tag{1.32}$$

where $j = \sqrt{-1}$ and $\vec{E}$ and $\vec{H}$ are *complex* vectors.

The constitutive relations remain the same.

The manner in which energy is transferred from one point to another by electromagnetic waves is described by the *Poynting vector*,

$$\vec{P} = \vec{E} \times \vec{H}\ \text{watts/m}^2, \tag{1.33}$$

which gives the peak rate of energy flow out through an arbitrary enclosing surface. If E and H are the complex vectors representing the electric and magnetic fields respectively, in the time periodic case, the average Poynting vector is

$$\vec{P}_{\text{avg.}} = \tfrac{1}{2}\mathrm{Re}\{\vec{E} \times \vec{H}^*\}\frac{\text{watts}}{\text{m}^2}, \tag{1.34}$$

where $\vec{H}^*$ = complex conjugate of $\vec{H}$.

$\vec{P}$ must be integrated over a closed surface to obtain power, or "rate of energy flow."

For an excellent derivation and discussion of the Poynting vector, see Ramo et al. (1965) Stratton (1941), and Wait (1985).

A "wave equation" may be obtained in its simplest form for a perfect dielectric containing no charge and no conduction currents (free space) by differentiating Maxwell's equations and using vector identities (see Jordan and Balmain 1968) to obtain the relations:

$$\begin{aligned}\nabla^2\vec{E} &= \mu\varepsilon\frac{\partial^2\vec{E}}{\partial t^2}\\ &= \mu\varepsilon\left\{\left[\frac{\partial^2}{\partial x^2} + \frac{\partial^2}{\partial y^2} + \frac{\partial^2}{\partial z^2}\right]E_x\vec{i}_z + \left[\frac{\partial^2}{\partial x^2} + \frac{\partial^2}{\partial y^2} + \frac{\partial^2}{\partial z^2}\right]E_y\vec{i}_y\right.\\ &\quad\left. + \left[\frac{\partial^2}{\partial x^2} + \frac{\partial^2}{\partial y^2} + \frac{\partial^2}{\partial z^2}\right]E_z\vec{i}_z\right\}\end{aligned} \tag{1.35}$$

and

$$\nabla^2\vec{H} = \mu\varepsilon\frac{\partial^2\vec{H}}{\partial t^2}. \tag{1.36}$$

The general solution of Eq. (1.35) is

$$E = f_1(x - v_o t) + f_2(x + v_o t), \tag{1.37}$$

where:

$v = 1/\sqrt{\mu\varepsilon}$

f_1 = any function $(x - v_o t)$

f_2 = any function of $(x + v_o t)$,

which may be recognized as the equations describing the behavior of voltage or current along a lossless transmission line.

If an electromagnetic wave propagates in a region where the conductivity is not zero and where conduction currents may exist, we may substitute Ohm's law into Maxwell's equations and obtain – after vector calculus manipulation – the wave equation for conducting media (see the derivation in Ramo et al. 1965):

$$\nabla^2 \vec{E} - \mu\varepsilon \ddot{\vec{E}} - \mu\sigma \dot{\vec{E}} = 0 \tag{1.38}$$

and for H:

$$\nabla^2 \vec{H} - \mu\varepsilon \ddot{\vec{H}} - \mu\sigma \dot{\vec{H}} = 0, \tag{1.39}$$

where:

$$\ddot{\vec{E}} \equiv \frac{\partial^2 \vec{E}}{\partial t^2}; \quad \dot{\vec{E}} \equiv \frac{\partial \vec{E}}{\partial t}$$

and

$$\ddot{\vec{H}} \equiv \frac{\partial^2 \vec{H}}{\partial t^2}; \quad \dot{\vec{H}} \equiv \frac{\partial \vec{H}}{\partial t}$$

or the conduction term

$$\nabla^2 \vec{E} = \mu\varepsilon \frac{\partial^2 \vec{H}}{\partial t^2} + \mu\sigma \frac{\partial \vec{E}}{\partial t}. \tag{1.40}$$

1.4.2 Some Characteristics of Propagating Plane Waves

In lossless linear, homogeneous, isotropic (LHI) media the solution of the scalar wave equation

$$\frac{\partial^2 \hat{E}_x}{\partial z^2} + \omega^2 \mu\varepsilon \hat{E}_x = 0, \tag{1.41}$$

where:

$\hat{E}_x$ is the complex value of E_x.

The positive z-traveling wave can be written as:

$$E_x^+(z, t) \equiv C_2[\cos(\omega t - kZ)], \tag{1.42}$$

where C_2 is a real constant amplitude coefficient.

The factor k from Eq. (1.42) is the "wave number" or a "phase-shift factor" which describes the amount of phase change that occurs per meter for this wave function. Then the following relation holds,

$$\lambda = 2\pi/k = \frac{2\pi}{\omega\sqrt{\mu\varepsilon}} \quad \text{or} \quad k = \frac{2\pi}{\lambda} = \omega\sqrt{\mu\varepsilon}. \tag{1.43}$$

The "phase velocity" of this wave is the speed of the traveling wave in the z-direction required so that the wave will experience no change in phase; therefore, surfaces of equal phase can be defined by setting:

$$(\omega t - kz) = \text{constant}. \tag{1.44}$$

The *phase velocity* can then be obtained by differentiating Eq. (1.44) with the result:

$$v_P = dz/dt = \omega/k = 1/\sqrt{\mu\varepsilon} \tag{1.45}$$

or:

$$v_P = f\lambda \tag{1.46}$$

A uniform, plane wave traveling in free space will have a phase velocity of

$$v_{P_o} = 1/\sqrt{\mu_o\varepsilon_o} = 3 \times 10^8\,\text{m/s}. \tag{1.47}$$

The phase velocity of this positive-z-traveling wave is then the velocity of equiphase planes for a sinusoidal steady-state traveling wave at a single angular frequency (ω). A signal or "modulation" will be carried by a wave at a different velocity, called the group velocity. The group velocity will equal the phase velocity only when k is directly proportional to F, i.e., $k = \omega\sqrt{\mu\varepsilon}$. The *general relation* for group velocity is

$$v_g = \left(\frac{\partial k}{\partial \omega}\right)^{-1} \text{m/s} \tag{1.48}$$

in the vicinity of the "carrier" frequency.

A "dispersive" medium is one where a traveling EM wave has different phase and group velocities. A concise description of the spreading of a pulse in a dispersive medium is given by Jackson (1975), and a complete treatment of the concept of "signal velocity" is contained in Brillouin's book (1960).

The concept of a complex wave-impedance ($\hat{\eta}$) for a uniform, plane traveling wave in a Cartesian coordinate system can be obtained by considering the positive- and negative-traveling z complex waves separately:

$$\hat{\eta} \equiv \frac{\hat{E}_x^+(z)}{\hat{H}_y^+(z)} \text{volts/amp (OHM)} \tag{1.49}$$

or

$$\hat{\eta} = \frac{\hat{E}_x^-(z)}{\hat{H}_y^-(z)}. \tag{1.50}$$

For free space this yields:

$$\eta = 120\pi = 377\,\text{ohms}.$$

Uniform plane waves traveling in a lossy unbounded medium can be described by:

$$E_x(z,t) = \mathrm{Re}[\hat{C}_1\, e^{j(\omega t + \hat{k}_c z)} + \hat{C}_2\, e^{j(\omega t - \hat{k}_c z)}]. \tag{1.51}$$

The complex propagation constant is then defined as

$$\begin{aligned}\gamma &\equiv j\hat{k}_c \equiv j\omega\sqrt{\mu\hat{\varepsilon}_c}\\ &\equiv j\omega\sqrt{\mu\varepsilon\left(1+\frac{\sigma}{j\omega\varepsilon}\right)} \quad \text{Note: } \mathrm{R\varepsilon}\{\gamma\} > 0\\ &\equiv \alpha + j\beta,\end{aligned} \tag{1.52}$$

with the attenuation and phase-shift constants (α and β respectively) given by:

$$\alpha \equiv \frac{\omega\sqrt{\mu\varepsilon}}{\sqrt{2}}\left[\sqrt{1+\frac{\sigma^2}{\omega^2\varepsilon^2}}-1\right]^{1/2} \text{(nepers/m)} \tag{1.53}$$

$$\beta = \frac{\omega\sqrt{\mu\varepsilon}}{\sqrt{2}}\left[\sqrt{1+\frac{\sigma^2}{\omega^2\varepsilon^2}}+1\right]^{1/2} \text{(radians/m).} \tag{1.54}$$

1.4.3 Properties of the Propagation Constant and Vector

Before proceeding further in our discussion of other properties of EM waves, some properties of the complex propagation constant, $\hat{\gamma}$, defined in Eq. (1.52) should be expanded. We can rewrite Eq. (1.52) as

$$\gamma \equiv \omega\sqrt{\mu\left(\varepsilon - j\frac{\sigma}{\omega}\right)} \tag{1.55}$$

and rewrite the expression for the complex wave impedance [Eq. (1.49)] as

$$\eta = \frac{\sqrt{\mu}}{\varepsilon - j\dfrac{\sigma}{\omega}}. \tag{1.56}$$

For every medium there is a frequency where the real and imaginary parts become equal

$$\varepsilon = \frac{\sigma}{\omega} \quad \text{or} \quad \omega = \frac{\sigma}{\varepsilon}. \tag{1.57}$$

This is called the transition or cross-over frequency (ω_{x-0}) and when $\omega \ll \omega_{x-0}$ or $\varepsilon = \dfrac{\sigma}{\omega_{x-0}} \ll \dfrac{\sigma}{\omega}$ the material behaves as a conductor. Conversely, when $\omega \gg \omega_{x-0}$ or $\varepsilon = \dfrac{\sigma}{\omega_{x-0}} \gg \dfrac{\sigma}{\omega}$, the material becomes a lossy dielectric. The values of σ and ε can be very frequency-sensitive and can vary widely for different media. The properties of dielectric and conducting materials are given in Table 1.3.

Table 1.3. Some propagation characteristic of dielectrics and conductors

Propagation characteristics	Dielectrics $\left(\varepsilon \gg \frac{\sigma}{\omega}\right)$	Conductors $\left(\varepsilon \ll \frac{\sigma}{\omega}\right)$
$\hat{\gamma}$	$\omega\sqrt{\mu\varepsilon}\left(1 - j\frac{\sigma}{2\omega\varepsilon}\right)$	$\sqrt{\frac{1}{2}\mu\sigma\omega}\,(1-j)$
β	$\omega\sqrt{\mu\varepsilon}$	$\sqrt{\frac{1}{2}\mu\sigma\omega}$
α	$\frac{1}{2}\sigma\hat{\eta}$	$\sqrt{\frac{1}{2}\mu\sigma\omega}$
αdB	$4.343\,\sigma\hat{\eta}$ dB/m	$8.686\sqrt{\frac{1}{2}\mu\alpha\omega}$
$\hat{\eta}$	$\sqrt{\mu/\varepsilon}\left(1 + j\frac{\alpha}{2\omega\varepsilon}\right) \simeq \sqrt{\mu/\varepsilon}$	$\sqrt{\mu\omega/2\sigma}(1+j)$
λ	$\lambda_0\frac{\beta_0}{\beta} = \lambda_0\sqrt{\mu\varepsilon_0/\mu_0\varepsilon}$	$\frac{2n}{\sqrt{\frac{1}{2}\mu\sigma\omega}}$
$\frac{\eta}{\eta_0}$	$\sqrt{\mu/\varepsilon\cdot\varepsilon_0/\mu_0}$	
$\frac{\alpha\text{dB}}{2\pi/\lambda}$	—	8.686 dB

As a practical application of the relations in Table 1.3, consider the propagation properties of land with the following typical values,

$$\varepsilon = 10\varepsilon_o$$

$$\varepsilon_o = 8.854 \times 10^{-12}\,\text{farads/m} = \left(\frac{10^{-9}}{36\pi}\right)\text{farads/m}$$

$$\mu = \mu_0 = 4\pi \times 10^{-7}\,\text{henries/m}$$

$$\sigma = 10^{-3}\,\text{siemens.}$$

These values give

$$\omega_{x-0} = \frac{\sigma}{\varepsilon} = \frac{10^{-3}}{10\varepsilon_o} = 36\pi \times 10^5\,\text{rad/s}$$

$$f_{x-0} = 18 \times 10^5 = 1.8\,\text{MHz.}$$

So for $f \gg 2$ MHz, land with the stated properties is a *lossy dielectric* with

$$\lambda = \frac{\lambda_o}{\sqrt{10}} = 0.3\,\lambda_o, \quad \eta = \frac{\eta_o}{\sqrt{10}} = 120\,\Omega, \quad \lambda_0 = \text{wavelength in free space}$$

and

$$\alpha = 4.343 \quad \sigma\eta = 0.5\,\text{dB/m.}$$

For $f \ll 2$ MHz, this land behaves like a *conductor*, with the following properties

$$\alpha = 8.686\sqrt{\tfrac{1}{2}(4\pi \times 10^{-7})(10^{-3})(2\pi f)} = 0.5\sqrt{f}$$

and a characteristic resistance of

$$\sqrt{\frac{1}{2}\left(\frac{4\pi \times 10^{-7}}{10^{-3}}\right)2\pi f} = 60\sqrt{f_{MHz}}\ \Omega.$$

As f decreases, the depth of penetration into the ground increases. Since the attenuation in a conductor is 8.686 dB per $(\lambda/2\pi)$, the increase in penetration is due to an increase in λ. The complex wave impedance is then given by

$$|\eta| = 60\sqrt{2}\, f_{MHz}^{1/2} = 89\sqrt{f_{MHz}}\ \Omega$$

so for $f < 2$ MHz, $|\eta| < \eta_o$
and

$$\frac{E^1}{E} \simeq -\frac{\eta_o}{\eta_o} = -1.$$

So a radio wave traveling in air at normal incidence upon the ground surface will be almost totally reflected. The foregoing example is illustrative of the reflecting properties of ground as a function of frequency in the HF portion of the spectrum, and is of considerable importance when designing HF antenna systems for ionospheric sounders and radars. In many regions of the earth the ground conductivity is so low and the water table or frozen ground is so variable in depth that the best solution is to construct a metallic ground screen under the antenna.

To recapitulate, we should now be quite familiar with the quantity defined in Eq. (1.35), the propagation, constant which we denote by k. For an ideal, nonlossy medium, k is real and is given by

$$k = \omega\sqrt{\mu\varepsilon}. \tag{1.58}$$

If the medium is conductive, and hence lossy, the propagation constant is complex and is given by rewriting Eq. (1.55) as

$$\gamma = \sqrt{j\mu\omega(\sigma + j\varepsilon\omega)} = \alpha + j\beta.$$

We see in any case that the propagation constant describes the properties of the medium and is dependent entirely on the medium through which the EM wave is propagating.

We now introduce a different concept, the propagation vector, $\vec{K}$, (defined only for plane waves) which relates not only to the medium, but also to the particular EM field whose propagation we are considering. The propagation vector is a vector having the magnitude of the propagation constant for the medium and the direction of the phase propagation of the plane wave. The direction of the propagation vector is related to the orientation of the planes of constant phase, and is taken to coincide with that of the normal to the planes of constant phase. In rectangular coordinates the general equation for the plane of constant phase in a

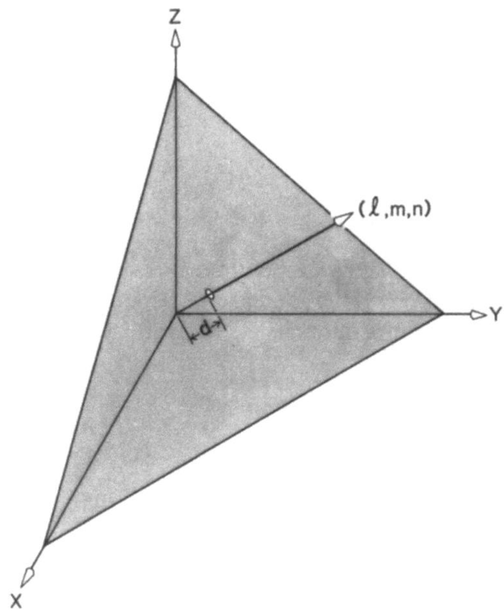

Fig. 1.17. Geometry of direction cosines for the propagation vector

lossless medium is of the form:

$$lx + my + nz = \text{constant}, \tag{1.59}$$

where l, m, and n are the direction cosines or direction numbers of the normal to the plane as shown in Fig. 1.17.

When l, m, and n are the direction consines, the constant is the distance, d, from the origin to the plane of constant phase. For a plane wave traveling in the direction of phase propagation (l, m, n), the time and space variation is of the form,

$$e^{j[\omega t - K(lx + my + nz)]}. \tag{1.60}$$

Having introduced the propagation vector, we can now describe some of its properties:

$$\vec{K} = \vec{K}(l, m, n) = (K_x, K_y, K_z) \tag{1.61}$$

hence,

$$\left(\frac{K_x}{K}\right)^2 + \left(\frac{K_y}{K}\right)^2 + \left(\frac{K_z}{K}\right)^2 = l^2 + m^2 + n^2 = 1.$$

In propagation problems we can then write the propagation vector as:

$$e^{j\{\omega t - (K_x x + K_y y + K_z z)\}}, \tag{1.62}$$

where

$$K_x^2 + K_y^2 + K_z^2 = K^2.$$

Usually the medium is given, so we know $\vec{K}$, and ω is also usually known. Then the direction of propagation may be chosen in the coordinate system. Assume in this case that propagation is along the z-axis, then $K = K_z$ and $K_x = K_y = 0$, and $e^{j(\omega t - K_z)}$ so we have determined the value of $\vec{K}$ in the exponential term. If we assume no conductivity, so that K is real $(K = \omega\sqrt{\mu\varepsilon})$, the planes of constant phase are then moving in the direction of the positive z-axis with a velocity of propagation

$$v_{ph} = \omega/K = dz/dt, \tag{1.63}$$

which follows from the relations

$$(\omega t - K_z) = \text{constant and } \omega dt - K d_z = 0; \text{ so } dz/dt = (\omega/K).$$

In the general form, the phase velocities along the x, y, and z axes are

$$V_x = \frac{\omega}{K_x};\quad V_y = \frac{\omega}{K_y};\quad V_z = \frac{\omega}{K_z},$$

each of which is greater than v. In free space

$$v_{ph} = c, \quad \text{so} \quad \frac{\omega}{K_x}, \ \frac{\omega}{K_y} \ \text{and} \ \frac{\omega}{K_z} \ \text{are all greater than c.}$$

We will conclude the discussion of the propagation vector by examining the properties of plane waves when the components of $\vec{K}$ are real or complex. The behavior of an EM wave may be described by the propagation vector in a general orthogonal coordinate system as follows, with the time and spatial variation given by

$$e^{j(\omega t - \vec{K}\cdot\vec{r})} \tag{1.64}$$

and the propagation vector as

$$\vec{K} = \vec{\beta} - j\vec{\alpha}, \tag{1.65}$$

where

$$K_1 = \beta_1 - j\alpha_1, \quad K_2 = \beta_2 - j\alpha_2, \text{ etc.}$$

squaring Eq. (1.61).

$$K^2 = K_x^2 + K_y^2 + K_z^2 = \omega^2\mu\left(\varepsilon - j\frac{\sigma}{\omega}\right) \tag{1.66}$$

or

$$(\beta - j\alpha)^2 = (\beta_1 - j\alpha_1)^2 + (\beta_2 - j\alpha_2)^2 + (\beta_3 - j\alpha_3)^2, \tag{1.67}$$

then the real part of K^2 can be written

$$\beta^2 - \alpha^2 = (\beta_1^2 + \beta_2^2 + \beta_3^2) - (\alpha_1^2 + \alpha_2^2 + \alpha_3^2), \tag{1.68}$$

and the imaginary part of K^2 is

$$\alpha\beta = \alpha_1\beta_1 + \alpha_2\beta_2 + \alpha_3\beta_3. \tag{1.69}$$

We can then examine the characteristics of three distinct types of waves which arise when we specify the real and complex nature of the propagation vector.

Case 1. If $\vec{K}$ is real and K_x, K_y, K_z are real then we have the "ordinary plane wave." For this wave the propagation vector is given by

$$K^2 = \beta_1^2 + \beta_2^2 + \beta_3^2 \tag{1.70}$$

and

$$\alpha_1 = \alpha_2 = \alpha_3 = 0.$$

The phase velocity is given by

$$v_{ph} = \frac{\omega}{K}, \tag{1.71}$$

and the planes of constant phase coincide with the planes of constant amplitude. Also,

$$\frac{\omega}{K_x}, \frac{\omega}{K_y}, \frac{\omega}{K_z} \text{ are all} > \frac{\omega}{K}.$$

Since K_x, K_y, K_z are all $< K$, the quantities above are *not* the components of a vector. However, $\left(\frac{1}{v_{ph}}\right)$ or $\left(\frac{\vec{K}}{\omega}\right)$ *is* a vector.

Case 2. If $\vec{K}^2$ is real and K_x, K_y, K_z are complex, but $\frac{K_x}{K_y}, \frac{K_y}{K_z}$ are real, then we have an "evanescent wave."

Since K^2 is real, then $\alpha_1\beta_1 + \alpha_2\beta_2 + \alpha_3\beta_3 = 0$ and the planes of constant phase are perpendicular to the planes of constant amplitude, and the amplitude falls off exponentially in the direction of propagation. The evanescent wave must be supported by a boundary to avoid exponential increase in the direction opposite to that of propagation. Another property of this wave is that while the wave progresses in the direction *away* from the boundary, there is also oscillation of energy back and forth *across* the boundary. This oscillation of energy signifies that we have *a reactive field* associated with the boundary. Examples of evanescent waves are those which exist in a waveguide below the cutoff frequency and do not propagate, and waves which travel from "free space" into a "totally reflecting" ionospheric layer. The word "evanescence" describes the spatial decay of a wave which occurs for kinematic reasons, not because of absorption processes.

Case 3. For $\vec{K}$ complex; K_x, K_y, K_z complex and $\frac{K_x}{K_y}$ and $\frac{K_y}{K_z}$ complex $\longrightarrow$ "exponential wave."

For K^2 complex and $\alpha_1\beta_1 + \alpha_2\beta_2 + \alpha_3\beta_3 \neq 0$
We have

$$e^{-(\alpha_1 x + \alpha_2 y + \alpha_3 z)} e^{-j[\omega t - (\beta_1 x + \beta_2 y + \beta_3 z)]} \tag{1.72}$$

or

$$e^{-\vec{\alpha}\cdot\vec{r}}e^{-j(\omega t-\vec{\beta}\cdot\vec{r})}$$

then the attenuation constant is given by

$$\alpha=\sqrt{\alpha_1^2+\alpha_2^2+\alpha_3^2} \tag{1.73}$$

and the propagation constant is

$$\beta=\sqrt{\beta_1^2+\beta_2^2+\beta_3^2}. \tag{1.74}$$

In general, the planes of constant phase are *neither* parallel *nor* perpendicular to the planes of constant amplitude. An exponential wave can decrease because of absorption processes (see Sect. 2.3.2), or increase because of plasma instability processes. For both increasing and decreasing amplitudes; $\nabla\cdot\vec{P}\neq 0$.

1.4.4 Polarization of an Electromagnetic Wave in Free Space

In addition to frequency, wavelength, phase, and amplitude, another fundamental property of uniform plane EM waves is the polarization – which can be qualitatively defined as the combined temporal and spatial behavior of the electric vector of the wave. A propagating uniform plane wave has the property that the electric vector, the magnetic vector, and the Poynting vector comprise an orthogonal set at any point along the direction of travel of a wave in an LHI medium. An unpolarized wave is defined as a combination of plane waves all propagating in the same direction, with random phases and arbitrary amplitudes.

An example of nearly unpolarized EM waves is direct sunlight observed on a cloudless day.

In their simplest form, vertically and horizontally polarized waves are shown in Fig. 1.18.

A "horizontally polarized" linear EM wave's electric vector lies in a plane parallel to the earth's surface, and a vertically polarized wave's electric vector lies in a plane perpendicular to the earth's surface.

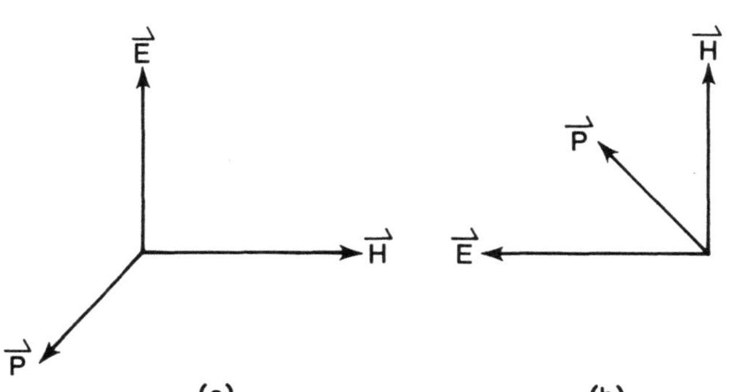

Fig. 1.18 a, b. Idealized representation of a vertical (**a**) and a horizontal (**b**) polarized plane wave

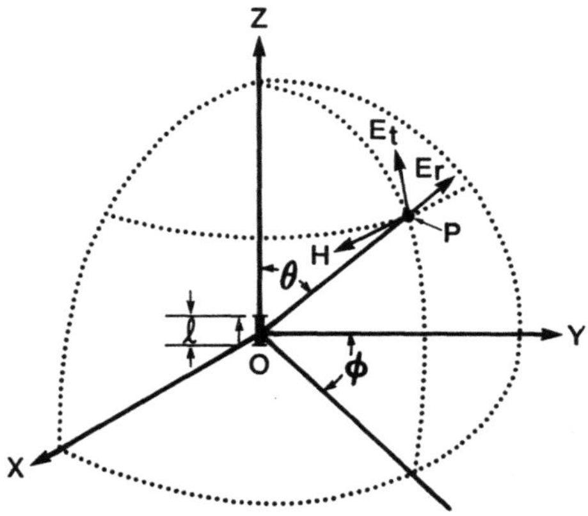

Fig. 1.19. Spherical coordinate system for analysis of the polarization of a plane wave propagating in a linear, homogeneous, isotopic (LHI) medium

In order to develop the concepts of wave polarization further, we will describe the behavior of uniform plane waves generated by an electric dipole propagating in an LHI medium in the spherical coordinate system shown in Fig. 1.19

where:

l = length of dipole
I = Current in dipole (amperes)
r = distance from origin, 0, to point P on surface of sphere
θ = angle measured from the z-axis to the line OP
λ = wavelength
f = frequency
$\omega = 2\pi f = 2\pi/\lambda$
$v = (\omega t - \alpha r)$.

Let us consider the wave field as a distance $r > 5\lambda$, then

$$E_r = 0$$
$$E_t = -(60\pi I l/r)\sin\theta\cos(\omega t - \alpha r) \tag{1.75}$$
$$H = +E_t/120\pi.$$

The wave produced by an ideal, linear antenna located at the origin of the coordinate system in Fig. 1.19 could be accurately described in free space and in the "far-field" region (the approximate boundary between the near-field and far-field regions is $D = 2r^2/\lambda$) as a uniform, vertically polarized EM wave. Most linear antennas and antenna arrays produce a uniform, linear EM wave in their far-field regions, but antennas can be configured to produce a circularly polarized wave such as those illustrated in Fig. 1.20. Two cases are possible according to the positive or negative rotation of the electric vector about the O-axis; right-hand circular (RHC) polarization and left-hand circular (LHC) polarization. RHC polarization is described as clockwise rotation of E when viewed in a direction opposite to the direction of propagation (looking toward the source).

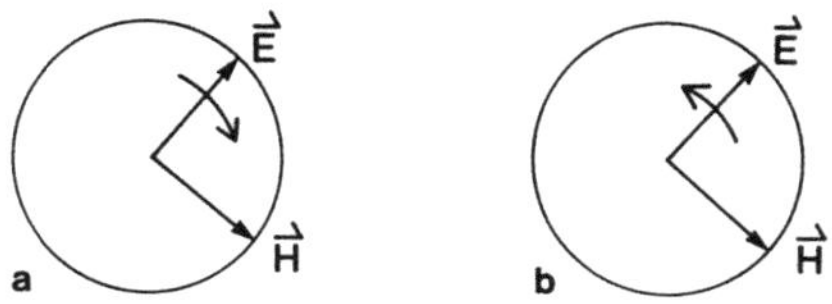

Fig. 1.20. **a** Right-hand circularly (RHC) polarized wave (looking toward the source). **b** Left-hand circularly (LHC) polarized wave

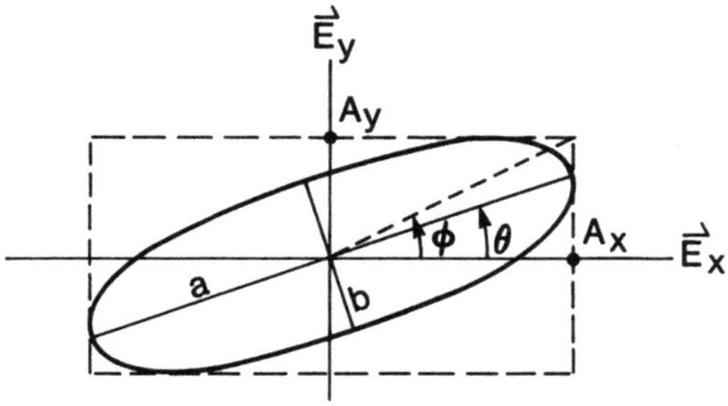

Fig. 1.21. Polarization ellipse in rectangular coordinates

As a linear or circularly polarized wave progresses in the terrestrial propagation environment, processes such as dispersion, refraction, reflection, and absorption alter the original polarization state – resulting in an elliptically polarized EM wave. We can therefore regard the elliptical polarization state to be the general case for EM waves propagated in the terrestrial atmosphere, and consider linear and circular polarization to be special cases of the elliptical case. Figure 1.21 presents the polarization ellipse in rectangular coordinates.

The polarization of the ellipse is then specified by the axial ratio (a/b), the orientation of the major axis, and the sense of rotation of this ellipse. The derivation of these parameters is given in detail in Born and Wolf (1970) and in Jordan and Balmain (1968), and is outlined below.

Consider two dipoles aligned with the x and y axes radiating a wave in the z-direction of the form

$$E_x = A_x e^{j\theta_x}$$

and (1.76)

$$E_y = A_y e^{j\theta_y},$$

where: A_x and A_y are real, positive amplitudes and θ_x and θ_y are the phase angles associated with the field components.

The complex quantity p describes the polarization ratio and is given by

$$p = E_y/E_x = |p| e^{j\theta}$$

and (1.77)

$$|p| = A_y/A_x.$$

The orientation of the tilt angle is finally given by

$$\tan 2\theta = \tan 2\Phi \cos \theta. \tag{1.78}$$

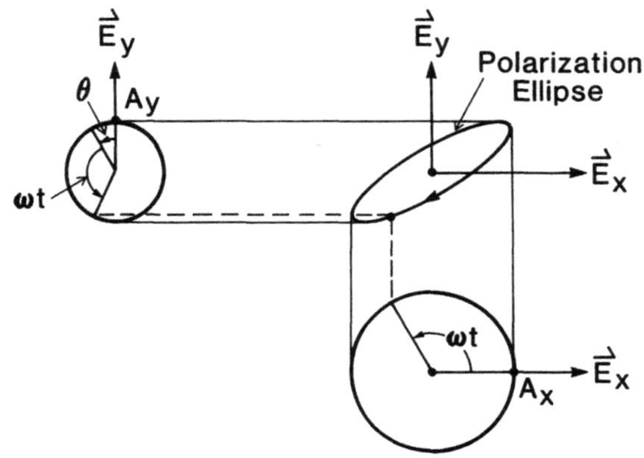

Fig. 1.22. Relation between amplitude and phase of rectangular field components and the polarization ellipse

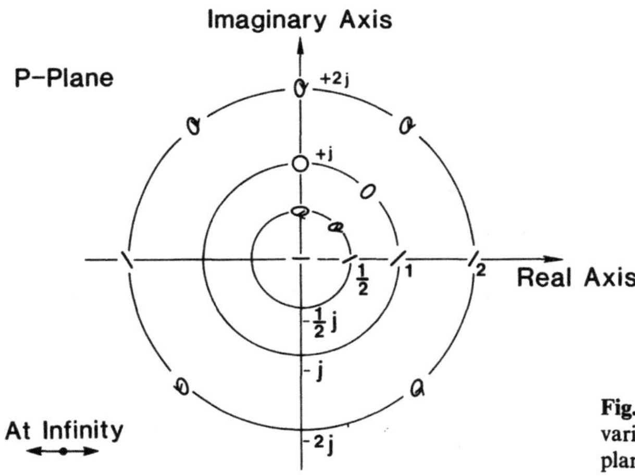

Fig. 1.23. Polarization ellipses at various points on the complex P plane

Thus, if we can specify the amplitude and phase of the rectangular field components, we can construct the ellipse from the foregoing relations. We can also employ a graphical method to specify the polarization ellipse by considering the sinusoidally varying E_x and E_y components as shown in Fig. 1.22.

Utilizing the graphical method of Fig. 1.22 and Eq. (1.77), we can obtain polarization ellipses for any point in the complex P plane as shown in Fig. 1.23.

If we imagine a screw turning clockwise and advancing in the direction of propagation in the positive z direction (into the page in Fig. 1.23) then the upper half-plane represents the left-handed sense of polarization, and the bottom half-plane includes the right-hand sense of polarization. Further inspection of Fig. 1.23 reveals that the linear polarization cases lie on the real axis, and that the LHC and RHC polarizations occur at the points $+j$ and $-j$ respectively.

It is also possible to obtain the polarization ratio in a rotating coordinate system (Jordan and Balmain 1968), which is particularly useful when considering polarization analysis of waves from helical antennas or the transmission of waves

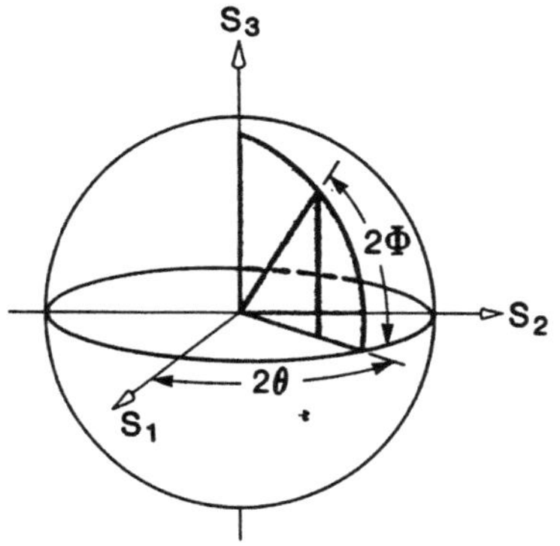

Fig. 1.24. Geometry of the Poincaré sphere

between elliptically polarized antennas, but this topic is beyond the scope of this book.

In order to completely characterize the polarization state of an EM wave including the polarization sense, we need to obtain some additional defining relations. Equation (1.78) is the first relation, and may be written as

$$\tan 2\Phi = \frac{2A_xA_y}{A_x^2 - A_y^2}\cos\theta. \tag{1.79}$$

The second and third relations are derived in Born and Wolf (1970) and may be written as

$$a^2 + b^2 = A_x^2 + A_y^2 \tag{1.80}$$

and

$$\sin 2\theta = \pm\frac{2ab}{a^2 + b^2}. \tag{1.81}$$

A powerful technique for analysis of partially polarized EM waves was developed by Stokes (see Born and Wolf 1970) and for single frequency radiation, the parameters can be written as

$$s_0 = A_x^2 + A_y^2 \tag{1.82}$$

$$s_1 = A_x^2 - A_y^2 \tag{1.83}$$

$$s_2 = 2A_xA_y\cos\theta \tag{1.84}$$

$$s_3 = 2A_xA_y\sin\theta. \tag{1.85}$$

These four parameters also satisfy the relation

$$s_0^2 = s_1^2 + s_2^2 + s_3^2. \tag{1.86}$$

Examination of Eq. (1.86) reveals that it describes a sphere of radius s. In this particular application, this is known as the Poincaré sphere and Eqs. (1.79)–(1.81) may be applied to the Poincaré sphere shown in Fig. 1.24 to obtain the following relations.

Where:

$$s_1 = s_0 \cos 2\theta \cos 2\Phi \qquad (1.87)$$

$$s_2 = s_0 \cos 2\theta \cos 2\Phi \qquad (1.88)$$

$$s_3 = s_0 \sin 2\theta. \qquad (1.89)$$

A careful examination of Fig. 1.24 and Eqs. (1.87)–(1.89) leads to the conclusion that all the points on the surface of the sphere represent all the possible polarization conditions, and that the top hemisphere represents the left-hand sense of polarization and the bottom hemisphere shows the right-hand polarization cases. It is rather difficult to illustrate the various polarization conditions pictorially on the surface of the Poincaré sphere, so various schemes are used to map significant points onto planes – as discussed in considerable detail in Kelso (1964), and in the handbook edited by Jordan (1985). Further discussions of antenna polarization will be found in books by Mott (1986) and Beckmann (1968).

1.4.5 The Radar Equation

We will conclude our discussion of the behavior of EM waves with a brief discussion of the "radar equation", starting with the derivation given by Skolnik (1980). This fundamental equation describes the basic principle of operation of many of the systems described in this book, including incoherent scatter, HF backscatter, and auroral radars – although the simple form of the equation must be considerably modified when applied to a magnetoionic medium. Figure 1.25 illustrates the basic geometry of a simple pulsed radar system.

We start by considering the power radiated by an isotropic antenna, P. The power density from this isotropic antenna (S) at a distance R from the radar antenna will be equal to the transmitter power divided by the surface area of a sphere of radius R,

$$S = \frac{P_t}{4\pi R^2}. \qquad (1.90)$$

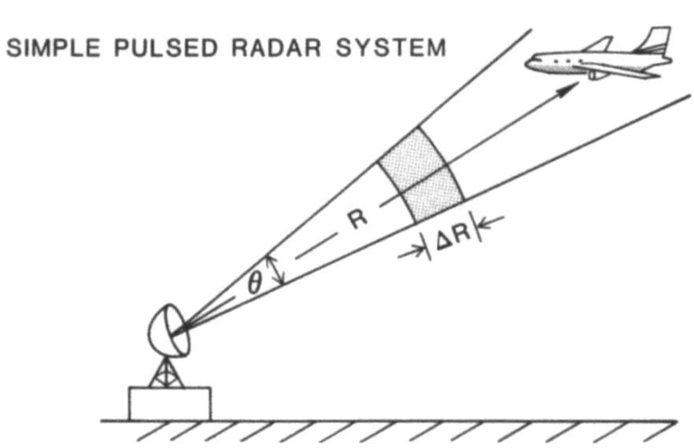

Fig. 1.25. Idealized geometry for the radar equation

Next, we must utilize the concept of the gain, G, of an actual antenna compared to an isotropic source. This is defined as the ratio of power required from an isotropic radiator to produce the given intensity in the desired direction to that required from an actual (practical) antenna, or

$$G = \frac{4\pi R^2 P_r}{W_i}, \tag{1.91}$$

where:

P_r = The power radiated in a given direction (usually in the direction of maximum radiated power) from the actual antenna
W_i = The power radiated in the same direction by an isotropic antenna with the same power input as the actual antenna.

Then the power density at the radar target from the radar antenna with gain, G, is

$$S_t = \frac{P_t G}{4\pi R^2}. \tag{1.92}$$

The radar target, however, will intercept only a portion of the incident power, then reradiates this power in various directions, depending on its reflectivity and geometry. The radar target cross-section, σ, is a measure of the amount of incident power that the target reradiates back in the direction of the radar antenna. The power density of the signal from the target radiated back to the radar antenna (P_{rad}) is related to the target cross-section as follows,

$$P_{rad} = \frac{P_t G A_e}{(4\pi)^2 R^4} \sigma. \tag{1.93}$$

The radar antenna, in turn, captures a part of the energy reflected back to it by the target. If we denote the effective area of the radar antenna as A_e, then the power, P_r, received by the radar is

$$P_r = \frac{P_t G}{4\pi R^2} \cdot \frac{\sigma}{4\pi R^2} A_e \tag{1.94}$$

or

$$P_r = \frac{P_t G A_e}{(4\pi)^2 R^4} \sigma.$$

The distance beyond which the radar target cannot be detected is called the maximum radar range, R_{max}, and it occurs when the received echo power, P_r, just equals the minimum detectable signal, S_{min}. We can then write the expression for R_{max} as

$$R_{max} = \left[\frac{P_t G A_e \sigma}{(4\pi)^2 S_{min}} \right]^{1/4}. \tag{1.95}$$

Equation (1.95) is the most common formulation of the radar equation.

Another form of Eq. (1.91) relates the gain of an antenna to its effective area as

$$G = \frac{4\pi A_e}{\lambda^2} \tag{1.96}$$

or

$$A_e = \frac{\lambda^2 G}{4\pi}. \tag{1.97}$$

Alternate forms of the basic radar equation may be obtained by substituting Eqs. (1.96) and (1.97) into Eq. (1.95) to yield:

$$R_{max} = \left[\frac{P_t G^2 \lambda^2 \sigma}{(4\pi)^3 S_{min}}\right]^{1/4} \tag{1.98}$$

or

$$R_{max} = \left[\frac{P_t A_e^2 \sigma}{4\pi \lambda^2 S_{min}}\right]^{1/4}. \tag{1.99}$$

A careful inspection of the expressions for the radar equation given by relations (1.95), (1.98), and (1.99) shows how easy it is to misinterpret this "simple" euqation, since R_{max} is shown as:

$$R_{max} \propto [A_e]^{1/4} \quad \text{and} \quad R_{max} \propto [A_e]^{1/2},$$

while

$$R_{max} \propto [\lambda]^0, \quad R_{max} \propto [\lambda]^{1/2} \quad \text{and} \quad R_{max} \propto [\lambda]^{-1/2}$$

and Eq. (1.95) shows R_{max} *to be independent of* λ! The correct interpretation of the radar equation depends on whether we assume that the *antenna gain* is constant, or whether the *effective area* is constant with wavelength. It should be emphasized that the forms of the radar equation given in Eqs. (1.95), (1.98), and (1.99) represent the simplest formulation of this relationship, and that many other complications arise when one considers a "practical system."

For a pulsed monostatic radar system with pulse length, τ, and a receiver signal-to-noise ratio defined (related to S_{min}) as follows,

$$S_{min} = N(S/N)_{min}, \tag{1.100}$$

where:

N = receiver noise power = KTB_nF
K = Boltzmann's constant
T = absolute temperature
B_n = noise bandwidth = $1/\tau$
F_n = receiver noise figure
τ = transmitter pulse length.

Introducing a system loss factor, L, which represents losses in the transmission lines, junctions, etc., we can then write an expression for the maximum range a target is detectable using a pulsed, monostatic radar (Flock 1979) as,

$$R_{max} = \left[\frac{LW_t \tau G A \sigma}{(4\pi)^2 KTF_n \left(\frac{S}{N}\right)_{min}}\right]^{1/4}. \quad (1.101)$$

A = effective area of radar antenna
W = energy of the transmitted pulse.

Equation (1.101) applies to discrete or "hard", non-beam-filling targets with the same caveats as for Eqs. (1.95), (1.98), and (1.99). An expression of the radar equation for targets which are distributed over a spatial volume which the radar beam intercepts is derived by Flock (1979) as

$$P_r = \frac{P_t A c \tau \sigma N}{32 R^2} \quad (1.102)$$

and

$$R_{max} = \left[\frac{P_t A c \tau \sigma \bar{N}}{32 S_{min}}\right]^{1/2}, \quad (1.103)$$

where:

N = the number of scatterers per unit volume
σ = cross-section of a single scatterer,
C = free-space velocity of e–m waves.

Notice in Eq. (1.102) that the backscattered power is now proportional to R^2 instead of R^4, since this is a "filled-beam" application of radar backscatter.

So far, we have considered the simple radar equation, a monostatic pulsed radar system, and idealized radar reflection from discrete and distributed targets.

The complexity of the radar equation increases when it is applied to more sophisticated systems such as continuous wave (CW) and frequency-modulated (FM) systems, moving-target indicators (MTI), pulse Doppler, synthetic-aperture radars (SAR's), and other esoteric systems. Descriptions of the appropriateness of the radar equation to these systems are contained in standard radar reference books such as Skolnik (1980). It is hoped that this brief review of the radar equation concept will serve to remind users of ionospheric radio techniques of the fundamental part it plays in describing the scatter of EM waves by matter.

The application of the radar equation to techniques used in ionospheric investigations is greatly complicated by the magnetoionic nature of the "targets" and the medium through which the radar beam travels. Specific radar equation adaptations for ionospheric probing will be described in Chapter 2.

Chapter 2 Radio Propagation in a Plasma

The behavior of radio waves in the terrestrial ionosphere has been meticulously described in several well-known texts and monographs, such as those by Rawer (1952), Mitra (1952), Ratcliffe (1959), Kelso (1964), Davies (1965, 1969, 1990), Wait (1970), Yeh and Liu (1972), Budden (1985), Kelley (1989). The purpose of this chapter is to review the fundamental processes whereby radio waves interact with the ionosphere, thereby laying a foundation for our understanding of the various radio techniques to be described in following chapters. Detailed derivations will not be given, since they are contained in the books cited above. Some of the most important aspects of the historical development of magnetoionic theory can be found in the works of Gillmor (1984). Salient points in the development of theories which describe the propagation of radio waves in the ionosphere will be included as each new topic is introduced. A very interesting bit of historical research on the origin of the term "ionosphere" was reported by Gillmor (1976), who concluded that Sir Robert Watson-Watt coined the term in a letter to the secretary of the British Radio Research Board in 1926. The acceptance of the term was aided by H. Plendl as "die Ionosphăre" and it came into wide use during the period 1932–1934 when Watson-Watt, Appleton, Ratcliffe, Dieminger, Ranzi, Martyn, Mitra, and others used the term in papers and books. Before the term "ionosphere" was accepted, it was variously called the Kennelly-Heaviside layer, the upper conducting layer, the ionized upper atmosphere, and other rather ponderous names!

2.1 Radio Wave Propagation in an Unmagnetized Plasma

The simplest approach in understanding radio wave propagation in the terrestrial ionosphere is to make the assumption of an unmagnetized plasma with no collisions between electrons and other particles, and develop the equations which describe this behavior. The results obtained using these assumptions describe the gross behavior of high frequency (HF) waves vertical and oblique propagation. The "fine-structure" or "second-order" behavior of radio wave propagation in the ionosphere must be analyzed using the magnetoionic theory developed by Appleton and others in the 1930's, which will be presented in Section 2.2.

2.1.1 Vertical and Oblique Propagation

Consider an ordinary plane wave, as described in Section 1.4.2, with the z-axis chosen to be the wave normal; then the definition of a progressing wave requires that all EM field quantities will contain z only, as in the exponent $e^{-K_z nz}$, where n is a constant and K_z is the z-component of the propagation vector. In free space n = 1, and in other media it can have other values and may even be complex. If v is the velocity of the wave, then $n = \frac{c}{v}$, where n is the refractive index and c is the velocity of an EM wave propagating in free space.

A *qualitative* approach for defining the refractive index is to consider a plane EM wave propagating in an unmagnetized plasma where collisions can be ignored. The E field sets the free electrons into motion, setting up a "convection" electric current which affects the behavior of the wave. This *convection* current is in *antiphase* with the *displacement* current and partly cancels the latter out, so that the effective rates of change of the resulting magnetic and electric fields of the wave are altered. This alteration results in the phase of the wave at a given point in the ionized medium being shifted, making the wave behave as though it had been speeded up. Therefore, as soon as the wave sets up a convection current, its phase velocity ($^+V_{ph}$) increases. Since ion density increases with altitude up to the ionospheric layer maximum density, as the wave penetrates farther into the ionized layer, convection current effects become more pronounced and the wave velocity increases.

As discussed in Section 1.4, the phase velocity (see Sect. 1.4.2) is not the speed at which a signal propagates. The wave propagates at the group velocity, in this simple case, (V_g) – which is always *less* than the phase velocity (V_{ph}) – and within the ionosphere V_g is actually reduced. For a simple ionosphere

$$V_g = cn, \tag{2.1}$$

where c = the speed of an EM wave in free space
= 2.99798×10^8 m/s.

(Note: at VLF V_g can be greater or less than V_p (but less than c), depending on f/f_H.)

As the EM wave travels from the nonionized atmosphere into the ionosphere, the group velocity (V_g) gradually decreases (since n = 1 for air an n < 1 for the ionosphere). Furthermore, the greater the wave velocity, the smaller is the group velocity, and so the signal is retarded. If the wave is propagating vertically, then $V_g \to 0$ when $n \to 0$. After reflection the wave travels downward with increasing velocity until it enters the un-ionized atmosphere where V = c.

The first mathematical description of the foregoing EM wave behavior was given by Eccles (1912), who started with an electric field described by

$$e(t) = E_0 \sin(2\pi ft), \tag{2.2}$$

where:

E_0 = maximum amplitude of the wave
f = frequency of the wave in Hz.

Then one can equate the force on an electron due to this EM wave to Newton's second law,

$$\text{force} = eE = m_e a = m_e \frac{\partial v}{\partial t}, \tag{2.3}$$

where:

m_e = mass of the electron = 9.1×10^{-31} kg
e = magnitude of the charge of the electron = 1.6×10^{-19} coulombs.

Integrating Eq. (2.3) yields the velocity of the electron,

$$v = -\frac{E_0 e}{2\pi m_e f} \cos(2\pi f t). \tag{2.4}$$

This electron motion produces a convection current (I_c) of density Nev, where N = number of electrons/unit volume and, at the same point, also a displacement current (I_d) due to the changing electric field, thus the total current is given by

$$I_t = I_d + I_c = \left[K - \frac{Ne^2}{\pi m_e f^2}\right] \frac{E_0 f}{2} \cos(2\pi f t), \tag{2.5}$$

where K is the dielectric constant of the medium without electrons.

The effect of the electrons is to reduce the dielectric constant of the medium from K to $(K - N_e^2/\pi m f^2)$ and since K = 1 for free space, for the ionosphere the dielectric constant is given by

$$K_i = 1 - \left(\frac{Ne^2}{\pi m_e f^2}\right), \tag{2.6}$$

and the refractive index μ[1] is given by

$$\mu = \sqrt{K_i} = \sqrt{1 - \frac{Ne^2}{\pi m_e f^2}}. \tag{2.7}$$

(Note: μ is the real part of the complex refractive index n)
Then the phase velocity is

$$v_{ph} = \frac{c}{\mu} = \frac{c}{\sqrt{1 - \left(\frac{Ne^2}{\pi m_e f^2}\right)}} \tag{2.8}$$

and the group velocity

$$v_g = c\mu = c\sqrt{1 - \left(\frac{Ne^2}{\pi m_e f^2}\right)}. \tag{2.9}$$

μ is also used to denote magnetic permeability in Chapter 1.

We can deduce the following information from Eq. (2.9):

1. If a radio frequency pulse of frequency f enters an ionospheric layer at vertical incidence, it will experience a decrease in group velocity as soon as the electron density increases above zero. As the pulse progresses vertically in the layer, the group velocity will decrease. If the layer electron density is sufficiently high, the wave of frequency f will eventually reach a point where the group velocity is reduced to zero and it will be reflected.

2. The condition for *vertical incidence reflection* is $n = 0$, or

$$1 - \frac{Ne^2}{\pi m_e f^2} = 0 \tag{2.10}$$

   or,

$$N = \frac{m_e \pi f^2}{e^2}$$

$$= 1.24 \times 10^4 f^2 \text{ electrons/cm}^3$$

$$= 1.24 \times 10^{10} f^2 \text{ electrons/m}^3 \text{ (in SI units).} \tag{2.11}$$

   Equation (2.11) then gives the electron density required for vertical incidence reflection at the frequency f in megahertz.

3. Equation (2.7) can describe ion *or* electron density in the ionosphere. However, since the electrons are much less massive than the ions (1/32000 of an oxygen ion) it is the *electron density* which determines the retardation and reflection of the wave.

4. We can also write Eq. (2.7) in a different form,

$$\mu = \sqrt{1 - \frac{f_n^2}{f^2}}, \tag{2.12}$$

   where f_n is the "plasma frequency" defined by

$$f_n = \sqrt{\frac{N_e^2}{\pi m}}, \quad \text{or} \quad f_n(\text{KHz}) = 9\sqrt{N_e(\text{cm}^{-3})}$$

$$\text{or} \quad f_n(\text{KHz}) = 9 \times 10^{-6}\sqrt{N_e(\text{cm}^{-3})}. \tag{2.13}$$

This plasma frequency is the frequency at which this particular plasma oscillates and radiates "plasma waves" and its value for the terrestrial ionosphere is given by Eq. (2.11). We can now define the condition for reflection of a wave at vertical incidence as $f = f_N$ from Eq. (2.12). The equations derived in Section 2.1.1 can also be obtained somewhat more rigorously by considering the volume polarization p which a wave of electric field strength E produces in the medium, as is done by Davies (1969) and Yeh and Liu (1972). In all cases, we assume that the medium is unmagnetized, isotropic, and uniform, with charge neutrality – it is a "cold plasma" (ignoring thermal motion of the particles) – and there is no Lorentz-polarization term.

2.1.2 The Virtual Height Concept

In the previous paragraphs we have developed the concepts of refractive index and plasma frequency and the reflection of a pulse of radio frequency energy at vertical incidence. Before we can utilize these relations to measure the height of a reflecting layer, we must introduce the concept of virtual height (h'). If we assume that the rf pulse is propagating vertically upward at the speed of light, $v = c$, then the time required for this pulse to be reflected from an ionospheric layer and return to earth is

$$t = 2/c \int_{0}^{h} dz/\mu, \tag{2.14}$$

then the *virtual height, h'*, can be found by

$$\begin{aligned} h'(f) &= 1/2\, c \cdot t \\ &= \int_{0}^{h} dz/\sqrt{1 - f_n^2/f^2}. \end{aligned} \tag{2.15}$$

Since the pulse always travels more slowly *in the layer* than in free space, the virtual height of a layer is always greater than the true height. The true height and virtual height are related by the integral equation

$$h'(f) = \int_{0}^{Z_{max}} dz/\mu(f, z), \tag{2.16}$$

where z is the true height, Z_{max} is the maximum height reached by frequency f, and n is the refractive index at Z_{max} for the frequency f. This can also be written as

$$h'(f) = \int_{0}^{Z_{max}} dz/(1 - f_n^2/f^2)^{1/2}. \tag{2.17}$$

To find the *true height* of reflection of a radio wave of frequency f, ignoring the geomagnetic field, one starts by obtain an inversion of Abel's integral equation of the form,

$$h(f_v) = 2/\pi \int_{0}^{f_v} h'(f)\, df/(f_v^2 - f^2)^{1/2}. \tag{2.18}$$

A good discussion of the solution of Abel's equation, and of the determination of ionospheric true heights is given by Kelso (1964).

2.1.3 Vertical and Oblique Propagation

Before considering the behavior of a radio signal in a magnetoionic medium, we will state the results of three theorems which relate oblique and vertical incidence propagation as depicted in Fig. 2.1.

The first relation is the *Secant law*, which relates the vertical-incidence frequency f_v, reflected at B to the oblique frequency f_{ob}, which is reflected at

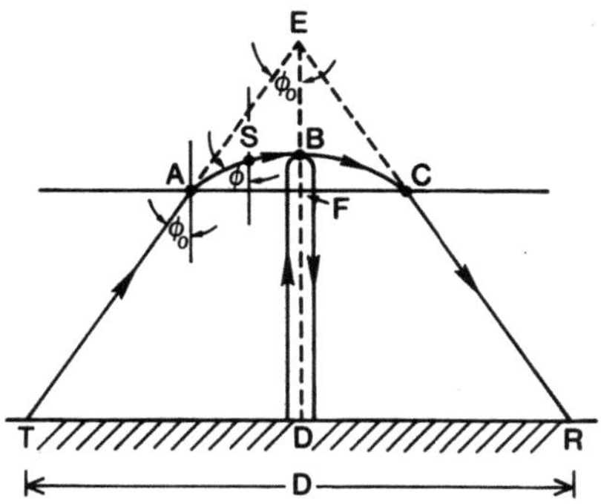

Fig. 2.1. Geometry describing vertical and oblique ionospheric propagation

this same true height. A typical derivation of this relation is given by Davies (1969), and it is usually written as

$$f_{ob} = f_v \operatorname{Sec} \phi_o. \tag{2.19}$$

The Secant law then, relates the two frequencies f_v and f_{ob} reflected from the same true height (the distance BD in Fig. 2.1).

In order to determine sec ϕ and f_{ob} values from vertical incidence soundings (which measure the virtual height, h′), we need the results of two more theorems. *Breit and Tuve's* theorem states that the time taken to traverse the actual curved path TABCR in Fig. 2.1 at the group velocity, V_g, equals the time necessary to travel over the straight-line path TER at the free-space velocity, c. Referring to the geometry shown in Fig. 2.1, we can write the expression

$$\begin{aligned} t &= 1/c \int_{TER} dx/\sin \phi_o \\ &= D/c \sin \phi_o \\ &= (TE + ER)/c. \end{aligned} \tag{2.20}$$

Martyn's theorem states that if f_v and f_{ob} are the vertical and oblique frequencies respectively, reflected from the same true height, h, then the virtual height at which the frequency f_{ob} is reflected equals the height of the equivalent triangular path for the frequency f_v. Referring to Fig. 2.1 and defining the equivalent path at oblique incidence for frequency f_{ob} as

$$P'_{ob} = 2\,TE, \tag{2.21}$$

we obtain

$$\begin{aligned} P'_v &= \cos \phi_o P'_{ob} \\ &= 2\,DE. \end{aligned} \tag{2.22}$$

Martyn's theorem may be written more concisely as

$$h'_{ob} = h'_v. \tag{2.23}$$

Newbern Smith (1939) devised a set of logarithmic transmission curves parametric in range for curved earth and ionosphere shown in Fig. 2.2 which are

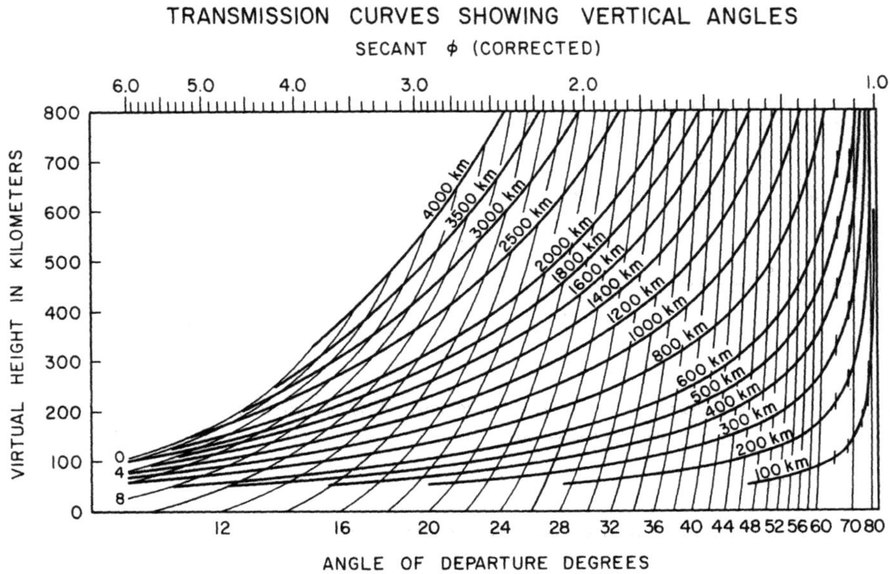

Fig. 2.2. Newbern Smith's logarithmic transmission curves for curved earth and ionosphere

sufficiently accurate for the distances shown. Details concerning the use of these curves to obtain f for distances from 100 to 4000 km may be found in Davies (1969) and in the *URSI Handbook of Ionogram Analysis* (1972).

2.2 Radio Propagation in a Magnetized Plasma

The effects of the terrestrial ionosphere on the propagation of radio waves cannot be described by the simple dispersion Eq. (2.7); we must employ the magnetoionic theory, which is best represented by the *Appleton* equations. Preceding the development of the magnetoionic theory, one "second-order" effect of the ionosphere on radio wave propagation was noted by Eccles in 1912 and Larmor in 1924, who independently pointed out that *collisional* effects would produce some *absorption* of radio waves.

It appears that Appleton was the first to consider the effects of the earth's magnetic field when he pointed out at a meeting at Cambridge in November 1924 that a plane polarized wave would be split into two oppositely rotating circularly polarized waves by the magnetized plasma. Appleton published these results in 1925 in the Proceedings of the Physical Society (London), and similar results were also published by Bell Telephone Lab scientists Nichols and Schelling (1925). A very complete description of the development of the magnetoionic theory is given by Gillmor (1982), who emphasizes the significant (but unpublished and unrecognized, at the time) work of Wilhelm Altar. Altar's contribution is considered by some people to be so fundamental (see Gillmor 1982) that the "magnetoionic" equations *should* be called the Appleton-Altar equations. The standard ray treatment of magnetoionic theory is given by Ratcliffe (1959) and a complete mathematical treatment of ray theory and of the "full-wave" theory is presented by Budden (1985).

2.2.1 Dispersion and Polarization

Before proceeding with a discussion of the Appleton (magnetoionic) equations, we need to define two quantities contained explicitly in the equation. The first is ν, the collision frequency between electrons and heavier particles (ions and neutrals) in collisions per second. Another quantity, the gyromagnetic frequency, is the natural frequency (Hz) of gyration of an ion or electron in a magnetic field, of strength B_o (webers/m) and is given by

$$f_H = \frac{|e|}{2\pi m_e} B_o \approx 2.80 \times 10^{10} B_o, \tag{2.24}$$

and the *angular* gyrofrequency is given by

$$\omega_H = \frac{|e|}{m} B_o \approx 1.76 \times 10^{11} B_o. \tag{2.25}$$

Since electrons are much less massive than ions, the *electron* gyrofrequency affects the propagation of HF waves in the ionosphere more than *ion* gyrofrequencies. For example, since $B \approx 0.5 \times 10^{-4}$ Wb/m, the electron gyrofrequency is ≈ 1.40 MHz, which falls in the upper end of the standard broadcast band.

We can perhaps begin to appreciate the difficulty of adequately describing the complete behavior of radio waves in the ionosphere by realizing that in the worst case, the terrestrial ionosphere is a partially ionized, spherically stratified plasma with a wide spectrum of nonuniformly spaced irregularities, upon which is imposed a nonuniform magnetic field – which is itself distorted by perturbations in the solar wind! The most difficult problems may be solved only by imposing the valid approximations and boundary conditions and obtaining full-wave solutions of the coupled differential equations using rather complex computer codes (Budden 1985 and Wait 1970).

Using the recommended URSI notation, the magnetoionic dispersion equation is given by

$$n^2 = 1 - \frac{X}{(1 - jZ) - \left[\dfrac{Y_T^2}{2(1 - X - jZ)}\right] \pm \left[\dfrac{Y_T^4}{4(1 - X - jZ)^2} + Y_L^2\right]^{1/2}}, \tag{2.26}$$

where:

n = *complex* refractive index $= (\mu - j\chi)^2$

ω = angular frequency of the "exploring wave" in radians per second

ω_N = plasma frequency

ω_H = angular gyrofrequency $= \dfrac{B_0|e|}{m}$ in rad/s

$$\omega_L = \text{longitudinal angular gyrofrequency} = \frac{B_0|e|}{m_e}\cos\theta$$

$$\omega_T = \text{transverse angular gyrofrequency} = \frac{B_0|e|}{m_e}\sin\theta$$

$$X = \omega_N^2/\omega^2$$

$$Y = \omega_H/\omega$$

$$Y_L = \omega_L/\omega,\ Y_T = \frac{\omega_T}{\omega}$$

$$Z = \nu/\omega.$$

The magnetoionic equation as stated in Eq. (2.26) describes the *dispersion* of a radio wave in a homogeneous, partially absorbing ionized gas upon which a constant magnetic field is impressed.

In order to calculate the effects of this anisotropic medium on the *polarization* of a radio wave traversing this region, it is convenient to define the polarization ratio, R, as

$$R = -H_y/H_x = E_x/E_y. \tag{2.27}$$

Then we can obtain the magnetoionic *polarization* equation (See Kelso 1964 and Ratcliffe 1959)

$$R = \left(-\frac{j}{Y_L}\right)\{\tfrac{1}{2}Y_T^2/(1-X-jZ) \mp [\tfrac{1}{4}Y_T^4/(1-X-jZ)^2 + Y_L^2]^{1/2}\}. \tag{2.28}$$

In the upper F-region where the electron-ion collision frequency is very low, we may simplify the Appleton-Hartree dispersion and polarization equations by dropping the Z term (since $\nu \approx 0$). Equations (2.26) and (2.28) then become

$$n^2 = 1 - \frac{2X(1-X)}{2(1-X) - Y_T^2 \pm \sqrt{Y_T^4 + 4Y_L^2(1-X)^2}} \tag{2.29}$$

(no absorption)

and

$$R = -\frac{H_y}{H_x} = -\frac{j}{Y_L}\left[1 + \frac{X}{n^2-1}\right]. \tag{2.30}$$

For a discussion of these equations and the implications of dropping the collision term, see Kelso (1964).

If we further simplify Eq. (2.29) by dropping the Y terms (no magnetic field), then we obtain

$$n^2 = 1 - X,$$

which is equivalent to Eq. (2.12).

According to magnetoionic theory, a plane polarized electromagnetic wave traveling in a medium like the terrestrial ionosphere will be split into two "characteristic waves." The wave which most closely approximates the behavior of a wave propagating in this medium without an imposed magnetic field is called the *ordinary wave* and the other is called the *extraordinary wave*. The foregoing terms are borrowed from the double refraction phenomenon in optics (Born and Wolf 1970, although the magnetoionic phenomenon is more complicated than the optics phenomenon. The ordinary wave is represented by the upper sign in the polarization Eq. (2.30), except when the wave-normal is exactly along the direction of the magnetic field.

If we recast Eq. (2.30) into the form

$$R = -j\left\{\frac{1}{(\omega/\omega_c)(1-X)-j/\omega_c} \mp \left[\left\{\frac{1}{\left(\frac{\omega}{\omega_c}\right)(1-X)-\frac{j}{\omega_c}}\right\}^2 + 1\right]^{1/2}\right\}, \tag{2.31}$$

where:

$$\omega_c = \left(B_0 \frac{|e|}{m}\right) f(\theta)$$

$$f(\theta) = \tfrac{1}{2}\sin^2\theta/\cos\theta$$

$$Y_T^2/2Y_L = \omega_c/\omega,$$

then it will be seen to describe an *ellipse* (see Fig. 1.14 in Sect. 1.4.4). The quantities $f(\theta)$ and ω_c play an important part in the description of the polarization behavior of waves in magnetoionic theory, as will be seen. The *magnitude* of ω_c is independent of frequency, but varies with the angle between the wave normal and the magnetic field, θ, whereas the *sign* of ω_c depends upon the sign of the charge, e, and on the direction of the magnetic field. For *longitudinal* propagation ω_c equals 0 and for *transverse* propagation ω_c approaches ∞. In the case when $X = 1$, the quantity ω_c primarily determines the polarization of the wave. An interesting discussion of R as a function of X and of the variation of the polarization ellipse is given by Ratcliffe (1959, pp. 71–74).

A better understanding of the behavior of EM waves in the terrestrial ionosphere may be obtained by employing two approximations. The *quasi-longitudinal* (QL) approximation is applicable when the wave is propagating nearly *parallel* to the geomagnetic field, and the *quasi-transverse* (QT) approximation applies when the wave propagates in a direction nearly *normal* to the geomagnetic field direction. The conditions for which they hold are

$$\text{QL:}\quad Y_T^4/4Y_L^2 \ll |(1-X-jZ)^2| \tag{2.32}$$

$$\text{QT:}\quad Y_T^4/4Y_L^2 \gg |(1-X-jZ)^2|. \tag{2.33}$$

When the relations (2.32) and (2.33) are valid, then Eqs. (2.29) and (2.30) are reduced to

$$\text{QL:}\quad n_{QL}^2 \simeq 1 - \frac{X}{(1-jZ \pm |Y_L|)} \tag{2.34}$$

$$R_{QL} \simeq \mp j \tag{2.35}$$

QT: (upper sign):

$$n_{QT}^2 \simeq 1 - \frac{X}{\{1 - jZ + (1 - X - jZ)\cos^2\theta\}} \tag{2.36}$$

$$R_{QT} \simeq 0 \tag{2.37}$$

QT: (lower sign):

$$n_{QT}^2 \simeq 1 - \frac{X}{\{1 - jZ - Y_T^2/(1 - X - jZ)\}} \tag{2.38}$$

$$R_{QT} \simeq \infty. \tag{2.39}$$

The negative sign in Eq. (2.35) describes a wave propagating downward through the ionosphere in the northern hemisphere.

As Ratcliffe (1959) points out, the QL approximation depends on the local electron density and the collision frequency, as well as the direction of propagation of the wave with respect to the magnetic field. In fact, it is possible that the QL approximation can hold even when the wave-normal departs significantly from the direction of the field.

When $X = 1$ and $\nu = 1/3\,|\omega_c|$, Ratcliffe (1959) shows that the *QTapproximation* is appropriate for

$$\omega_c^2/\omega^2 \gg (1 - X)^2, \tag{2.40}$$

or, equivalently

$$Y|f(\theta)| > 3|(1 - X)|, \tag{2.41}$$

and the *QL approximation* holds when

$$Y|f(\theta)| < 1/3|(1 - X)| \tag{2.42}$$

now, for the important condition $X = 1 \pm Y$

$$\tan^2\theta < 2/3 \tag{2.43}$$

or

$$\theta < 40°, \tag{2.44}$$

so in other words, if $\theta < 40°$, it is always safe to assume that the QL approximation is valid when $X = 1 \pm Y$ (see Ratcliffe 1959, p. 78). The validity of the QL and QT approximations for various ionospheric models is discussed in Ratcliffe (1959) and Budden (1985).

2.2.2 Absorption of Radio Waves in the Ionosphere

The refractive index (n) is modified when one introduces collisions between the electrons and heavy particles, and the wave experience absorption – which physically is due to the conversion of ordered momentum into random motion of the particles after collision. For each collision, some energy is transferred from the electromagnetic wave to the neutral molecules and appears as thermal energy.

Ratcliffe (1959) discusses the *microscopic* processes involved in the absorption process in considerable detail.

We will follow the standard treatment of absorption of radio waves in the ionosphere presented by Kelso (1964) and Davies (1969) by first considering absorption in the absence of a magnetic field, then introducing the magnetoionic case.

For the propagation of an electromagnetic wave in an unmagnetized plasma, we can define the absorption index (or coefficient) as

$$K = \frac{\omega}{c}\chi, \tag{2.45}$$

where χ = the imaginary part of the refractive index, n. And from the Appleton-Hartree equation without collisions, we can write

$$K = \frac{e^2}{2\varepsilon_o m_e c\mu} \cdot \frac{N\nu}{\omega^2 + \nu^2}. \tag{2.46}$$

Based on Eq. (2.46), we can conveniently divide absorption into two limiting types, commonly called *nondeviative* absorption and *deviative* absorption. *Nondeviative* absorption occurs in regions where the product $N\nu$ is large and $\mu \simeq 1$, and is characterized by the absorption of HF waves in the D region. *Deviative* absorption, on the other hand, occurs near the top of the ray trajectory or anywhere else on the raypath where significant bending takes place (for small $N\nu$ and μ).

When the refractive index is ≈ 1, there is essentially no bending of the ray and we can write

$$K \approx 4.6 \times 10^{-2}\left(\frac{N\nu}{\omega^2 + \nu^2}\right) \text{ (db/km)}. \tag{2.47}$$

We can further simplify Eq. (2.47) for the VHF case, since $\omega^2 \gg \nu^2$, as

$$K \simeq 1.15 \times 10^{-3}\frac{N\nu}{f^2} \text{ (db/km)}. \tag{2.48}$$

In the *medium and high frequency* (MF and HF) case, Eq. (2.47) may be written as

$$K \simeq 4.6 \times 10^{-2}\frac{N}{\nu} \text{ (db/km)}. \tag{2.49}$$

Unlike nondeviative absorption, deviative absorption occurs when the wave experiences significant group retardation and consequently spends a relatively long time in the absorbing layer and there is considerable curvature of the raypath. The general expression for the absorption index in a deviating region of a nonmagnetic plasma is

$$K = \frac{\nu}{2c\mu}(1 - \mu^2 - X^2). \tag{2.50}$$

In the *ionosphere*, Eq. (2.51) reduces approximately to

$$K \simeq \frac{\nu}{2c} \mu', \tag{2.51}$$

where μ' = the group refractive index.

For large values of μ', we can write the preceding equation as

$$K = \frac{\nu}{2c} \cdot \frac{X}{\sqrt{1-X}}. \tag{2.52}$$

We should remember that the concepts of deviative and nondeviative absorption are *limiting* cases, and that as a wave approaches the reflecting level, ray theory breaks down, so we must employ the wave theory to obtain a complete description of the behavior of the wave.

A practical approach for including the effects of the magnetic field upon the absorption index is to separately examine the QL and QT approximations. For the QL approximation, nondeviative absorption, and for $\mu^2 \gg X^2$, we can write

$$K \simeq \frac{e^2}{2\varepsilon_0 mc} \cdot \frac{N\nu}{(\omega \pm |\omega_L|^2) + \nu^2}. \tag{2.53}$$

If we further assume the VHF case, that is if $(\omega \pm |\omega_L|^2) \gg \nu^2$, the absorption index becomes

$$K \simeq \frac{e^2}{2\varepsilon_0 m_e c} \cdot \frac{N\nu}{(\omega \pm |\omega_L|)^2}. \tag{2.54}$$

Comparing these relations with Eq. (2.45), it appears that the absorption index of the ordinary wave is less in the magnetoionic medium than in the unmagnetized plasma. This is because the energy taken from the incident wave per cycle is smaller in the magnetized plasma due to smaller electron orbits. It should also be remembered that absorption becomes very large and ray theory breaks down as the frequency of the extraordinary wave approaches the electron gyrofrequency.

The absorption index for *deviative* absorption (QL case) for the *extraordinary* wave for the conditions $\mu \ll 1$ and $X^2 \ll 1 - \mu^2$ is

$$K \simeq \frac{\nu}{2c(1 \pm Y)} \left(\frac{1}{\mu} - \mu \right), \tag{2.55}$$

which does not hold for the ordinary wave since the QT approximation holds near its reflection level.

When the QT approximation is valid, $(X \cos^2 \theta) \ll 1$, and we can express the *nondeviative absorption for the ordinary wave* as

$$K \simeq \frac{e^2}{2\varepsilon_0 mc\mu} \cdot \frac{N\nu \sin^2 \theta}{\omega^2 + \nu^2}. \tag{2.56}$$

For the conditions one obtains near the ordinary wave reflection level, the absorption index can be written as

$$K \simeq \frac{1}{2} \frac{\nu}{c} \csc^2 \theta \left(\frac{1}{\mu} - \mu \right). \tag{2.57}$$

In order to obtain the total absorption experienced by a wave traversing an ionospheric path, we must integrate k over the path s. We then define the *total absorption* (loss or attenuation) for isotropic media as

$$L = 8.68 \int_s K \, ds (dB/m), \tag{2.58}$$

where K is in nepers per unit length.

The *absorption index* is defined in the direction of propagation (the wave normal direction), and the absorption can be expressed as

$$L = 8.86 \int_s K \cos \alpha \, ds, \tag{2.59}$$

where α is the angle the wave normal makes with the ray path.

One may obtain some idea how total absorption varies with different assumed profiles of electron density and collision frequency by referring to Table 6.1 in Davies (1969).

If we ignore magnetic field effects in the terrestrial ionosphere, we may also express the integrated absorption in terms of group height or phase height by integrating the equation for deviative absorption to yield the following relations for the case of *reflection*.

$$L = \int K \, ds = \frac{\bar{\nu}}{c} \int (\mu' - \mu) \, dh = \frac{\bar{\nu}}{c} (h' - h_p), \tag{2.60}$$

where:

$\bar{\nu}$ = an *average* collision frequency
h_p = group height
h' = group height
μ = group refractive index.

This equation describes the basis of the technique for obtaining collision frequencies near the maximum electron density of the D-layer (see Davies 1969) and will be discussed in more detail in later sections of this book.

2.3 Ionospheric Scattering of Radio Waves

In the previous sections of this book we have discussed the effects of the terrestrial ionosphere on the attenuation (absorption), dispersion, and polarization of a radio wave propagating in this medium. The phenomenon of *scattering* of a radio wave in the ionosphere will be the subject of this section as it applies to coherent and incoherent backscatter used to probe the ionosphere. The subject of "ionoscatter" as a *communications* mode will not be discussed, since this is not the subject of this book.

One could *qualitatively* describe scattering to be either *strong or weak* in terms of the received signal strength of the scattered signal at the receiving radar antenna. An example of the former is VHF/UHF backscatter echoes received from *electron density gradients* in the auroral E-region, and of the latter is

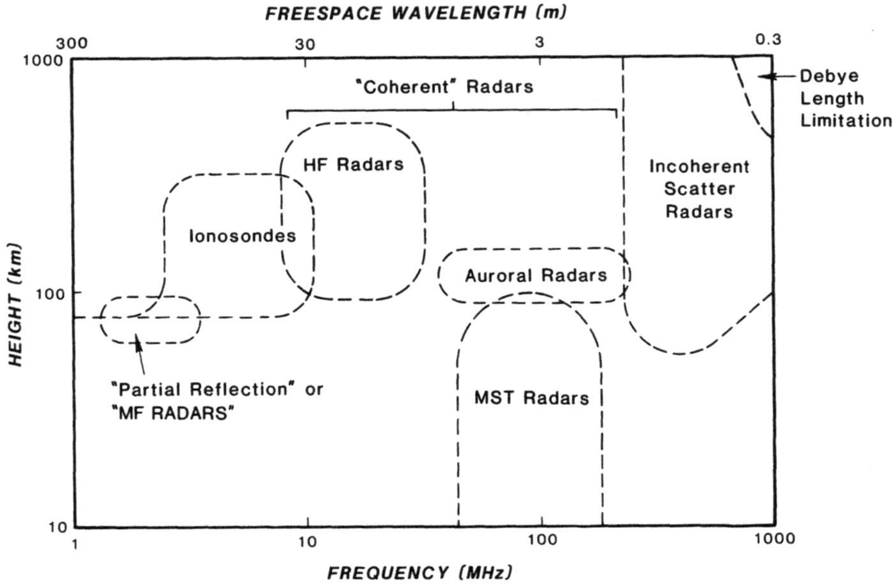

Fig. 2.3. Height-frequency regime of various ionospheric radar probes (Schlegel 1954). (Two ISRs in current use actually operate at 50 MHz)

incoherent backscatter received by a VHF/UHF radar from the undisturbed E- or F-layer.

Another way of classifying scattered echoes is in terms of their *backscatter cross-section* (σ) using pulsed radar system, and their *temporal stability*. A *coherent echo* exhibits a statistical correlation of the amplitude and phase from one pulse to another and emanates from quasi-deterministic gradients in electron density which have correlation times usually greater than 1 ms, which corresponds to a spectral width of the radar echo of less than 1000 Hz (sometimes less than 100 Hz). It also has a backscatter cross-section 10^4 to 10^9 greater than that from an *incoherent scatter radar echo*. Other important considerations in the case of *coherent* backscatter are the relation between the scattering irregularity *size* relative to the backscatter sounder freespace wavelength, the mean fractional deviation in electron density of the scatterer and the aspect angle between the radar line-of-sight, and the major axis of the irregularity.

On the other hand, an *incoherent* scatter echo arises from random thermal fluctuations in the ionosphere which have typical correlation times of $\approx 20\,\mu s$, corresponding to a radar echo spectral width of ≈ 50 kHz.

Figure 2.3 (Schlegel 1984) summarizes the height-frequency regimes of the "generic radars" used for ionospheric probing and may be of some help in selecting a particular technique for probing a particular region.

2.3.1 Coherent Scatter from Ionospheric Irregularities

The first quantitative description of coherent backscatter from ionospheric irregularities was published by Booker (1956a) [an extension of the Booker–

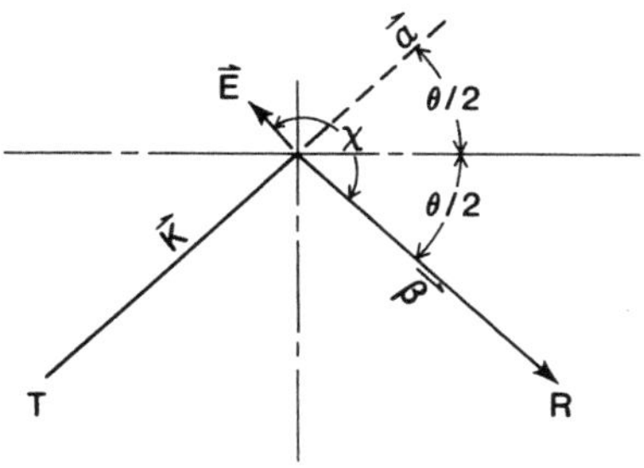

Fig. 2.4. Geometry for scatter from irregularities

Gordon (1950) tropospheric scatter theory], when he developed a theory which described backscatter from field-aligned irregularities in the auroral E-region. The results are also applicable to backscatter from F-region irregularities.

The geometry of scatter from an ionospheric irregularity is shown in Fig. 2.4. Based on this geometry, we can obtain one form of the Booker ionospheric irregularity scatter equation

$$\sigma(\theta, \chi) = \overline{\left(\frac{\Delta N}{N}\right)^2} \frac{\pi^2 \sin^2 \chi}{\lambda_N^4} P\{\vec{k} \cdot (\vec{\beta} - \vec{\alpha})\}, \tag{2.61}$$

where:

$\sigma(\theta, \chi)$ = backscatter cross-section of the irregularity

$\overline{\left(\frac{\Delta N}{N}\right)^2}$ = mean square fractional deviation in electron density

λ_N = wavelength of plasma oscillation

$\vec{k}$ = wavenumber of transmitted signal

(α, β, θ) = defined in Fig. 2.4

$P\{\vec{k} \cdot (\vec{\beta} - \vec{\alpha})\}$ = Power spectrum of the irregularity.

Equation (2.61) can also be expressed in terms more closely related to ionospheric parameters as

$$\sigma(\theta, \chi) = \overline{\left(\frac{\Delta N}{N}\right)^2} \frac{(2\pi L/\lambda_N)^2 \sin^2 \chi}{\lambda_N \{1 + (4\pi L/\lambda)^2 \sin^2(\theta/2)\}^2}, \tag{2.62}$$

where L = scale size of the irregularity along B.

Assuming the Booker–Gordon (1950) distribution

$$\rho(x, y, z) = \exp[-1/2(x^2/a^2 + y^2/b^2 + z^2/c^2)] \tag{2.63}$$

and the Gaussian spherical symmetrical function

$$\rho(x, y, z) = \exp\left[-\frac{\sqrt{(x^2 + y^2 + z^2)}}{L}\right], \tag{2.64}$$

we can rewrite Eq. (2.63) for the case of small irregularities ($L \ll /2\pi$) as

$$\sigma(\theta, \chi) = \cdot \overline{\left(\frac{\Delta N}{N}\right)^2} \frac{1}{\lambda_N} \left(\frac{2\pi L}{\lambda_N}\right)^3 \sin^2 \chi, \tag{2.65}$$

which indicates that the small ionospheric irregularity acts as an infinitesimal dipole re-radiating the incident radio pulse. If, on the other hand, the irregularity is *large* – ($L > \lambda/2\pi$), then

$$\sigma(\theta, \chi) = \overline{\left(\frac{\Delta N}{N}\right)^2} \frac{(2\pi L/\lambda_N)^3 \sin^2 \chi}{\lambda_N (4\pi L/\lambda)^4 \sin^4(\theta/2)} \tag{2.66}$$

for the special *backscatter* case (large irregularities), we can write

$$\sigma_B(\pi, \pi/2) = \overline{\left(\frac{\Delta N}{N}\right)^2} \frac{(2\pi L/\lambda_N)^3}{(4\pi L/\lambda)^4}, \tag{2.67}$$

which reduces to

$$\sigma_{B,\,large}(\pi, \pi/2) = \frac{1}{32\pi L}\left(\frac{\lambda^4}{\lambda_N}\right)\overline{\left(\frac{\Delta N}{N}\right)^2}. \tag{2.68}$$

Similarly, backscatter from *small* $\left(L \ll \dfrac{\lambda}{4\pi}\right)$ irregularities can be described by

$$\sigma_{B,\,small}(\pi, \pi/2) = \overline{\left(\frac{\Delta N}{N}\right)^2} \frac{(2\pi L/\lambda_N)^3}{\lambda_N \{1 + (4\pi L/\lambda)^2\}^2}. \tag{2.69}$$

In a recent paper, Walker et al. (1987) start with Booker's expression for backscatter cross-section

$$\sigma_B = r_e^2 \langle \Delta N^2 \rangle P_3(2kl, 2km, 2kn), \tag{2.70}$$

where:

$\langle \Delta N^2 \rangle$ = the mean square fluctuation level of the electron density
r_e = the classical electron radius $= 2.813 \times 10^{-15}$ m
$P_3(k_\xi, l_\eta, m_\zeta)$ = the three-dimensional power spectrum of the irregularities
$k = 2\pi/\lambda$ = wave number of the transmitted signal,

and derives a general expression for the backscattered power (P_r) at the receiver as

$$P_r = r_e^2 \langle \Delta N^2 \rangle f_2(\theta_l) \frac{(l_p A_r G_t P_t g_t g_r \Delta\Phi_B)}{4\pi r'} \int_{\Delta\Psi} P_3(2k, 0, 2k\Psi)\, d\Psi, \tag{2.71}$$

where:

$f_2(\theta_l) = \sin\theta / [\mu_p^2 f_1(\theta_l) \sin\{\theta_I + (\alpha - \alpha_o)\}]$
$f_1(\theta_l) = \dfrac{PQ}{P'Q'}$, the correction factor for focusing
l_p = spatial length of radar pulse
A_r = effective area of receiving antenna
G_t = gain of transmitting antenna
P_t = transmitter power output
μ_p = refractive index at the reflection level
g_r = normalized receiving antenna pattern
g_t = normalized transmitting antenna pattern

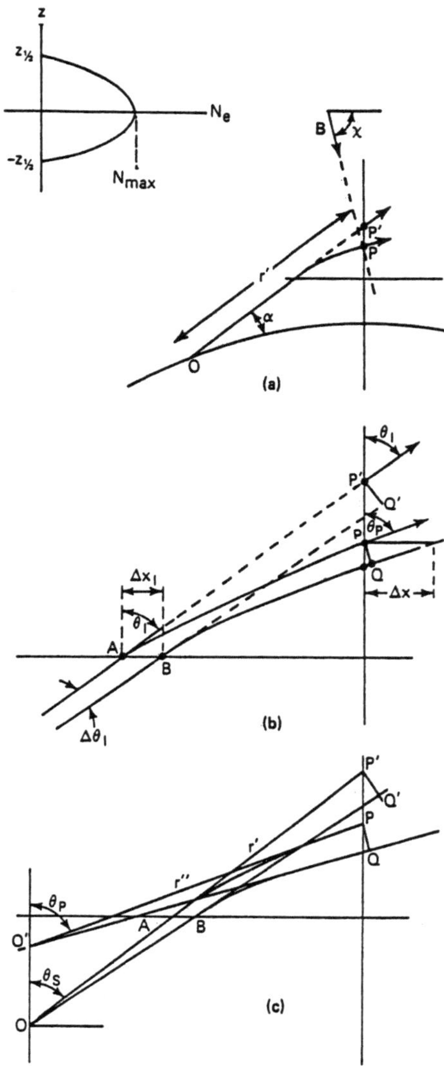

Fig. 2.5 a–c. Geometry and nomenclature for backscatter from irregularities. (Walker et al. 1984)

Ψ = elevation angle of radar ray
$= \chi = \theta$
θ = angle between the ray and vertical
χ = magnetic dip angle.

Other symbols are defined in Fig. 2.5.

Walker et al. (1987) give a special case of Eq. (2.71) for VHF/UHF backscatter when ionospheric refraction does not play a significant role as

$$P_r = r_e^2 \frac{l_p A_r G_t P_t}{4\pi r^2} \int_{\Delta\Omega} g_t g_r \langle \Delta N^2 \rangle P_3(2kl, 2km, 2kn)\, d\Omega. \tag{2.72}$$

Walker et al. (1987) also obtained numerical results which permit the comparison of specific radar system sensitivities such as the HF radar in Goose Bay, Greenland, the VHF radar at Jicamarca, Peru, and the STARE radar in Scandinavia. Some of these specific results will be discussed in Chapter 4 of this book.

2.3.2 Incoherent Scatter from the Ionosphere

The development of the incoherent scatter radar technique as a very powerful method for investigating the ionosphere is an excellent example of how, in some instances, theory precedes practical application by many years. Two excellent review papers by Evans (1969, 1972) and the references therein present the essentials of incoherent scatter theory and practice, and elegant and rigorous derivations of the salient equations are given by Krall and Trivelpiece (1973).

The basic theory of scatter of electromagnetic waves from free electrons was developed by the discoverer of the electron, J.J. Thomson, who in 1906 showed that the energy scattered by a single electron is

$$W = (r_e \sin \Psi)^2, \tag{2.73}$$

where:

W = energy scattered by a single electron into unit solid angle per unit of incident EM flux ($1\ \mathrm{W/m^2}$)
r_e = classical electron radius $= e^2/\varepsilon_0 m_e c^2 = 2.82 \times 10^{-15}$ m
Ψ = the angle between the direction of the incident electric field and the direction to the observer.

The "radar cross-section" of an individual electron would then be

$$\sigma_e = 4\pi(r_e \sin \Psi)^2 \approx 10^{-28} \sin^2 \Psi\, (\mathrm{m}^2) \tag{2.74}$$

and for *backscatter* ($\Psi = \pi/2$),

$$\sigma_e = 4\pi r_e^2. \tag{2.75}$$

Fejer (1960) showed that the *radar cross-section per unit volume* is simply

$$\sigma = N\sigma_e, \tag{2.76}$$

where:

N = electron density,

and Buneman (1962) showed that the effective radar cross-section (σ_{eff}) can be written as

$$\sigma_{eff} = \frac{\sigma_e}{(1+\alpha^2)(1+T_e/T_i+\alpha^2)} \tag{2.77}$$

for $T_e/T_i < 3.0$ and

T_e = electron temperature
T_i = ion temperature

$\alpha = 4\pi D/\lambda$
D = Debye length = $6.9[T_e/N_e]^{1/2}$ in cm
λ = freespace wavelength of radar.

Since the electrons are in *random* thermal motion, they will scatter signals whose phases are varying with time and are not related to one another. At the radar-receiving antenna the signal *powers* will add so that, on the average, the cross-section per unit volume is that given by Eq. (2.76), giving rise to the name "incoherent scatter".

Advances in radar technology during and shortly following the second World War prompted W.E. Gordon of Cornell University in 1958 to suggest that it should be possible to detect this incoherent scatter from the terrestrial ionosphere. He predicted that, due to the random thermal motion of the electrons, energy would be scattered with a wide variety of Doppler shifts. He also predicted that the scattered signal spectrum should have a Gaussian distribution with a half-power width of 0.71 Δf_e where:

$$\Delta f_e = 1/\lambda(8\,KT_e/m_e)^{1/2}\ \text{Hz} \tag{2.78}$$

and

Δf_e = the Doppler shift of an electron *approaching* the radar at the mean thermal speed
T_e = electron temperature (Kelvin)
K = Boltzmann's constant
λ = radar wavelength (m),

or, expressing the wavelength in meters

$$\Delta f_e = 11(T_e)^{1/2}/\lambda\,(\text{kHz}). \tag{2.79}$$

The foregoing equations predicted that the incoherent scatter echo would be widely Doppler spread (i.e., at f = 430 MHz, the Doppler spread would be $\simeq$ 713 kHz for T_e = 2000 K), so that a very large antenna would be required to make this technique feasible. These results were incorporated in the design of the large antenna systems at Arecibo, Puerto Rico, and Jicamarca, Peru, which comprised the "first generation" of incoherent scatter radars (ISR's).

After hearing from Gordon a description of the preceding results at the URSI fall meeting at Pennsylvania State University in October, 1958, Kenneth Bowles of the U.S. National Bureau of Standards (NBS) calculated on "the back of an envelope" whilst on the airplane en route to his lab (perhaps a somewhat apocryphal story!) that a newly completed meteor scatter radar located in Long Branch, Illinois, had the capability to detect this incoherent scatter – and though his calculations were not precise, because of the "ion component" of incoherent scatter, the experiment was successful. Within a few weeks, Bowles utilized this NBS radar to obtain the first incoherent backscatter from the ionosphere in October 1958 – a typical example from these early experiments is shown in Fig. 2.6 – and these results were reported by Bowles (1958, 1961). The Long Branch radar operated at 40.92 MHz, with approximately one megawatt of pulse power output into an antenna consisting of 1024 halfwave dipoles fed in phase in

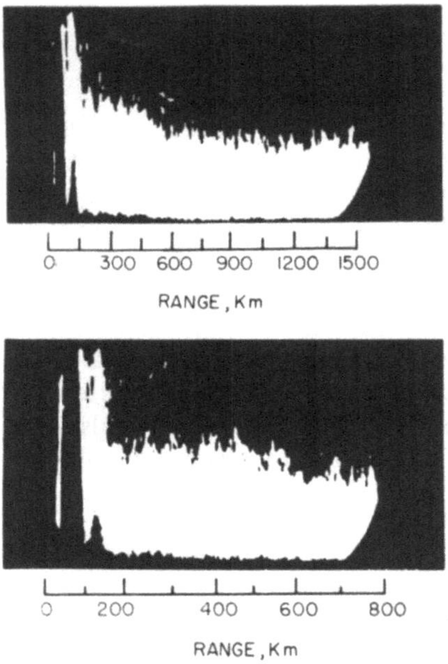

Fig. 2.6. Typical example of data obtained on Bowles' first incoherent scatter radar experiment at Long Branch, Illinois. (Bowles 1958)

a square array with 3.1° halfpower beamwidth. A standard communications receiver using a preamplifier with a noise factor of 2.0 was modified for pulse reception and used in conjunction with a "multiple deck mercury jet commutator" type of integrator which gave a 20-dB improvement in signal-to-noise ratio. An example of the output of this integrator is shown in Fig. 2.7 and one of the first electron density profiles of the terrestrial ionosphere is shown in Fig. 2.8.

Results obtained by Bowles indicated that the actual Doppler broadening of the echo spectrum was two orders of magnitude less than that predicted by Eq. (2.77), and he correctly interpreted this to be due to the effect of ions. Theoretical developments in the early 1960's (Dougherty and Farley 1960) explained these observational results in terms of the relation of the radar wavelength, λ, compared to the electron Debye length in the plasma, which is given by

$$D_e = 6.9(T_e/N_e)^{1/2}\,(\text{cm}). \tag{2.80}$$

A simplified explanation of the Debye length is that electrons within a sphere of radius D centered on the positive ion are influenced by *that ion* and cannot act independently. In general, the Debye length is a measure of the sphere of influence of a given test charge in a plasma and depends on the speed of the test charge relative to the plasma.

Incoherent scattering occurs from fluctuations in electron density having a scale of D_e. The backscatter, then, is actually due to *local fluctuations in electron density* instead of purely scatter from free electrons, and more correctly should be called quasi-incoherent scatter, but the term *incoherent scatter* has persisted. Some authors continue to refer to the incoherent scatter phenomenon as

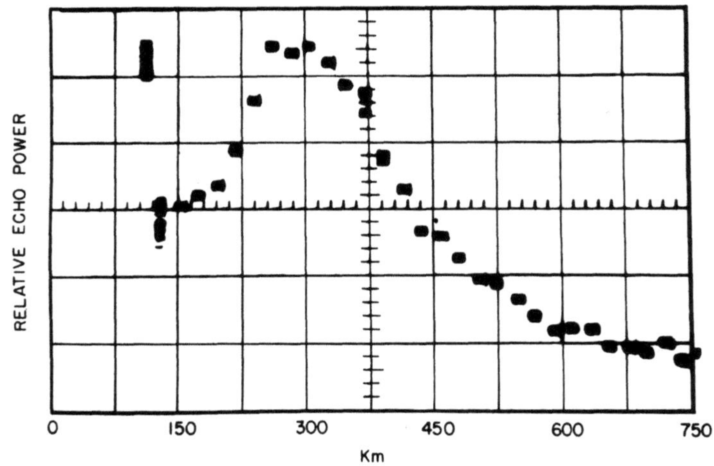

Fig. 2.7. Output of integrator used in Bowles' ISR experiment. (Bowles 1958)

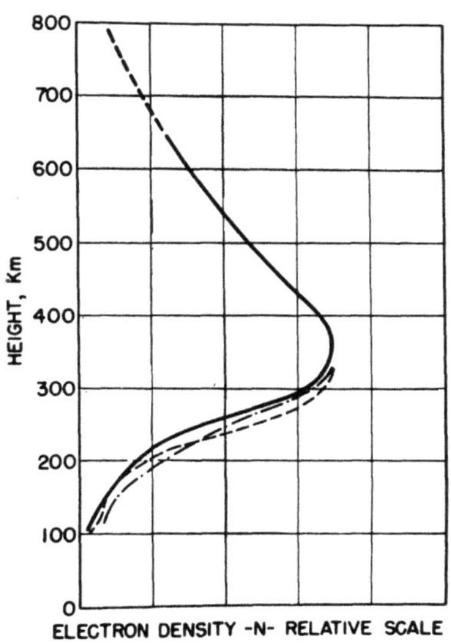

Fig. 2.8. One of the first electron density profiles of the ionosphere obtained using the ISR technique. (Bowles 1958)

Thomson scatter for a variety of reasons; however, the term Thomson scatter is normally reserved for situations of scattering from free electrons without influence from ions.

In practice, this incoherent scatter is detected from the ionosphere principally when $\lambda \gg D$, although experiments at Arecibo have detected incoherent scatter

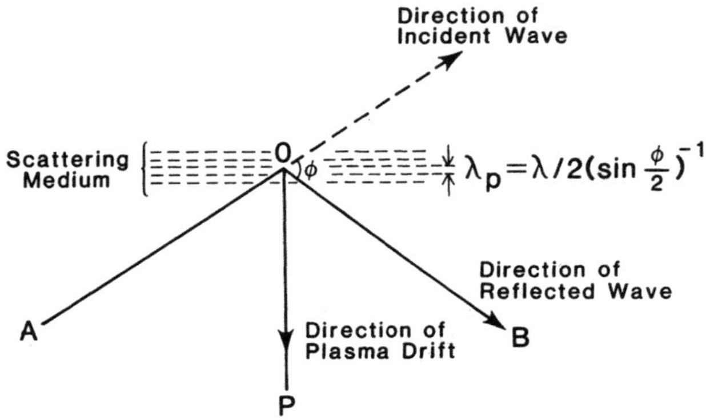

Fig. 2.9. Geometry for determining the spatial scale of irregularities producing "incoherent scatter"

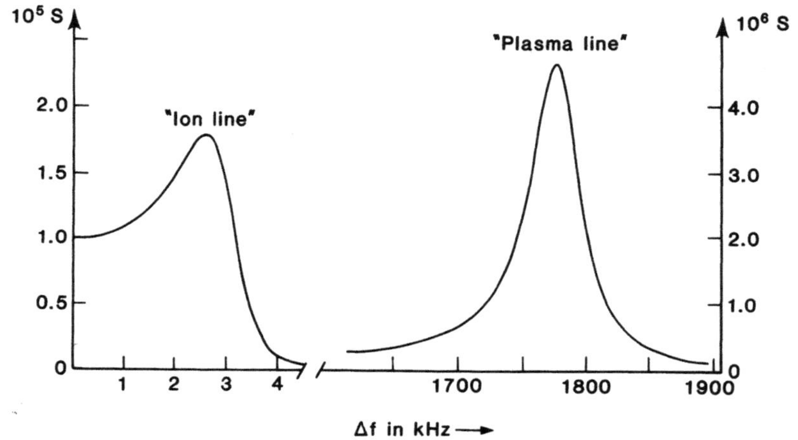

Fig. 2.10. Idealized spectrum of incoherent scatter echo

when $\lambda \simeq D$ (Hagen and Behnke 1976). The spatial scale of the irregularities λ_p is given by the Bragg formula,

$$\lambda_p = \frac{\lambda}{2\sin(\Psi/2)}, \tag{2.81}$$

with the geometry as shown in Fig. 2.9.

The *spectrum* of an incoherent scatter echo is very rich in information about the magnetoplasma which it is probing. A few of these plasma properties are easy to obtain; most require only straightforward data analysis techniques; and some require complex processing using specific plasma models.

Figure 2.10 is an idealized sketch of the spectrum of an incoherent scatter echo from the ionosphere, showing the "ion line" on the left and the "plasma line" on the right. The *ion line* is centered on the operating frequency f_1 and is the energy backscattered by irregularities of scale characterized by the Debye length which are in random thermal motion, while the *plasma line* is centered on the plasma frequency f_N [Eq. (2.13)] and is due to the collisions of electrons with other electrons. The plasma line is a weak line, except when it is enhanced by "hot" photoelectrons; and when the line is enhanced, both local electron density and the characteristics of the photoelectron flux may be measured.

A first-order estimate of the electron density, N, may be obtained from the maximum power, P, of the ion line spectrum; the *peak-to-valley ratio*, T_r, gives the ratio of electron-to-ion temperature (T_e/T_i); the *frequency shift* (Δf) of the center of the ion line from the operating frequency f_1 is a measure of the line-of-sight plasma drift (+ or −) = toward or away; and the shape of the spectrum is related to the atmospheric composition T_e and T_i, in a rather complex manner at any given altitude.

As an example of how one ionospheric parameter, electron density (N), is obtained from incoherent scatter radar data we can start by expressing the radar equation in terms of incoherent scatter measurements (Baron et al. 1970) as

$$N_e = \underbrace{K(1/P_t\tau)}_{\substack{\text{numerical}\\ \text{constants}}} \quad \underbrace{(r^2P_r)}_{\substack{\text{fixed}\\ \text{values}\\ \text{for each}\\ \text{data run}}} \quad \underbrace{\{(1+\alpha^2+T_r)}_{\substack{\text{variables}\\ \text{which are}\\ \text{directly}\\ \text{measured}}} \quad \underbrace{(1+\alpha^2)/2\}}_{\substack{\text{variables}\\ \text{measured}\\ \text{from frequency}\\ \text{spectrum}}} \tag{2.82}$$

where:

α^2 = Debye length term = 13.92 T_e/T_i
N_e = electron density at range r (el/m^3)
K = system constant (includes backscatter cross-section of electron, antenna gain, transmitter wavelength, etc.)
P_t = transmitter pulse power (watts)
τ = radar pulse length (s)
t = time from start of transmitted pulse (s)
c = speed of light (m/s)
P_r = power received from "target" at range r (watts)
λ = radar wavelength (meters)
$T_r = T_e/T_i$
D = Debye length (m).

Table 2.1. Ionospheric and thermospheric parameters obtained for various height regions

Zone 1	E-region (85 to 120 km) $N_e, T_i, T_e, V_i, \nu_{in}$
Zone 2	E- to F-region (115 to 250 km) N_e, T_i, T_e, V_i, ion composition (O^+, NO^+, O_2^+)
Zone 3	F-region (200 to 700 km) N_e, T_i, T_e, V_i
Zone 4	Upper F-region (500 to 1200 km) N_e, T_i, T_e, V_i, ion composition (O^+, H_e^+, H^+)

In this example T_e/T_e is derived from analysis of the spectrum shape. The system constant, K, is found by calibration with an ionosonde. Typically, the following ionospheric and thermospheric parameters are obtained for the various height regions in Table 2.1.

To conclude this section on incoherent scatter, the Faraday rotation effect will be briefly described. A much more complete description of this effect as it pertains to the determination of total electron content and a discussion of how Faraday rotation affects ISR measurements are contained in Evans (1969).

Most simply stated, a plane EM wave of frequency $f \gg f_N$ propagating through the ionosphere will split into two circularly polarized components with opposite senses of rotation and different phase velocities. At any time, t, these two components may be combined to produce a wave which will have been rotated with respect to its original plane of rotation by the angle Ω),
where:

$$\Omega = \left(\frac{1}{4\pi}\right)^3 e\mu_o/(m_e f^2 c\varepsilon_o) \int_o^R N_r H_r \cos\theta_r \, dr(\text{rotations}) \tag{2.83}$$

and:

H_r = magnetic field intensity (ampere-turns/meter)
θ_r = angle between the magnetic field and the ray at any point along the ray path, r
N = electron density at any point along the ray path

[all other quantities in Eq. (2.85) are defined in Chapter 1].

Millman et al. (1961) were the first to point out the effects of Faraday rotation on incoherent scatter data, consequently most ISR's avoid Faraday rotation effects by employing circular polarized antennas.

In Chapters 1 and 2 we have outlined the underlying theories upon which most of the radio techniques described in this book are based. The remaining chapters will concentrate on description of the principles of operation, some of the historical developments, the capabilities, and the limitations of each of the "generic" techniques.

Part II Earth-Based Radio Techniques

In Part II of this book we will classify "earth-based" radio techniques as those radar or sounder systems whose transmitters and receivers are located on the surface of the earth, or in the case of passive systems, earth-surface-located receivers monitoring either natural or man-made radio emissions from the earth's surface or from space. Rocket, satellite, and balloon-borne radio instrumentation will be covered in Part III.

Chapter 3 Vertical Sounders – the Ionosonde

3.1 History of Development and Basic Principles of Operation

One of the first measurements of ionospheric layer height was reported by Appleton and Barnett in 1926, who used a continuous wave transmitter and a receiver located from 65 to 130 km distant from the transmitter. The ground wave and the skywave arrived at the receiver with a relative phase difference. When the frequency of the transmitter and the receiver were simultaneously and rapidly changed by a small amount, the investigators obtained maxima and minima as the ground and skywave signals reinforced and canceled each other, respectively. Ionospheric layer heights were deduced from these data (see Kelso 1964, for a complete description of the technique and of the results in terms of both phase path and group path changes). It is interesting that G.W. Pierce in his 1910 textbook suggested that waves reflected from the ionosphere might interfere with the direct groundwave. As a matter of history, perhaps the *first* observations of this kind, followed by a correct deduction as to the causative mechanism, were made in 1912 by de Forest and Fuller, using continuous wave arc transmitters and receivers of the Federal Telegraph Company in California (Villard 1976). The first *vertical* sounding technique employed to probe the terrestrial ionosphere was the "pulse-echo" equipment used by Breit and Tuve to measure ionospheric layer height in 1925 (reported in 1926). This was the progenitor of the *ionosonde* – the most ubiquitous of all ionospheric measuring devices.

In its simplest form, an ionosonde works basically as follows: an oscilloscope sweep is initiated, and a short time later the transmitter sends a short pulse of rf energy at a given radio frequency upwards toward the ionosphere. After a time delay of a few milliseconds, the pulse reflected from an ionospheric layer returns to the receiver and is displayed on the same oscilloscope sweep. If the ground conductivity is high and ionospheric absorption is low, several echoes (or "multiples") may be observed. The "virtual height" (Sect. 2.1.2) of the layer may be directly deduced from the time delay between the transmitted and received pulses, assuming that the radio wave travels at the speed of light. If the transmitter and receiver frequency are then slowly varied together over a range of typically 1–20 MHz, an *ionogram* – a plot of virtual height versus frequency – is obtained.

Shortly after Breit and Tuve demonstrated the pulse-echo system, scientists in several different countries began developing various versions of the ionosonde. The need for sounding at more than one frequency was immediately apparent. First attempts consisted of multiple fixed-frequency sounders with automatic tracking between the sounders. It was soon appreciated that such frequency sampling was inadequate, and by the early 1930's a major effort to develop a sweep-

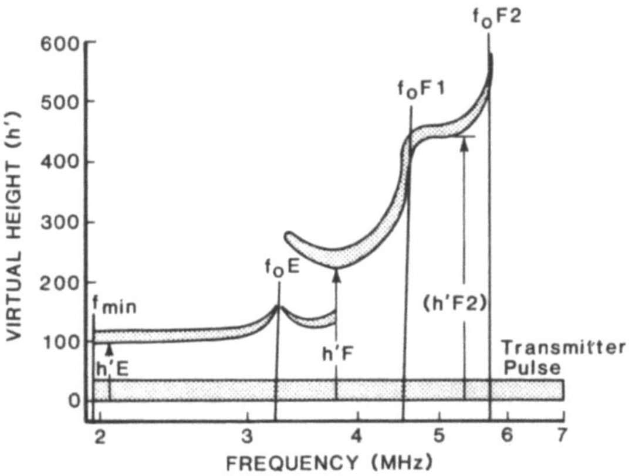

Fig. 3.1. Idealized ionogram. (After URSI Handbook 1972)

frequency sounder was in full swing. A major conceptual hangup in the early days was the idea that such sounders would sweep through the entire spectrum, thus interfering with every other user! Fortunately, the ionosphere usually changes slowly enough so that sweeping at 15-min intervals suffices, and experience soon showed that the interference was not serious in practice, except perhaps very near the transmitters.

An idealized ionogram is shown in Fig. 3.1, with some of the most-used parameters labeled. The virtual height-frequency regime covered by ionosondes is illustrated in Fig. 2.2 of Section 2.3.

The "scaling" or reduction of ionograms is both an art and a science and it is all but impossible to describe adequately in a book covering all the ionospheric techniques "how to" scale an ionogram. However, a description of the procedure and nomenclature for the analysis ("scaling") of ionograms is given in the *URSI Handbook of Ionogram Interpretation and Reduction* (1972), and in the URSI *Revision of Chapters 1–4* (1978). Another useful guide is the *High-Latitude Supplement to the URSI Handbook on Ionogram Interpretation and Reduction* (Piggott 1975). The latter two reports are available in *Microfiche* from *World Data Center A for Solar-Terrestrial Physics*, 325 Broadway, Boulder, C0 80303.

The historical development of ionospheric sounders by British, German, and American research organizations has been documented by Gladden (1959), Dieminger (1974, 1975), Villard (1976), Gardiner et al. (1982), Snyder and Bragaw (1986), respectively. Figures 3.2–3.5 illustrate some of the "first generation" ionospheric sounders.

3.2 Standard Ionosondes

Standardized ionosondes were developed and placed into service shortly after the end of the second world war. The term "standard" is used here to describe "analog" sweep-frequency pulsed systems which usually house the transmitter,

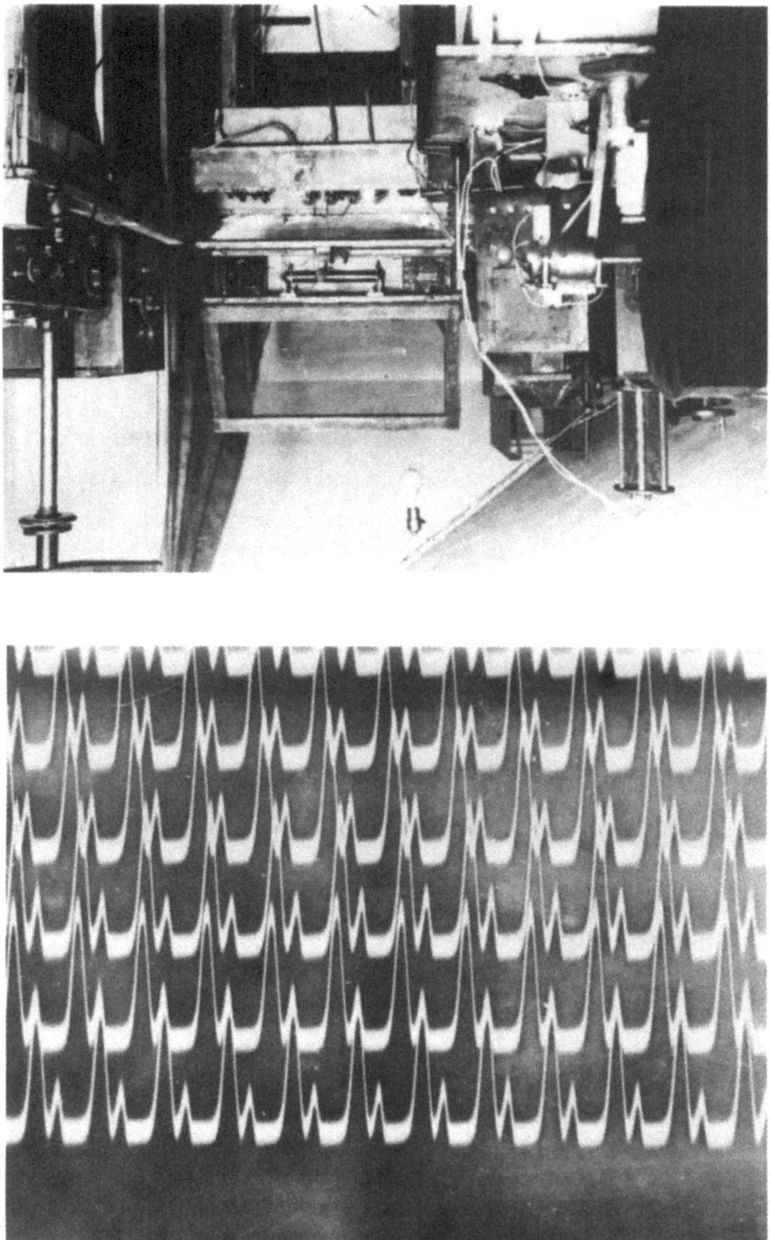

Fig. 3.2. "First-generation" ionospheric sounders

Fig. 3.3. "Frequency-change" transmitter operated at the National Physical Laboratories, Teddington, for routine soundings in 1931 under the auspices of the Radio Research Station (RRS) at Ditton Park, Slough, UK. (Courtesy of Dr. GE Gardiner)

receiver, and timing units in the same cabinet, and which can operate essentially unattended (except for film changes and maintenance). Anyone who operates one of these sounders, however, might debate the "unattended operation" statement! It is the experience of many operators that ionosondes need a dedicated technician or engineer who really understands the system, to lavish tender, loving care (TLC) on the equipment at unpredictable intervals (usually weekends and holidays!). These sounders incorporated the best technology available at the time, including many of the wartime radar advances. Standardization made possible the economies of mass production and greatly aided in processing the mountains of data produced. The essential difficulty was that early sounders were electromechanical devices. It was mainly the mechanical features that gave everyone problems. The modern all-electronic sounders are much more reliable.

Fig. 3.4. Pulse ionosonde receiver designed by LH Bainbridge-Bell in 1931 for RRS studies. (Courtesy of Dr. GW Gardiner)

Figure 3.6 portrays "one of that breed" – ionosonde engineer, Mr. John Pitts, lavishing TLC upon the C-3 sounder at the U.S. Department of Commerce field station at Boulder, Colorado, in 1974. Figure 3.7 shows another typical installation of the predecessor of the C-4 ionosonde – the NBS Model C-3 with its separate power supply.

These "standard" ionosondes usually consist of an accurate, stable oscillator capable of being swept in frequency, intermediate frequency multipliers and amplifiers, a wideband rf power amplifier capable of pulse outputs from 1 to 10 kW, a wideband pulse receiver, transmit/receive switch, etc., arranged as shown in the block diagram in Fig. 3.9.

Since many of these "standard" ionosondes are still in operation, Table 3.1 describes characteristics of several of the more commonly found types.

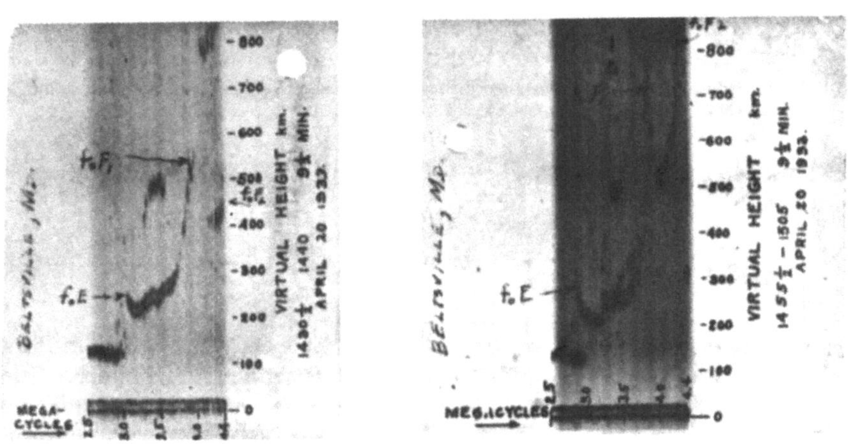

Fig. 3.5. *Above* First US National Bureau of Standards (NBS) multi-frequency ionosonde, designed by T.R. Gilliland in 1929. *Below* First NBS sweep-frequency ionogram obtained in 1923 using the sounder shown above

Fig. 3.7. NBS Model C3 ionosonde installation

Fig. 3.6. Mr. John Pitts, ionosonde engineer, attending a C3 sounder at the Boulder, Colorado field station in 1974

Fig. 3.8. Another post-WWII ionosonde, developed by RRS. Prototype of the Union Radio ionosonde, first commercially produced in 1948

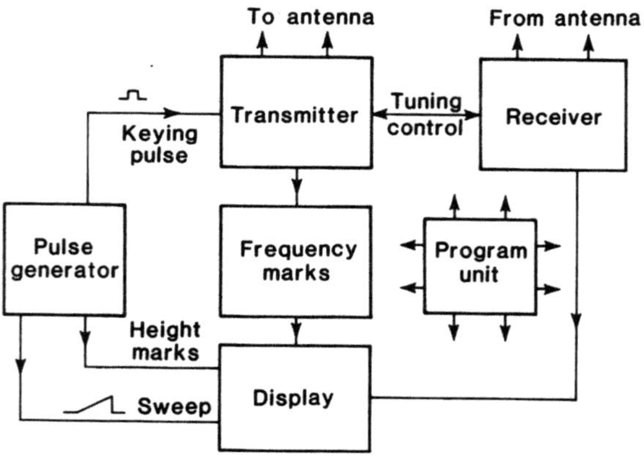

Fig. 3.9. Block diagram of a typical "standard" ionosonde

Table 3.1. Characteristics of some ionosondes

	NBS models C-2 and C-3 (USA)	Union Radio Mark II (UK)	LG-17 (Canada)
Peak power	10 kW	≈ 1 kW	5 to 10 kW
Frequency sweep	1-25 MHz logarithmic	0.55–17 MHz	0.6–20 MHz
Sweep duration	$7\frac{1}{2}$, 15, 30 s	5 min in 5 bands	4–320 s/ sweep
Pulse duration	50 to 100 µs		50 and 100 µs
Pulse-repetition rate (PRF)	10 to 90 per s	50 and 100 50 Hz	0–60 per s
Frequency marks	Every 1.0 MHz		100, 500, and 1000 kHz
Sounding schedule	5, 15, 30 or 60 min or continuous		
Height marks	50 and 100	50 km	50–500 km
Height range	0–500, 0–1000 or 0–4000 km	Continuously variable from 100 to 2000 km	500–4000 km
Tuning	Common xmtr rcvr oscillator		Common oscillator
Display	B-type 35 mm film	B-type 70 mm paper	B-type 35 mm film
Antennas	Orthogonal vertical deltas for rcvr and xmtr	Xmtr-vertical term. rhombic Rcvr-rhombic	Terminated vertical delta
Supply voltage	90 to 260 v.a.c. single phase		115 V
Supply frequency	50 to 60 Hz		60 Hz
Power required	30 amp at 115 V		2.8 kW

	Model SPIM (France)	Type 2 (Australia)	Type 6 (Japan)
Power output	2–10 kW	≈ 300 W	10 kW
Tuning	Ganged capacitors	Ganged capacitors	
Frequency sweep	1 to 20 MHz	1 to 16 MHz	1 to 20 MHz
Sweep duration	12 s	2 min	10, 20, 30, 60 s
Pulse duration	100 µs	80 µs	40 and 120 µs
PRF	66 Hz	50 Hz	50 or 60 Hz
Height range	1000 km		50–1000 km
Height marks	150 km	50 km	50 km
Display	Panoramic	A-type 35 mm	35 mm camera
Antenna			Deltas for both xmtr and rcvr

PANORAMIC (JAPAN)

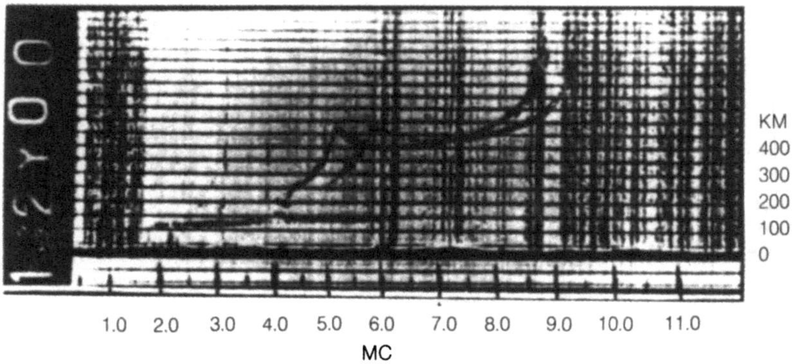

IONOSONDE: TYPE OR DESIGNATION: PANORAMIC
FREQUENCY RANGE: 1.0 MC TO 25.0 MC
HEIGHT RANGE: 1000 KM
POWER: 10 KW
SWEEP TIME: 20 SEC
LINEAR FREQUENCY SCALE

STATIONS: AKITA, JAPAN
DE BILT, NETHERLANDS
MACAO, CHINA SEA
PARAMARIBO, SURINAM
SYOWA BASE, ANTARCTICA
TOKYO (KOKOBUNJI), JAPAN
WAKKANI, JAPAN
*YAMAGAWA, JAPAN

SP-3 (U.S.S.R.)

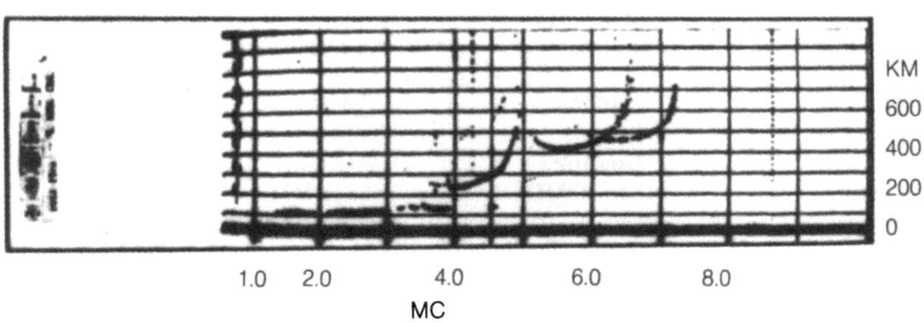

IONOSONDE: TYPE OR DESIGNATION: SP-3
FREQUENCY RANGE: 1.0 MC TO 20.0 MC
HEIGHT RANGE: 250, 750, 1500 KM
POWER: 20 KW
SWEEP TIME: 30 SEC
LINEAR FREQUENCY SCALE

STATIONS: AŞHKHABAD, U.S.S.R.
LENINGRAD, U.S.S.R.
*MOSCOW, U.S.S.R.

Fig. 3.10. Typical ionograms obtained with "standard" sounders. (Wright and Knecht 1957)

SP-35 (FRANCE)

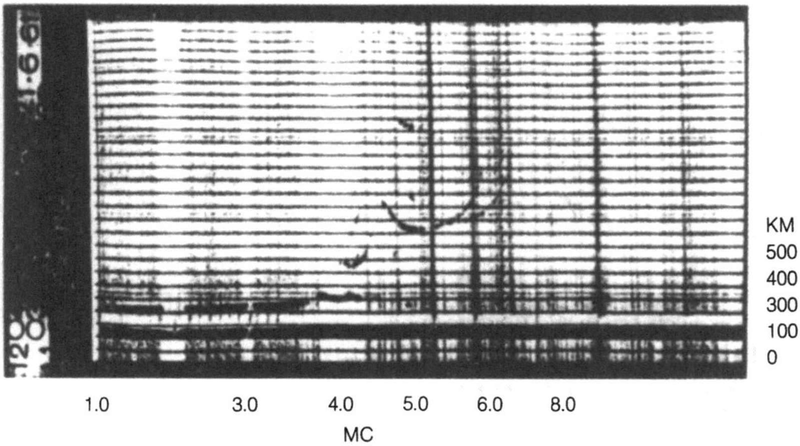

IONOSONDE: TYPE OR DESIGNATION: SP-35
FREQUENCY RANGE: 1.6 MC TO 17 MC
HEIGHT RANGE: 1000–1200 KM (ABOUT)
POWER: 2.6 KW
SWEEP TIME: 60 SEC
LOGARITHMIC FREQUENCY SCALE

STATIONS: *POITIERS, FRANCE
RABAT, MOROCO
TORTOSA, SPAIN

Fig. 3.10 (*Continued*)

Many examples of ionograms obtained with the standard ionosondes have been collected in *Atlases* published during the period $\approx$ 1957–1982 (two examples are those of Wright and Knecht (1957) and Shapley (1970). Figure 3.10 shows ionograms from Japan, the USSR, and France. A catalog of all ionospheric vertical sounding data held by World Data Center-A in Boulder, Colorado, has been published by Conkright and Brophy (1982), and should be consulted when specific ionograms are required.

3.3 Advanced Ionospheric Sounders

The sounders described in the preceding sections are also sometimes called "synoptic" ionosondes, analogous to the worldwide network of basic meteorological instruments which are capable of describing the global morphology of the lower atmosphere. Similarly, the network of ionosondes shown in Fig. 3.11 has been used to characterize the global morphology of the bottomside ionosphere (below the F2 peak) primarily as an aid in HF propagation predictions. It should also be emphasized, however, that the "synoptic" ionosondes have been utilized to provide a considerable amount of information on the

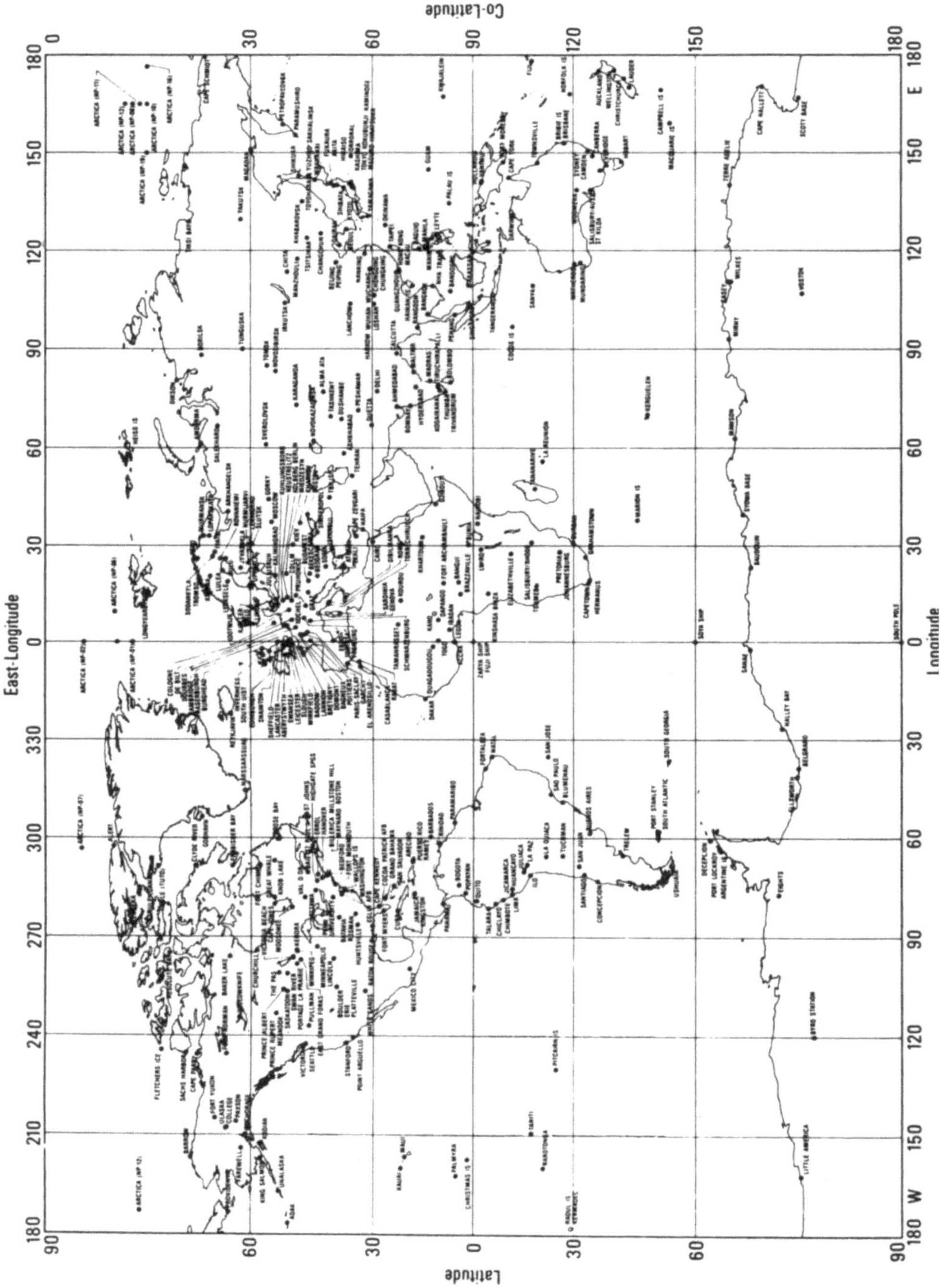
Co-Latitude
East-Longitude
Latitude
Longitude
E
W

physics of the ionosphere – within the limitations of the technique. Most of the sounders shown in Fig. 3.11 are the so-called standard ionosondes.

The basic standard ionosonde utilizes the *amplitude* of the return pulse to estimate delay time and makes no attempt to quantitatively measure the other characteristics of the returned rf pulse, i.e., the exact frequency, phase, polarization, and direction of arrival of the echo. As recently pointed out by Grubb (1986, pers. commun.), the concentration on range (virtual height) measurements by ionosondes was largely a US aberration. Early work at Slough by Watson-Watt et al. (1933) reported ionosonde operation in the "K-mode" (fixed-frequency versus time display), determination of angle-of-arrival of echoes, and echo polarization measurements.

In the years following World War II, investigations made with the aid of soundings became more sophisticated. One example was the need to study small traveling disturbances which led to a requirement for more sophisticated sounders. Beginning in the mid-1960's, various research groups started the development of a family of ionospheric sounders capable of measuring some or all of the characteristics of the returned pulse. These sounders were described by several names, i.e., Dynasondes, Digisondes, HF Radars, Kinesondes, etc., but we will use the generic term – Advanced Ionospheric Sounders (AIS) – coined by W. R. Piggott to describe them. This generation of sounders are much more than just ionosondes – by using suitable antenna systems they are capable of measuring Doppler shift of irregularities, location of echoes, polarization of echoes, etc., plus rejection of noise and interference. This in turn permitted less power to be used, so the sounders create less interference in an increasingly crowded spectrum. These AIS systems were made possible, of course, by technological advances in solid state devices, computers, and digital electronics. Some of the first of the new generation of AIS systems – the Dynasonde and Kinesonde – developed at the US Department of Commerce Labs in Boulder, Colorado, were described by Wright (1969, 1975, 1977), Wright et al. (1971, 1980), and Wright and Pitteway (1979).

The "production model" of the AIS system developed at the Boulder Labs was called the NOAA SEL HF Radar System, and was described by Grubb (1979). Software documentation for the NOAA AIS system has been published by Walden et al. (1984).

Six of the NOAA AIS systems were built and have been deployed since 1980 at various locations around the world, and the sounder has undergone a continual modification program to refine this measurement technique (see Dudeney and Jarvis 1986; Jarvis and Dudeney 1986; also Argo and Hindman 1987). Figure 3.12 shows the NOAA HF Radar AIS system and Figs. 3.13–3.15 are block diagrams of this system and illustrate one particular research organization's design philosophy for an AIS system. The NOAA AIS design philosophy was to make their sounder "software-controlled" to the maximum extent possible, so that it could be programmed to adapt it to different situations. This was a logical extension of Cohen's "flexible sounder" concept and Wright's Dynasonde, to

◄

Fig. 3.11. Map of all ionosphere vertical sounders known to have existed as of 1982. (Courtesy Ray Conkright 1982)

Fig. 3.12. NOAA Advanced Ionospheric Sounder (AIS)

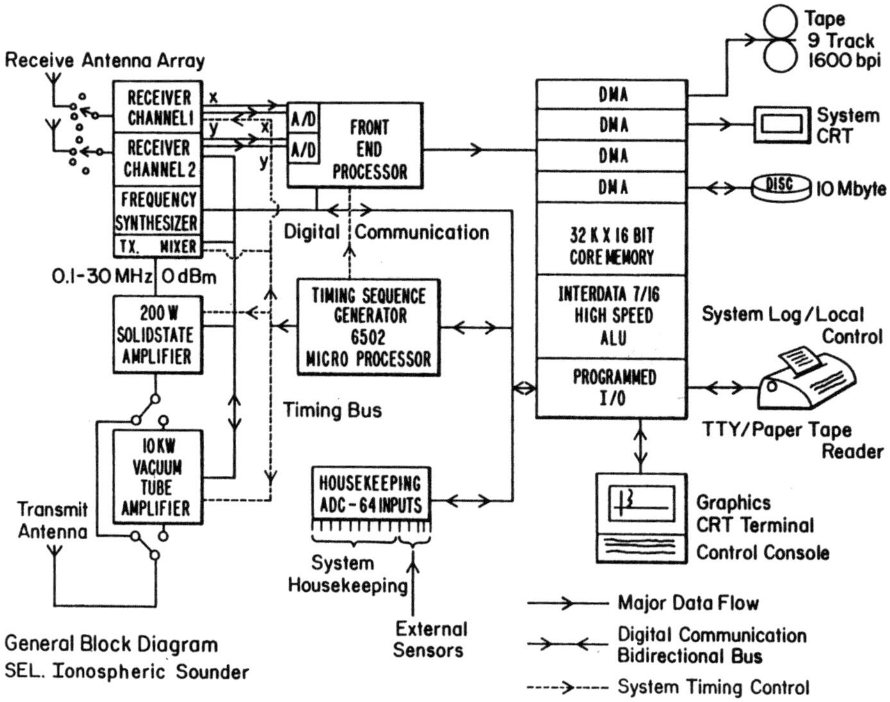

Fig. 3.13. Block diagram of the NOAA AIS/HF radar, Dynasonde system

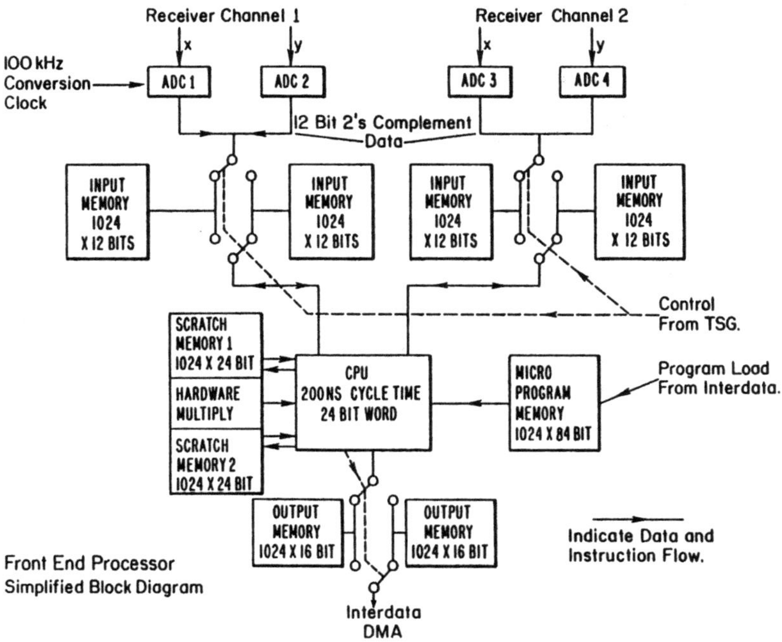

Fig. 3.14. Block diagram of NOAA AIS front-end processor

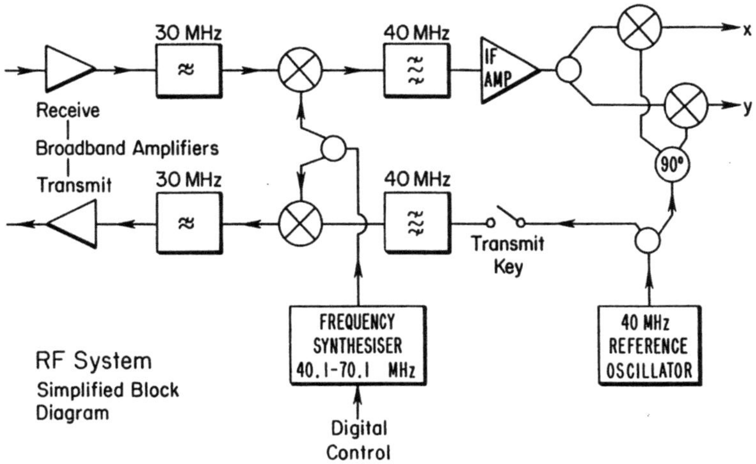

Fig. 3.15. Block diagram of NOAA AIS RF system

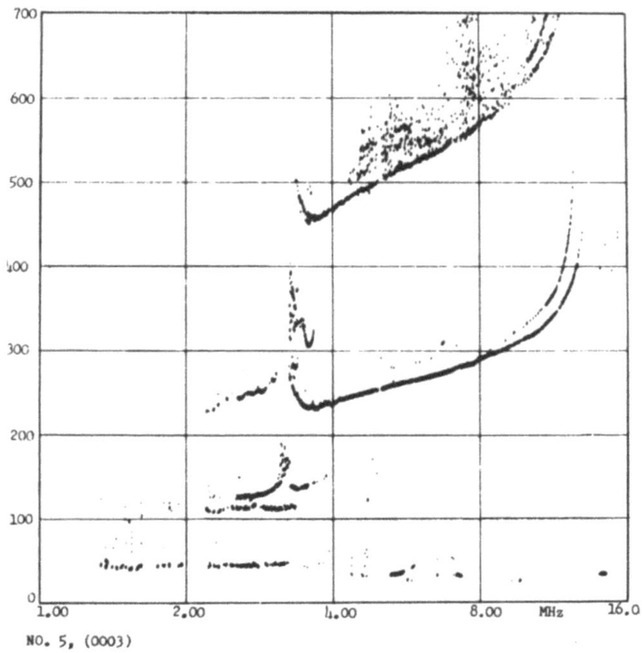

Fig. 3.16. Typical ionogram obtained with a NOAA AIS

extend the computer control from that of frequency and signal recognition by timing coincidence to full digital processing of the received echoes. Most users of the AIS have restricted themselves to the software emulation of the Wright Dynasonde mode of operation.

A parallel development of another AIS employing a different design philosophy also took place during the late 1960's and 1970's at the University of Lowell in Massachusetts, and has been reported by Bibl et al. (1970) and Phillips (1974). It culminated in the production model called the Digisonde 128 PS described by Bibl and Reinisch (1978). The latest model in this series of sounders (incorporating 1980's technology) is the Digisonde 256 (Reinisch 1986) shown in Fig. 3.17.

Figure 3.18 is a block diagram of the Digisonde 128 PS system and should be compared with Fig. 3.13 to illustrate how the two different groups approach the design of an AIS. A typical ionogram obtained with a Digisonde 128 PS is shown in Fig. 3.19. In retrospect, an analogy was offered which attempted to contrast the signal processing philosophy of the *early* Digisondes and Dynasondes. The potential problem with the Digisonde was that in trying to record the receiver output at all ranges without any echo detection, the system seriously undersampled the data and mixed the wanted signals irretrievably with interference by the use of a nonlinear signal averaging technique. The problem with the Dynasonde was the opposite. By using a very tightly constrained test for an echo, before accepting it for recording, it could easily reject echoes with unusual

Fig. 3.17. Digisonde 128 PS, University of Lowell Center for Atmospheric Research (ULCAR)

characteristics such as rapidly changing amplitude or range, or a large Doppler. Thus, at least at high latitudes it could "throw the baby (signal) out with the bathwater" (interference), whereas the Digisonde could "drown the baby" in the bathwater. Since both the NOAA and the Lowell AIS systems have evolved further, and are currently deployed globally acquiring quantitative ionospheric data, it is quite apparent that these technical problems were solved and the "baby" (data) was neither "drowned" nor "thrown out"!

Typical of further advances in AIS technology is the PC-Radar developed by Argo and Hindman (1987), which modified the basic NOAA HF radar design by using a "super microcomputer" which permits full digital processing of echoes. Further efforts to miniaturize and improve AIS systems using VLSI, powerful microcomputers and advanced digital signal processing (DSP) are underway at this time. Examples of research using AIS techniques have been published by

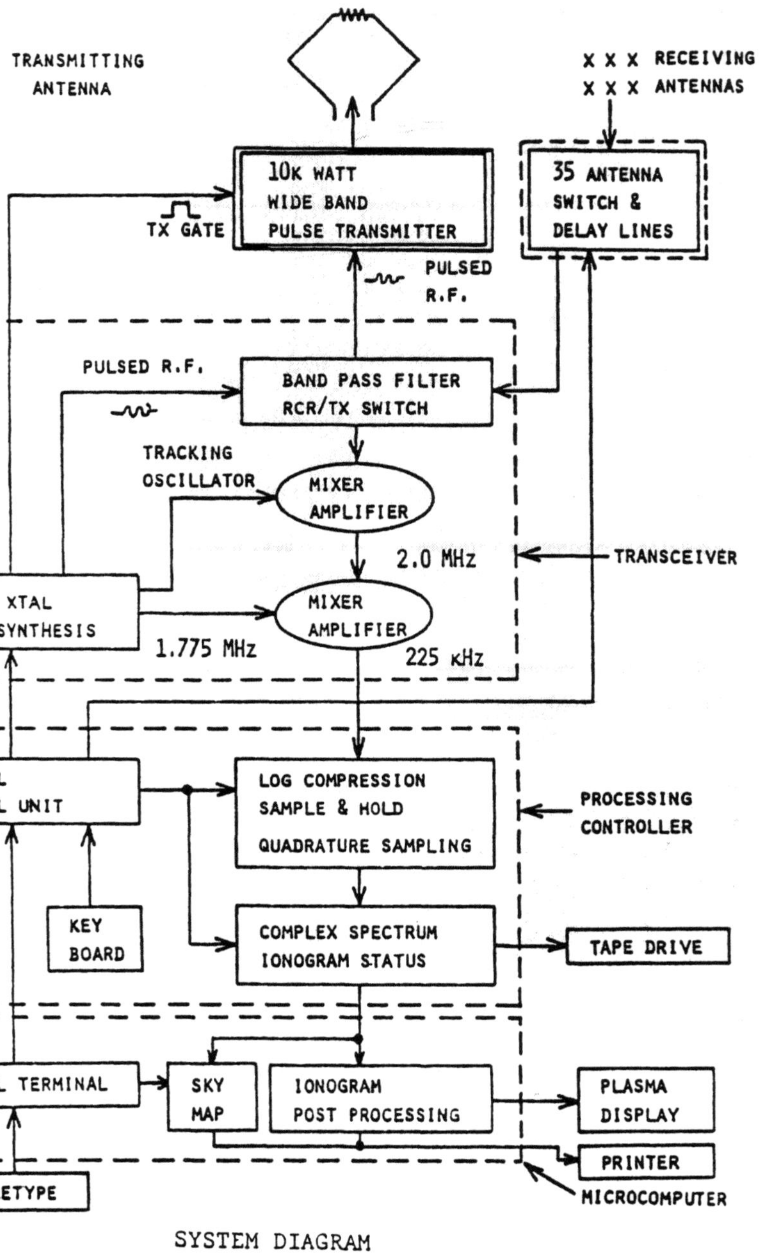

Fig. 3.18. Block diagram of Digisonde 128 PS

Fig. 3.19. Typical digital ionogram from D256 sounder and map of D256 network as of early 1990. (Courtesy of Bodo Reinisch)

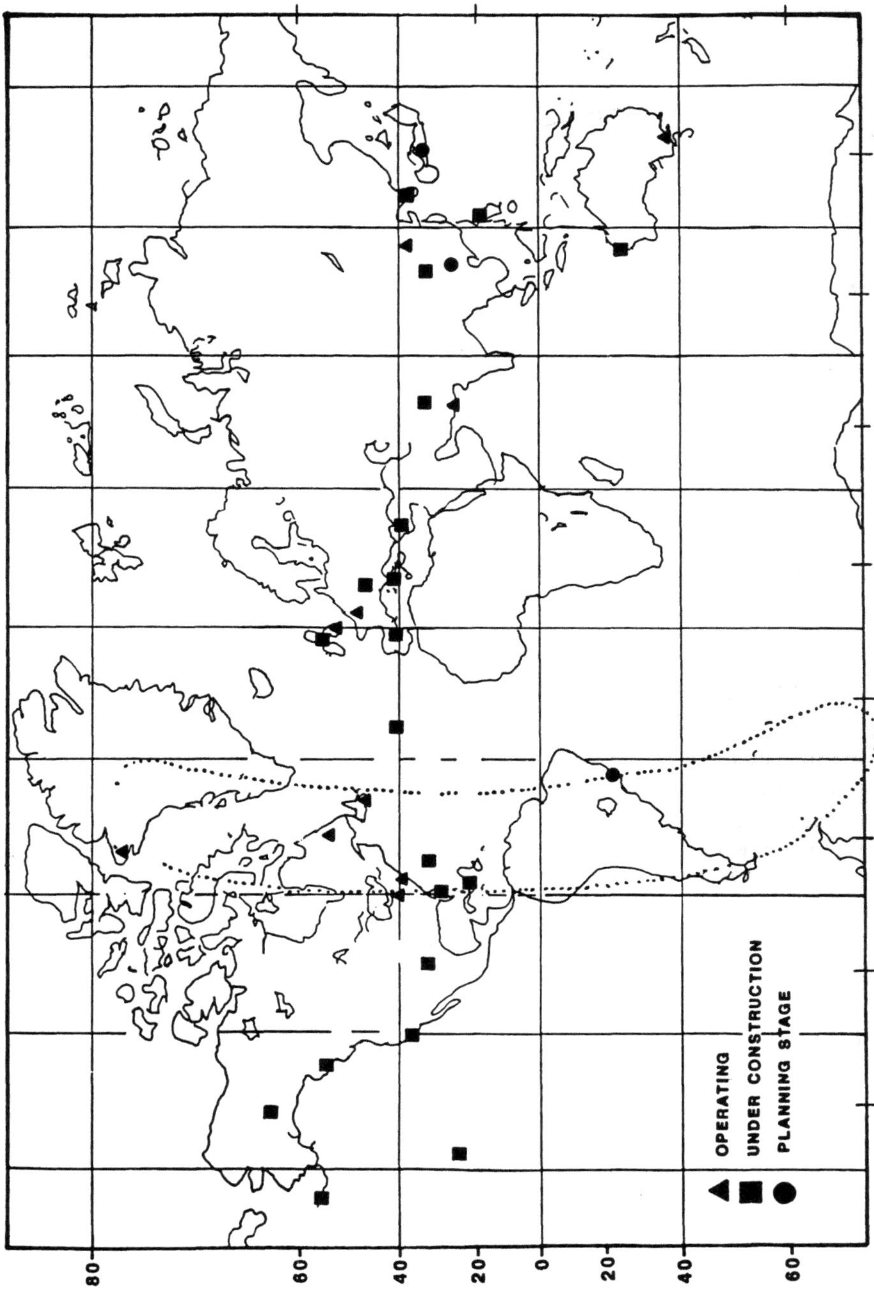
OPERATING
UNDER CONSTRUCTION
PLANNING STAGE
180
160
140
120
100
80
60
40
20
0
20
40
60
80
100
120
140
160
180
80
60
40
20
0
20
40
60

Wright and Hunsucker (1983), Argo and Hindman (1987), and Buchau et al. (1988). The latest developments in AIS systems are exemplified by the portable sounder described by Haines et al. (1989), the "PC Radar" (Argo and Hindman 1987), and the IPS-61 Digital ionosonde (Kelly 1990). Specifications for several of the more recent AIS systems are listed below:

Model: Digisonde 256

Organization: University of Lowell Center for Atmospheric Research (ULCAR) – Lowell, MA 01854

Selected specifications:

Frequency sweep: log and linear 0.5 to 30 MHz
Sweep duration: 20 s to several minutes
Frequency steps: log sweep, 20, 40, or 80 steps per octave; also linear and pseudo-random linear
Pulse repetition frequency: 50, 100, 200 Hz
Pulse width: 66 μs or 133 μs Hz
HF phase coding: interpulse and intrapulse pseudo-random
Peak pulse power: 10 kW
Receiver:
Bandwidth: 20 kHz
Signal to noise ratio: 6 dB for 2 μV input
Dynamic range: 64 dB + 42 dB switchable
Signal processing: sixteen data integration channels shared for Doppler, antenna configuration (incidence angle and polarization), and range doubling
Reference: URSI INAG Bulletin INAG 38, January 1983, Bibl et al. (1981), and Reinisch (1986).

Model 610M1 Digital Ionosonde

Organization: SRI-I, Radio Physics Lab, 333 Ravenswood, Menlo Park, CA 94025.

This sounder was designed for minimally attended, remotely controlled operation. Programming and data retrieval are via modem and voice-grade lines to a central facility that manages and processes data from up to four sounders.

Selected specifications:

The normal antenna is a crossed, terminated delta, one of which is used to transmit, and both of which are phased for circular polarization to assist in separation of ordinary and extraordinary echoes.

Frequency range: 0.5 to 30 MHz
Number of frequencies per sweep: 256

Peak pulse power: 2 to 7 kW
Pulse width: 50 μs
Pulse interval: 5.33 ms
Height resolution: 75 m
Maximum height: 800 km
Sounding intervals: 2, 5, 10, 15, 30, and 60 min (selectable)
Data interface: RS-232
Power requirements: 115/230 V, 60 Hz, 200 W for 5-min sounding intervals
Size: 30 in high, 19.5 in wide, 24 in deep
Weight: 120 lbs
Reference: URSI INAG Bulletin INAG 38, January 1983.

Model IPS-42

Organization: KEL Aerospace Pty Ltd, 12 Brennan Close, Asquith, New South Wales 2078, Australia.

Selected specifications:

Frequency range: 1.0–22.6 MHz
Pulse power: 5 kW

Fig. 3.20. Digisonde 256

Pulse width: 41.67 μs
Pulse intervals: 5.33 ms (3 per channel)
Number of channels: 576
Height range: 50–800 km
Height marker intervals: 100 km
Sweep: Logarithmic
Sweep time: 12 s
Film: 16 mm
Power supply: from power mains or 24V-DC, 4 amps (standby), 6 amps (transmitting)
Reference: INAG 38 and KEL brochures.

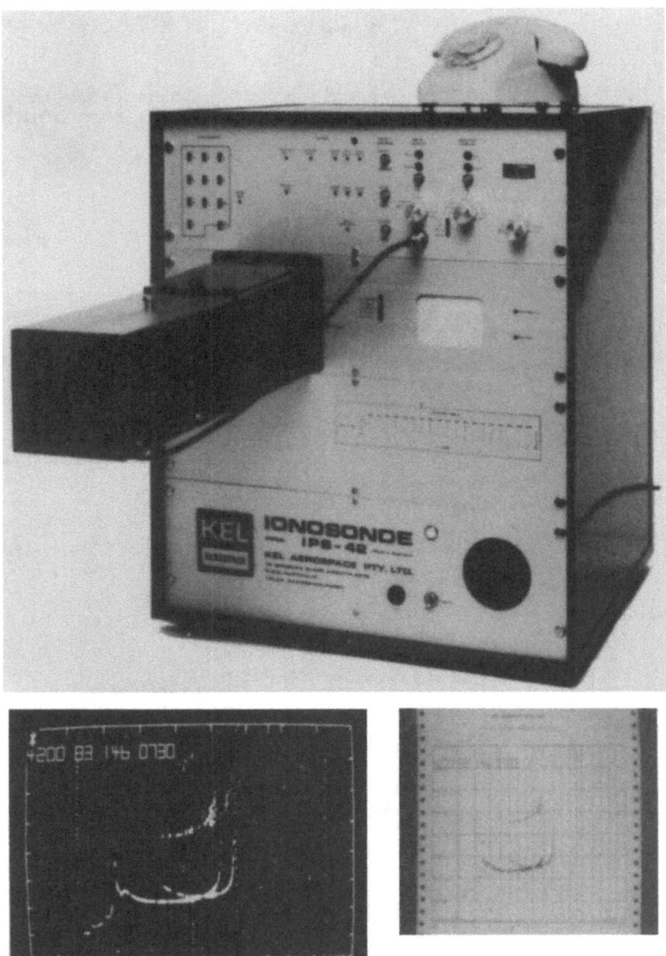

HIGH RESOLUTION
PRINTED IONOGRAM
FOR SCALING

REAL-TIME
VIDEO DISPLAY

Fig. 3.21. *Above* KEL IPS-42 ionosonde. *Below* KEL IPS-42 ionogram using DBD-43 receiver

Two of the presently available AIS systems are illustrated in Figs. 3.20 and 3.21 and should be compared with the "standard" ionosondes shown in Figs. 3.7 and 3.8 and described in Section 3.2.

Several other research groups have also developed their own versions of AIS systems in the last two decades, as described by Kasuya and Koseki (1969), Clarke (1971), Morgan and Pratt (1975), Poole (1985), Poole and Evans (1985), and Bowman et al. (1986).

Among the more novel approaches in the design of AIS systems were the Chirpsounders, FM-CW sounders developed by Barry Research Corp., and the "single-pulse" sounder developed at SRI-I (Watt et al. 1968). Photographs of the Chirpsounder system are shown in Fig. 3.22 as an illustration of the small physical size of some of the advanced sounder systems. The latter system employed a spark-switched impulse generator requiring 80 kV with a center frequency of 5.8 MHz which produced a 10–15 MW, 1 μs pulse. Figure 3.23 shows a block diagram of this system and Fig. 3.24 shows a record of ionospheric echoes as a function of frequencies within the passband of the system. This

Fig. 3.22. *Above* a barry chirpsounder receiver; *below* chirpsounder transmitter

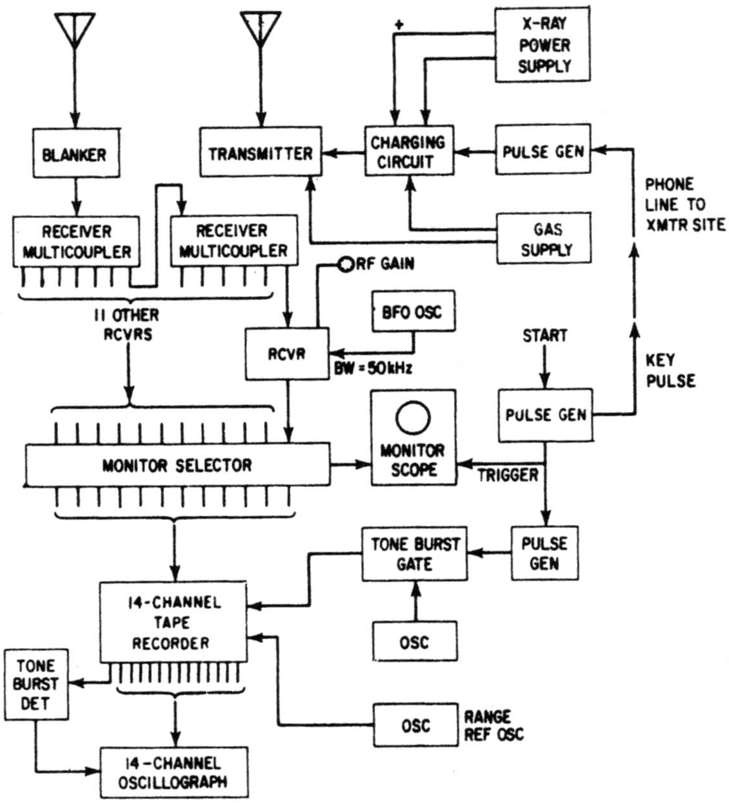

Fig. 3.23. Block diagram of the "single-pulse-sounding system"

sounding system permitted a resolution of ≈ 600 m in the E-region, but because of the RFI generated by this type of a transmitter when operated frequently, and the large physical size, the system was not practically feasible. It was, however, an innovative and rather "heroic" attempt to use an ancient technique (a *spark* transmitter!) to sound the ionosphere!

The vertical-incidence one-pulse sounder, developed and tested by SRI, was singly pulsed only once in 15 min, and did not result in interference complaints. However, similar units were used in some oblique tests with receivers in Florida, and were doubly pulsed at more frequent intervals. These *did* result in complaints.

3.4 Capabilities of Ionosondes

We can summarize the essential capabilities of ionosondes, generically speaking, as follows.

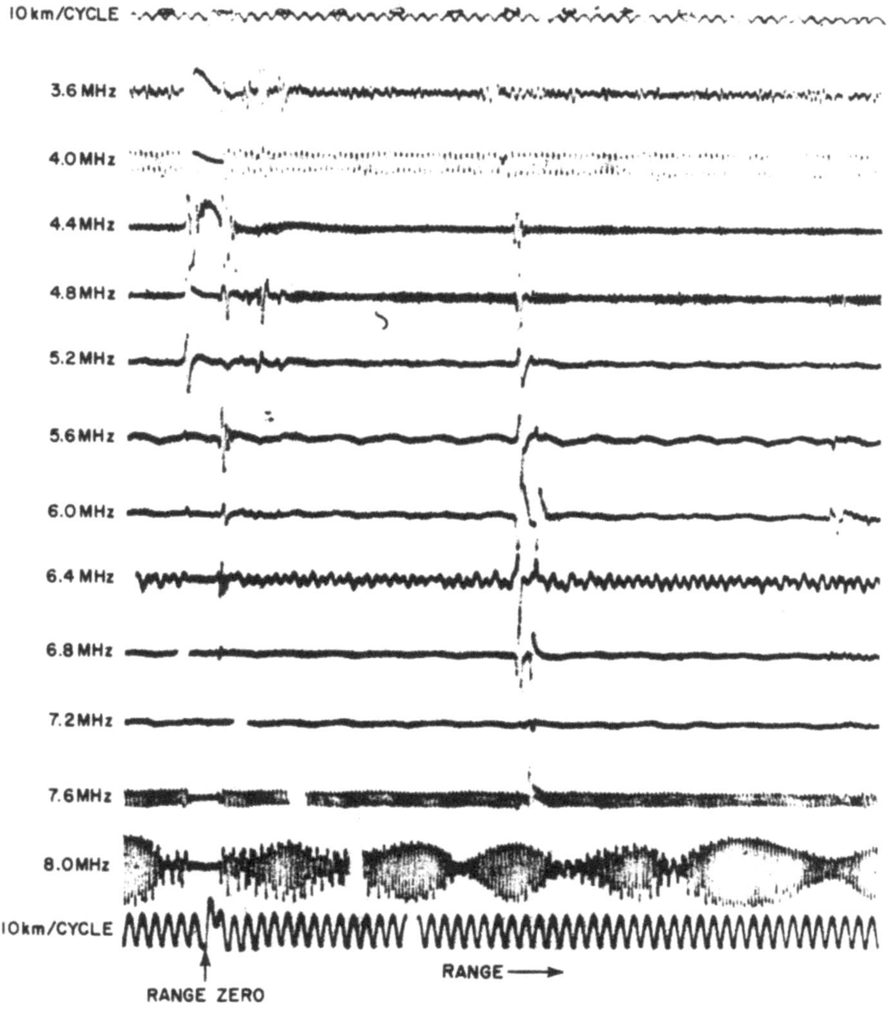

Fig. 3.24. Oscillographic ionogram of a single-pulse sounding at 1120 PST, December 5, 1968

"Standard" ionosondes (also called synoptic or analog ionosondes)

Directly measurable quantities:

- virtual height vs. plasma frequency (ionogram);
- critical frequency of layers;
- minimum frequency of layers;
- range and frequency spread of traces;
- magnetoionic splitting of traces.

Derivable quantities:

- electron density profile of bottomside ionosphere (for $h < hF\ max$)[2];
- estimates of the maximum usable frequency for various path lengths (for HF communications);
- an estimate (poor) of D-layer absorption (from f min);
- detection of TID's and an estimate of their horizontal speed;
- an approximate location of large ionospheric features, such as the edge of the main electron density trough and magnetospheric cusp signatures and an estimate of their motion.

Advanced ionospheric sounders

Directly measurable quantities:

- those listed under "synoptic" ionosondes plus;
- angles of arrival of echoes from ionospheric irregularities;
- line-of-sight Doppler shift of irregularities;
- amplitude of received signals;
- relative phases of received signals;
- polarization of received signals;
- spectral signatures of received signals.

Derivable quantities:

- electron density profiles;
- echo location;
- resolved horizontal velocities of irregularities.

In addition, AIS's can avoid sounding on frequencies already in use, thus cutting down interference to other services, and improving the quality of the records obtained.

3.5 Limitations

One limitation which applies to all techniques described in this chapter is that they can yield information on the ionosphere only up to the height of maximum ionization of the F2 layer (the "bottomside" of the ionosphere). Also, unless one extends the low-frequency end of the sweep and uses relatively high power, not much information can be obtained from the D-region. This is in contrast with the incoherent scatter technique, which, however, is much more expensive and definitely not as portable! Another limitation is that during intense E-region ionization episodes ("blanketing sporadic-E") it is not possible to obtain much information on the F-region. The "standard" ionosondes are also very susceptible to RFI from the broadcast band (550–1705 kHz). At certain locations and

[2] See Titheridge (1988) for a recent description of ionospheric real height analysis.

times, this MF broadcast interference can all but obliterate the ionospheric traces.

Cost can also be a limiting factor, with the "top-of-the-line" AIS systems (including antennas) available in 1991 requiring a capital expenditure of $\gtrsim$ \$ 300,000 US.

Additionally, some of the "derivable" quantities such as echo location can be obtained with confidence only if one pays very close attention to the design, construction, calibration, and phasing of the antenna system and to the assumptions one makes concerning the interaction of the radar pulse with ionospheric irregularities.

At auroral latitudes *all* ionosondes are subject to several rather severe limitations – namely, that during some of the most "interesting" times sporadic-E ionization or D-region absorption precludes the gathering of any ionospheric information on the layers above! These *interesting* times include magnetic storms and substorms and associated auroral and polar-cap absorption, intense auroral sporadic-E, and extreme spread-F conditions.

As an example, several decades of simultaneous operation of a C-3/C-4 type ionosonde and a 30 MHz vertical riometer at Fairbanks, Alaska have revealed that whenever the 30 MHz absorption exceeded ~ 1.2 dB, no ionospheric traces were observed on the ionosonde! Again, the incoherent scatter radar technique is not limited by such disturbance phenomena.

Chapter 4 Oblique Backscatter Sounders

Having discussed sounders which transmit and receive RF energy at *vertical incidence* to the ionosphere in the previous chapter, we now turn our attention to a class of sounders which direct RF energy at *oblique angles* to the ionosphere. By Oblique Backscatter Sounders (OBS) we include any technique which uses HF through UHF radio waves coherently backscattered from ionospheric irregularities either directly or via reflection from the surface of the earth to deduce properties of ionospheric irregularities. Examples of recently or currently deployed OBS systems are the STARE (Scandinavian Twin Auroral Radar Experiment, Greenwald et al. 1978) the SABRE VHF radar (Nielsen et al. 1983), the HF phased array radar located at Goose Bay, Labrador (Greenwald et al. 1985), the Cornell University CUPRI Radar (Riggin et al. 1986) Swedish/French SAFARI radar in Scandinavia (Hanuise et al. 1981) and the PACE (Goose Bay and Antarctic HF radars Baker et al. 1989). Table 4.4 lists all the known currently operating OBS systems.

We do not include in our discussions the class of sounders known as oblique ionosondes or synchronized step-sounders, which are used primarily for assessing propagation characteristics of the ionosphere for HF communication circuits. There is an important "subset" of OBS, most often referred to as over-the-horizon (OTH) radars, which are used by military services primarily for the detection of airplanes, ships and missiles. The hardware and software utilized in OTH systems is usually state-of-the-art and quite sophisticated, and the subject is highly classified, but it may be possible to acquire information about OTH systems from the unclassified literature which could be put to use in ionospheric research systems. Descriptions of some of the OTH radar systems and results are given by Barnum (1986), Brookner (1987), Maresca and Barnum (1982a, b), Earl and Ward (1987), Bourdillon et al. (1987), Boutacoff (1985), Trizna and McNeal (1985), in the special issue of the *IEEE* Journal on Oceanic Engineering of April 1986, and in a book by Kolosov et al. (1987).

We will first discuss the HF *groundscatter* technique which was discovered in 1926 and utilized mainly for mid-latitude ionospheric studies through the mid-1970's. Next, the *direct* backscatter technique (including VHF/UHF auroral radars and HF OBS's will be covered. Lastly, several examples of *currently deployed* OBS's will be described in some detail and the capabilities and limitations of all the OBS techniques will be evaluated.

Oblique backscatter sounders (OBS) are also referred to in the literature as coherent sounders and sometimes as HF, VHF, or UHF ionospheric radars, but we will use the term oblique backscatter sounder (OBS) in this book.

4.1 HF Pulsed Groundscatter Systems

A comprehensive review of HF groundscatter radar systems and analyses of some of the data obtained by these sounders up to the early 1970's has been published by Croft (1972), and an atlas of typical data acquired by these radars at mid-latitudes was published by Hunsucker (1970).

In the HF pulsed groundscatter OBS, a train of short pulses of radio frequency energy are refracted by the ionosphere and scattered nonuniformly at the surface of the earth. A small portion of the incident radio frequency energy is backscattered over approximately the same path, is amplified and detected at the receiver, and is displayed on a suitable cathode ray tube presentation. In this section we are primarily concerned with the *groundscatter* mode as opposed to

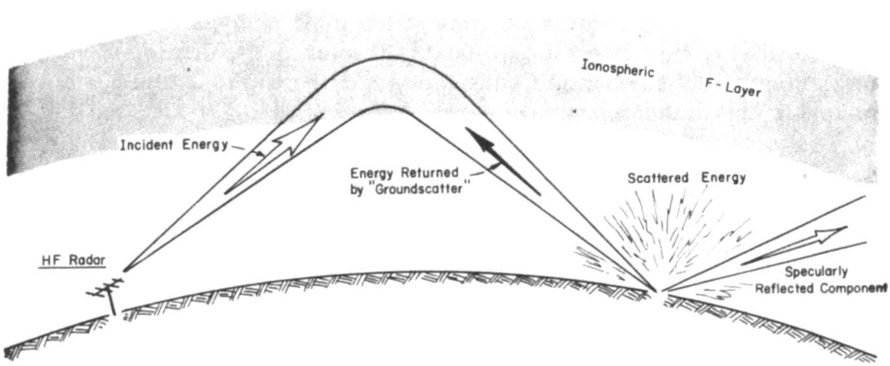

Fig. 4.1. Idealized sketch of the groundscatter mode

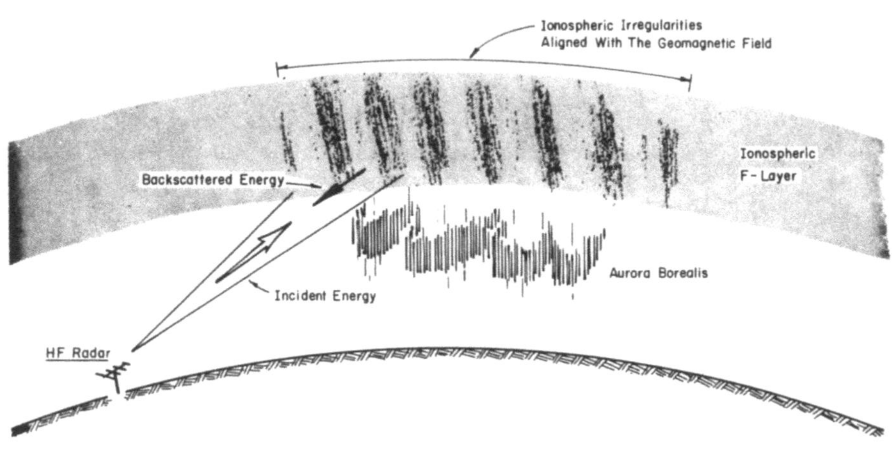

Fig. 4.2. Idealized sketch of direct backscatter from field-aligned ionospheric irregularities in the auroral oval. In reality, the HF raypath is refracted by the ionosphere to bring it to orthogonality with the field-aligned irregularity

direct backscatter from ionospheric irregularities. A comprehensive discussion and interpretation of *direct* HF backscatter echoes from ionospheric irregularities in the *auroral zone* is given in a series of papers by the late H.F. Bates (1959, 1961a; Bates et al. 1965) and in a report by Bartholomew (1966). Figures 4.1 and 4.2 are simplified representations of the groundscatter and direct backscatter modes.

4.1.1 The Development of Backscatter Sounding as a Remote Sensor of the Ionosphere

Ionospheric *groundscatter* was first observed some six decades ago independently by three investigators, Mögel (1930), and Taylor and Young (1928). It was not until 1951 that Dieminger described and explained Mögel's 1926 observations. By using highly directional antennas, these early investigators concluded that the scatter was, in fact, backscatter which occurred somewhere on the propagation path, but they could suggest no satisfactory theoretical explanation. Following the observations of Taylor and Young and Mögel, there were several attempts to explain the backscatter phenomenon. Quäck and Mögel (1929) suggested that the echoes were reflected from a height of 1500 km in the atmosphere, while Hoag and Andrew (1928) suggested that the echoes came from the polar ionosphere. During this period, Eckersley (1932, 1937, 1939, 1940) made a determined effort to solve the backscatter problem, but he erroneously believed that the backscatter was predominantly due to E-region irregularities.

During World War II only a few investigators were able to pursue HF ionospheric scatter problems. Pierce and Mimno (1940) suggested that traveling disturbances increased the curvature of the ionosphere to cause the scatter echoes. Edwards and Jansky (1941) attempted to determine the origin of the echoes by measuring echo time delay and direction of arrival, and postulated that some of the echoes were scattered from the ground. After the war, research to solve the HF scatter problem was vigorously resumed and much evidence was accumulated to support the hypothesis that the surface of the earth was the source of most of the scatter echoes as reported by Peterson (1949), Gates (1949), Silberstein (1949), and Hartsfield et al. (1950). These investigators concluded that the majority of the groundscatter echoes originated at or near the edge of the skip zone.

Conclusive experimental verification and a quantitative theoretical explanation of the groundscatter phenomenon was reported in the early 1950's by Dieminger (1951), Peterson (1951), Abel and Edwards (1951), and Silberstein (1954). The actual implementation and extensive use of the backscatter technique as a remote sensor of the ionosphere began during the International Geophysical Year (IGY). A worldwide network of 13 fixed-frequency backscatter sounders was operated during the IGY (Peterson et al. 1959), and a considerable amount of information on the phenomenology of the F2 layer was obtained.

From the early 1950's through the mid-1970's, many investigators utilized the HF groundscatter technique to measure various parameters of the ionosphere. Most of these investigations were concerned with the characteristics (height,

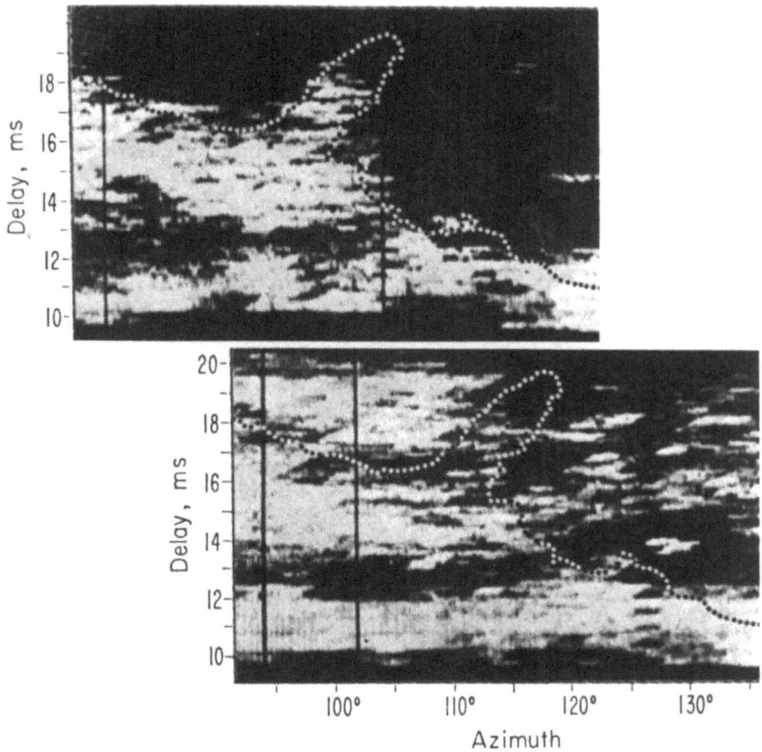

Fig. 4.3. A graphic demonstration of the difference between sea and land backscatter. The *dotted line* follows the United States coast, including the Florida peninsula. *Left* All sky wave backscatter echoes. *Right* Only the sky-wave backscatter that passes a narrow band filter; thus the sea return is largely eliminated

speed, direction, and structure) of patches of sporadic-E ionization (Villard et al. 1952; Clark and Peterson 1956; Shearman and Harwood 1958; Dueño 1962; Egan and Peterson 1962; Steele 1964. Other investigators utilized the technique to ascertain irregularity structure in the F-region (Villard and Peterson 1952; Silberstein 1954; Shearman 1956; Wilkins and Shearman 1957; Widdel 1957; Stein 1958; Valverde 1958; Ranzi and Dominici 1963; Tveten 1961; Davis et al. 1964; Wickersham 1964; Gilliland 1965; Croft 1967; Hunsucker and Tveten 1967; Croft 1968). An even more recent review on high latitude F-region irregularities has been published by Tsunoda (1988).

At this point, one might question whether some of the effects observed by an HF backscatter sounder might be due to inhomogeneities (either surface or subsurface) at the ground. The results of several experimental and theoretical studies of surface scatter effects at HF have been summarized by Hayden (1968), who concluded that, for the pulse lengths and antenna beamwidths normally employed in HF backscatter sounder investigations, the ionospheric focusing effects are much stronger than terrestrial terrain feature effects. However, if one employs narrow-beam antenna arrays, Doppler sounding, and long enough integration times, it is possible to detect the effects of some terrain features during periods when the ionosphere is relatively undisturbed. For example, Tveten (1967), using appropriate equipment parameters, has measured sea wave velocities from a distance of about 3000 km, and land-sea boundaries have also been mapped (Blair et al. 1969), as illustrated in Fig. 4.3.

4.1.2 Basic Ionospheric Groundscatter Theory

This section presents salient points in the quantitative theory of ionospheric ground backscatter developed by Peterson (1951). The starting point for this theoretical development is Eq. (2.19) from Chapter 2 – the secant law for the ionosphere

$$f_{ob} = f_v \sec \phi,$$

where:

f_{ob} = the oblique-incidence frequency (operating frequency of the backscatter sounder)

f_v = equivalent vertical-incidence frequency

$$= \frac{1}{2\pi}\sqrt{\frac{Ne^2}{m_e \varepsilon_o}} \approx 9\sqrt{N} \quad \text{(hertz)}.$$

The plane-earth, plane-ionosphere geometry as shown in Fig. 4.4 is sufficient to develop the first-order backscatter theory. For distances > 1000 km, it will be necessary to use a curved earth and concentric ionosphere, as will be done in a later section (see Dieminger 1951 for details of the derivation.

From Fig. 4.4 we start with

$$2r = \frac{2h'}{\cos \phi} \tag{4.1}$$

and

$$2d = 2r \sin \phi = 2h' \tan \phi, \tag{4.2}$$

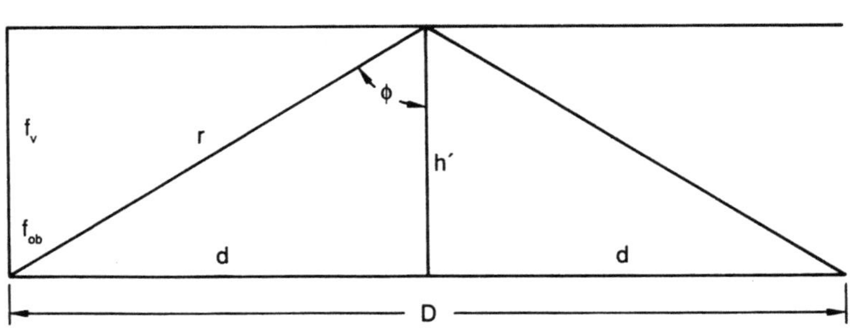

Fig. 4.4. Plane-earth, plane-ionosphere geometry for the oblique backscatter mode

where r/2 = the length of the oblique path from ground to ionosphere. Finally we obtain the equations for virtual height, path length, and distance.

$$h' = h_o + \frac{y_m}{2} \times \cos\phi_o \ln\left[\frac{1 + x\cos\phi_o}{1 - x\cos\phi_o}\right], \tag{4.3}$$

where ϕ_o = the angle of incidence at the bottom of the layer. Substituting Eq. (4.3) into Eqs. (4.1) and (4.2) yields

$$P = \frac{2h_o}{\cos\phi_o} + xy_m \ln\left[\frac{1 + x\cos\phi_o}{1 - x\cos\phi_o}\right] \tag{4.4}$$

and

$$D = 2h_o \tan\phi_o + xy_m \sin\phi_o \ln\left[\frac{1 + x\cos\phi_o}{1 - x\cos\phi_o}\right], \tag{4.5}$$

where:

$x = f_{ob}/f_c$
y_m = half thickness of the parabolic layer

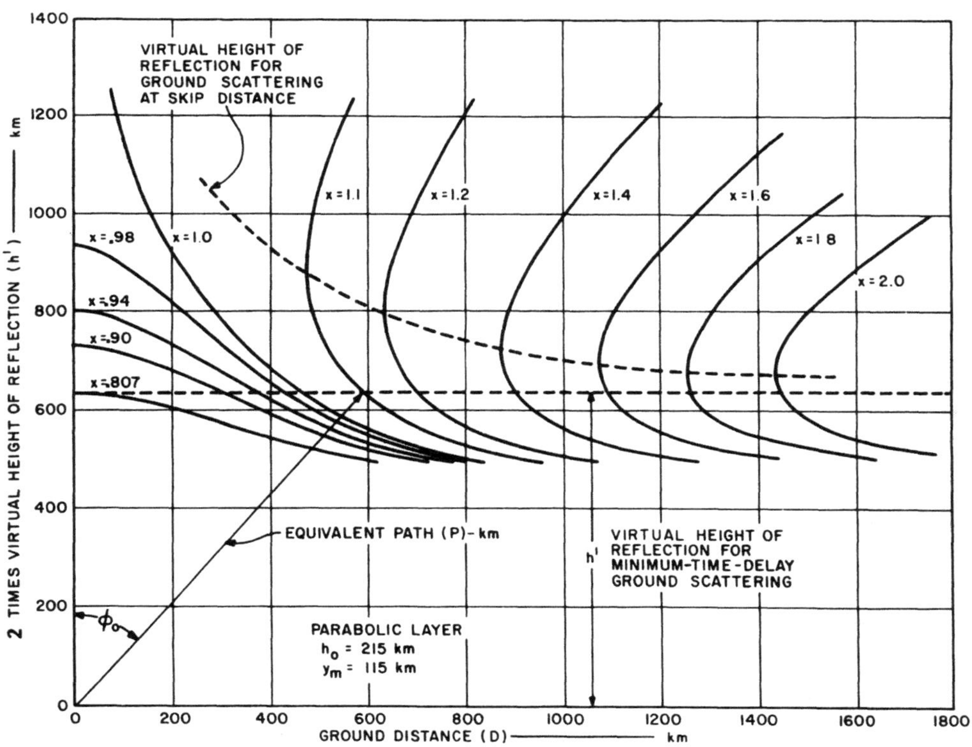

Fig. 4.5. Plot of virtual height (*h*), group path (*P*), ground distance (*D*), and takeoff angle (*o*), for a parabolic ionospheric layer. (Peterson 1951)

h_o = height to the bottom of the layer
ϕ_o = the angle of incidence on the bottom of the layer.

The relationship between the quantities h′, P, D, and ϕ is shown graphically in Fig. 4.5, for 11 values of x.

The skip distance for a given frequency may be determined by a projection normal to the ground-distance axis (D) and tangent to the x contour. A radial line from the origin to the x contour defines the virtual ray path, and the angle of incidence ϕ_o is the angle from the virtual height axis to the radial line. The minimum path distance P is equal to the minimum distance from the origin to the x contour, and these P_{min} points fall along a line of constant virtual height defining a ground distance greater than the skip distance. P_{min} is found by differentiating Eq. (4.4) with respect to ϕ and setting the resultant equal to zero.

$$\frac{dP}{d\phi_o} = 2h_o \sec\phi_o \tan\phi_o + xy_m\left(\frac{-x\sin\phi_o}{1 + x\cos\phi_o}\right) - xy_m\left(\frac{x\sin\phi_o}{1 - x\cos\phi_o}\right)$$

$$= 2h_o \sec\phi_o \tan\phi_o - \frac{2x^2 y_m \sin\phi_o}{1 - x^2\cos\phi_o} = 0$$

thus,

$$\cos\phi_o = \frac{1}{x}\left[\frac{h_o}{y_m + h_o}\right]^{1/2} \tag{4.6}$$

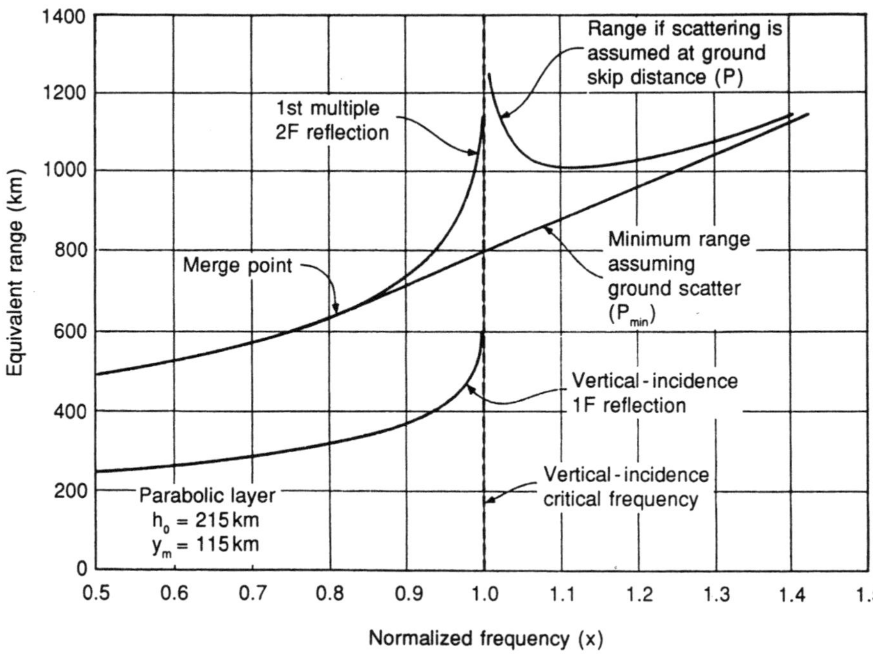

Fig. 4.6. Behavior of minimum range for groundscatter as a function of x. (After Peterson 1951)

Our final result is

$$P_{min} = x\left[2h_o\sqrt{\frac{y_m + h_o}{h_o}} + y_m \ln\frac{1+\sqrt{\dfrac{h_o}{y_m+h_o}}}{1-\sqrt{\dfrac{h_o}{y_m+h_o}}}\right]. \tag{4.7}$$

Figure 4.6 shows the behavior of P_{min} as a function of x for a given parabolic layer for plane-earth geometry. This mechanism is sometimes referred to in the literature as least-time focusing (LTF).

Figure 4.6 illustrates the fact that the focusing mechanism which enhances groundscatter is least-time focusing and *not* skip-distance focusing. Although Peterson (1951) and Dieminger (1951) demonstrated quite conclusively the validity of the least-time focused (LTF) groundscatter mechanism, there still seems to be some confusion concerning the determination of skip distance from groundscatter observations, as was pointed out by Bates et al. (1961). The error in determining skip distance for communications purposes from groundscatter observations is shown in Table 4.1.

Figure 4.7 is a Newbern Smith-type transmission curve parametric in backscatter delay instead of ground range (Tveten and Hunsucker 1969). If the abscissa of Fig. 4.6 were extended to the left to $x = 0$, one would observe that a line drawn from the origin and tangent to the second-order vertical-incidence trace would describe the minimum virtual ray path for each frequency propagated. This feature will be apparent on experimental sweep-frequency data included later in this chapter. Bates (1959) has compared the range-frequency behavior of groundscatter for three cases:

1. Flat earth.
2. Spherical earth.
3. Extended plane ionosphere-spherical earth.

Figure 4.8 illustrates these three cases, and it is apparent that serious divergence does not occur until $x \gtrsim 1.6$ or when the equivalent path >2000 km.

Table 4.1. Error in determining skip distance

$x = \frac{f_{ob}}{f_c}$	Difference between skip distance and LTF distance (km)
0.95	70
1.05	220
1.10	140
1.15	95
1.20	70
1.30	35
1.40	20

For $x > 1.4$ the skip distance and LTF groundscatter coincide for all practical purposes.

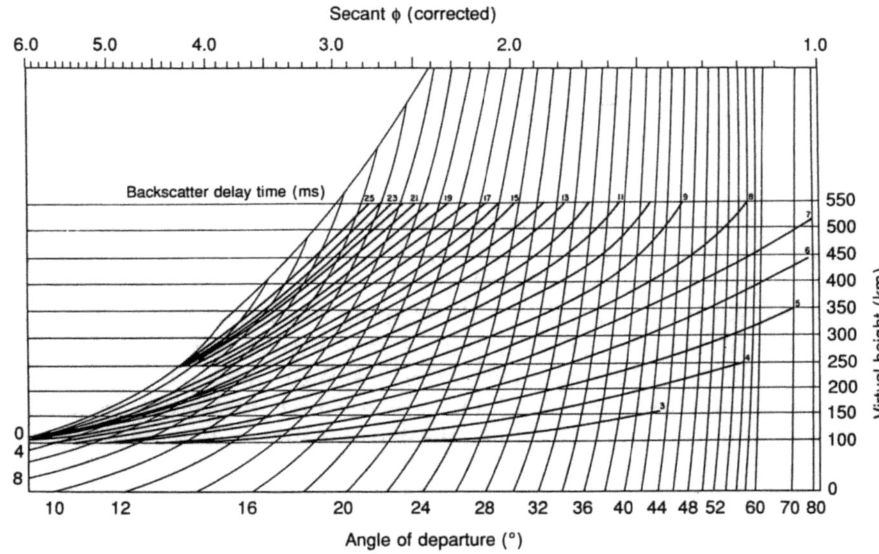

Fig. 4.7. Backscatter delay time vs. angle of departure. (Tveten and Hunsucker 1969)

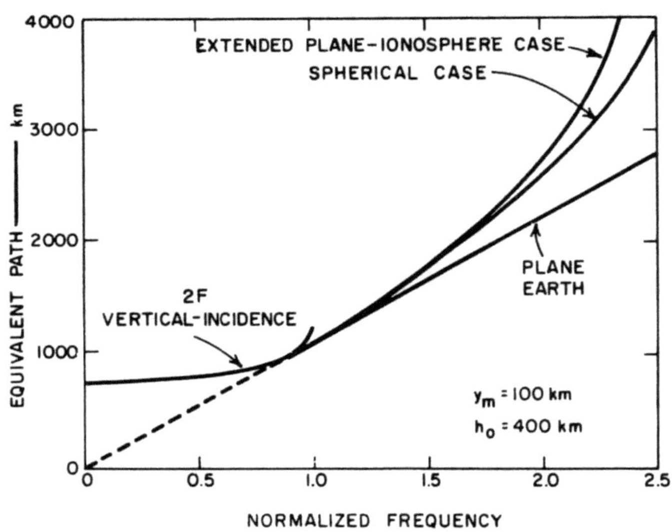

Fig. 4.8. Behavior of equivalent path (P) as a function of normalized frequency (x) for three geometries. (Bates 1961)

Other studies by Bates (1969) have shown that the shape of the ionospheric layer in the vicinity of the layer maximum is not critical for the LTF groundscatter mechanism. That is to say, the actual electron density distribution (which is best described by a Chapman layer) may be approximated by using a quasi-parabolic layer without serious error in the predicted groundscatter time delays. Figure 4.9 shows the types of layer shapes considered, and Fig. 4.10 compares the behavior of the groundscatter time delay as a function of frequency for various layer heights. The error is seen to be rather insignificant considering the experimental accuracies involved.

4.1.3 Various Techniques Employed in Groundscatter Sounding

There are four or five backscatter sounding techniques which have been utilized to provide most of the observations of ionospheric phenomena reported in the literature up until the early 1980's. Each of these techniques is briefly described in Table 4.2 and typical data are shown in Fig. 4.11, 4.12, 4.13, and 4.14 for the four most commonly used techniques.

4.2 Characteristics of a Typical Scanned Groundscatter System with Variable Frequency

4.2.1 The ITS/ESSA OBS System (1960's Technology)

An azimuth and elevation scanned HF groundscatter system developed in the US in the 1960's is described in this section as an example of a fairly sophisticated sounder. Characteristics of the high-resolution ground backscatter sounder developed at the Institute for Telecommunication Sciences of the ESSA research laboratories in Boulder in the 1960's are described in Table 4.3. As may be seen in Table 4.3, this system combined high angular resolution with the azimuth scan capability to permit observation of a large area of the ionosphere.

In 1967 a team from Stanford Research Institute of Menlo Park, California, employed their airborne system to measure the radiation patterns of the ITS/ESSA antenna array. This technique is described in detail by Barnes (1965). Antenna patterns measured at several typical operating frequencies and the results of these full-scale measurements indicated that the actual patterns are very close to the theoretical patterns. *It should be emphasized that when a full-scale sophisticated antenna array is constructed (especially if a ground screen is not used), it is quite important to validate the pattern predicted by modeling with actual measurements.*

4.3 Direct Ionospheric Backscatter

Although pulsed *groundscatter* OBS systems have been used to obtain much information on the regular ionospheric layer behavior and on irregularity structure (Sect. 4.1), that technique has been largely superseded by the *direct backscatter* technique since the early 1970's. At auroral and equatorial latitudes *most* irregularities are aligned with the geomagnetic field, both in the E-region and in the F-region, which requires rather stringent siting criteria for direct backscatter sounders.

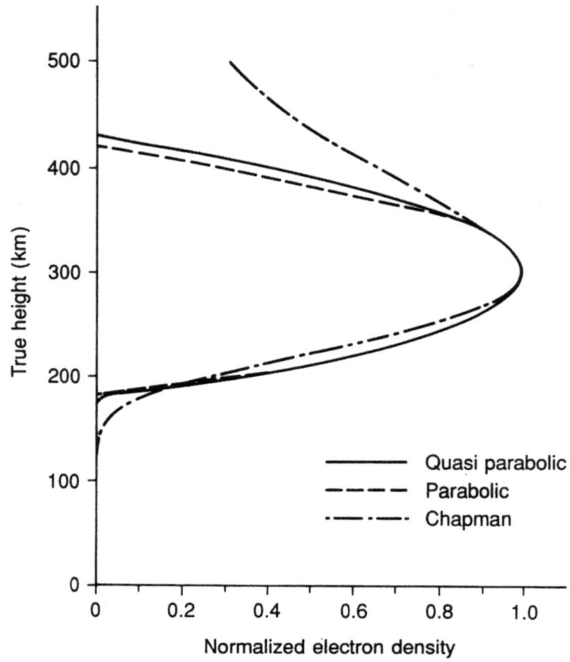

Fig. 4.9. Three different electron density profiles used in computing groundscatter behavior

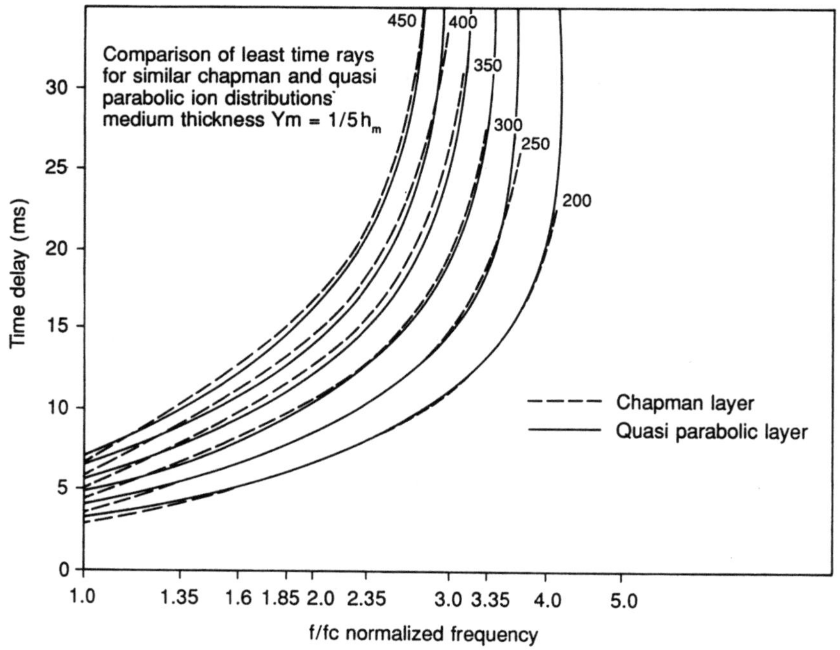

Fig. 4.10. Comparison of groundscatter rays for Chapman quasi-parabolic electron density distributions at different layer heights

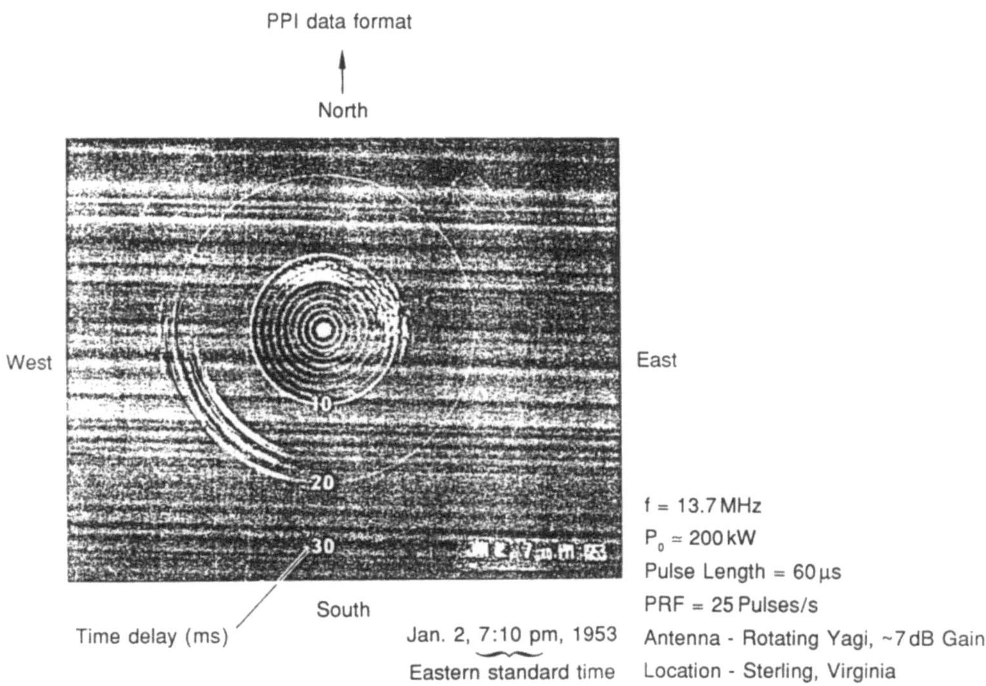

Fig. 4.11. Typical rotating antenna fixed-frequency (PPI) display of groundscatter echoes propagated by the F-region. (Hunsucker 1970)

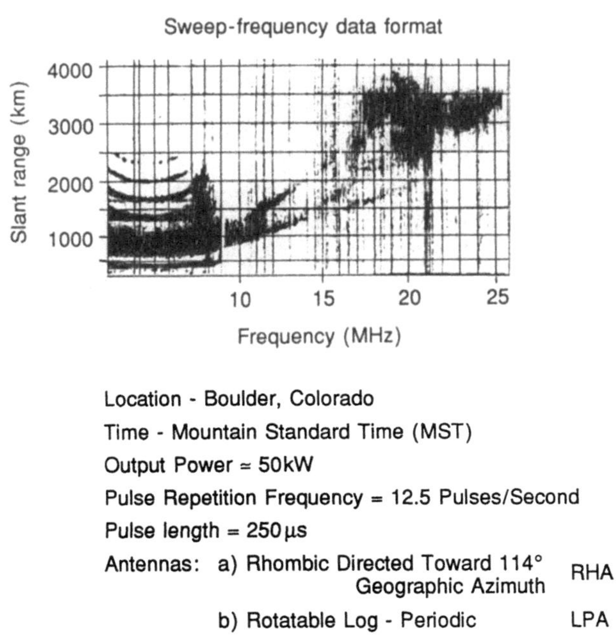

Fig. 4.12. Typical sweep-frequency backscatter record. The groundscatter trace branches off the second-order vertical incidence trace at the lower *left*. (Bates 1961)

Table 4.2. Ionospheric groundscatter techniques

Technique	Typical frequency range (MHz)	Typical antenna half-power beamwidths (degrees)		Advantages	Limitations	Reference and typical data
		Azimuth (deg.)	Elevation (deg.)			
Range-azimuth Range-elevation scan	12–25	3–1.5	4.2–2.1	High angular resolution	Limited scan width	Hunsucker and Tveten (1967); Tveten and Hunsucker (1969); Fig. 4.14
Rangetime	Fixed frequency (HF)	20–30 ≈ 40	— —	Good range resolution	Poor angular resolution	Tveten (1961); Ranzi and Dominici (1963); Fig. 4.13
Sweep or step frequency	3–25 4–64	≈ 20 105	— ≈ 60	Good resolution in frequency domain	Very poor angular resolution	Silberstein (1954); Bates (1961); Fig. 4.12
Rotating antenna, PPI display	Fixed frequency (HF)	≈ 40	≈ 30	Large area surveillance	Poor angular resolution	Peterson et al. (1959); Fig. 4.11
Rotatable fixed frequency	Fixed frequency (16 MHz)	8	≈ 15	Quite good angular resolution	Long rotation time (2–3 minutes)	Thomas and McNichol (1960)

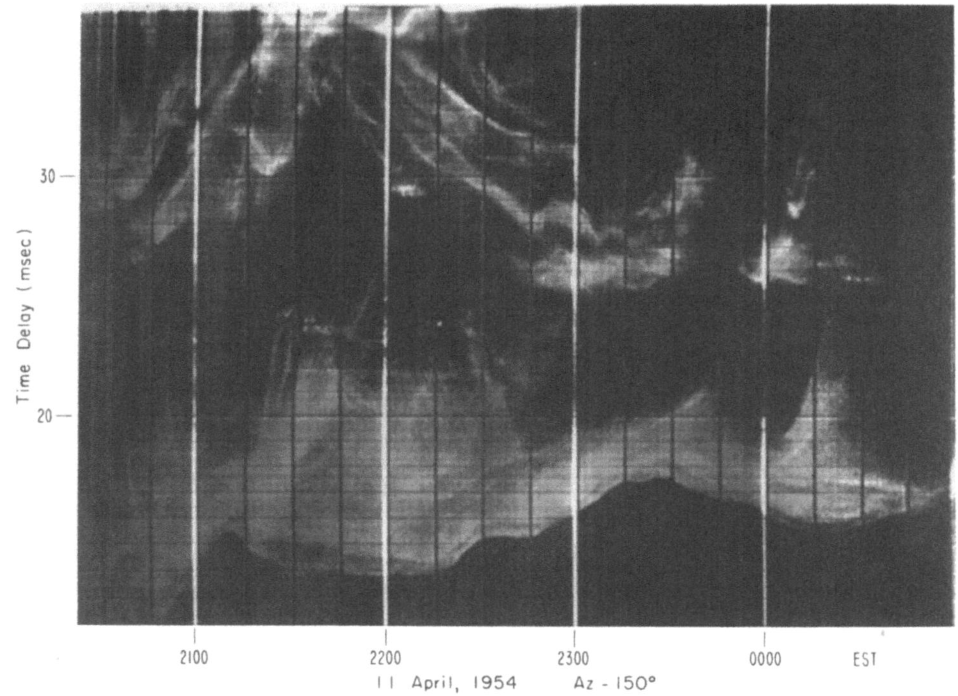

Fig. 4.13. Fixed-frequency, range-time backscatter record obtained by L.H. Tveten at Sterling, Virginia. (Hunsucker 1970)

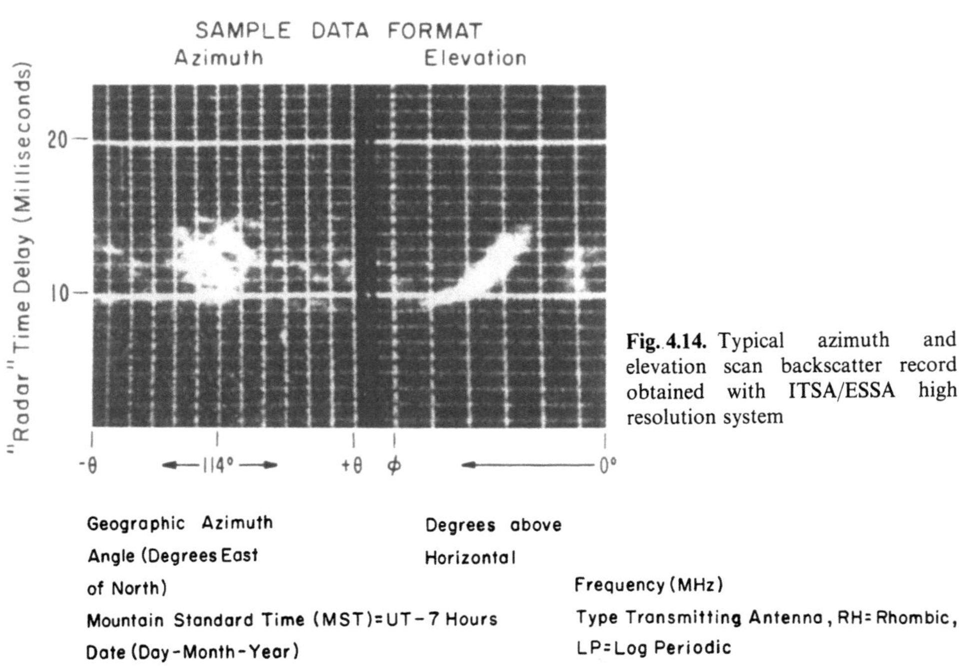

Fig. 4.14. Typical azimuth and elevation scan backscatter record obtained with ITSA/ESSA high resolution system

Table 4.3. ITS/ESSA backscatter sounder. System parameters

Azimuth array (425 m aperture)

Elements:	Log periodic horizontally polarized transposed dipoles (13 elements) Frequency, 12–25 MHz Gain, 5 dB above isotropic Front-to-back ratio, 20 dB E-plane beamwidth, 72° H-plane beamwidth, 115°
Element spacing:	17.67 m
Element height above ground:	24.38 m
Boresight azimuth:	114° true bearing

Horizontal antenna

Frequency (MHz)	Horizontal beamwidth (deg.)	Main lobe elevation (deg.)	Width of sector scan (deg.)
12	3.0	14.8	90
15	2.3	11.8	69
18	2.0	9.8	48
21	1.7	8.4	48
25	1.4	7.1	40

Elevation array (305 m effective aperture)

Elements:	Same as for azimuth array
Element spacing:	First element 8 m aboveground and remaining ones successively spaced at 16 m
Boresight azimuth:	114° true bearing

Vertical antenna

Frequency (MHz)	Vertical beamwidth (deg.)	Effective elevation sector scanned (deg.)
12	4.2	3–52
15	3.2	2.6–39
18	2.8	2.3–32
21	2.4	1.9–26
25	2.0	1.5–22

4.3.1 History of Development of the Technique

The purpose of this section is to trace the evolution of various radio techniques employed to investigate *ionospheric irregularities*, and to see how the direct backscatter method has been optimized to yield the best results. Perhaps the first use of *radio techniques* to study irregularities was Berkner and Wells' use of an *ionosonde* to study spread F (1934) and sporadic E (1937) in the equatorial region. The ionosonde can be regarded in this case to be an HF swept frequency radar which obtains reflections from ionospheric regions where the operating frequency is near the electron plasma frequency.

A comprehensive review of both coherent and incoherent direct scatter from ionospheric irregularities, including backscatter from irregularities associated with the equatorial electrojet, was given by Farley (1971). An outline of the theory of *coherent* scattering of radio waves from ionospheric irregularities as initiated by Booker (1956b) and extended by Walker et al. (1986) has been presented in Section 2.3.1 of this book. The analysis of scattering in terms of *plasma waves* developed by Farley (1963), Buneman (1963), Rogister and D'Angelo (1970), and others is summarized and compared with experimental results by Fejer and Kelley (1980) in a very comprehensive review of ionospheric irregularities. An even more recent review has been published by Tsunoda (1988). The Fejer–Kelley (F–K) review contains such a wealth of material on theory, techniques, and experimental results up to 1980 that it will be found to be extremely useful to anyone desiring information on radio techniques for investigating ionospheric irregularities.

As mentioned in a preceding section, ionospheric irregularities are, in most cases, aligned with the geomagnetic field, and the raypaths from the sounders must be nearly normally incident to the irregularities in order to produce a usable backscatter echo. Backscatter sounders utilized to investigate irregularities usually operate in the VHF/UHF region for E-region studies, and in the HF bands for F-region studies, except in the equatorial region, where VHF/UHF has also been used to study F-region irregularities.

The use of techniques other than ionosondes to investigate ionospheric E-region irregularities probably started with investigations of the radio aurora, which term was introduced by Collins and Forsyth (1959). Harang and Stoffregen (1938, 1961) were the first to observe and describe the scattering of VHF radio waves (on communication circuits), while Lovell et al. (1947) were the first to study irregularities using radio techniques. Radio amateurs also noted that VHF signals were influenced by the aurora, and even utilized auroral ionization poleward of their stations to communicate with each other over much greater than line-of-sight range (Moore 1951). Summaries of the early work have been published by Kaiser (1956) and by Little et al. (1956), and reviews of radio aurora research up until 1960 were presented by Booker (1960) and by Chamberlain (1961), while ionospheric radio scattering studies up to 1971 were reviewed by Farley (1971), and more recently by Haldoupis (1989).

Two basic system configurations are used to carry out direct backscatter studies of the ionosphere: the monostatic (pulsed) system and the multistatic (pulsed or CW) system – the respective geometries being discussed in detail in the F–K paper.

4.3.2 Direct Backscatter from the E-Region

The term radio aurora was introduced by Collins and Forsyth (1959) and has been accepted as describing the ionization associated with the visual aurora which produces certain types of radio reflection in the VHF and UHF frequency bands.

The two most plausible causative mechanisms for the auroral echoes appear to be the two-stream plasma instability (Buneman 1963; Farley 1963) and/or the gradient-drift mechanism (Rogister and D'Angelo 1970); and the phenomenon of auroral backscatter may be interpreted in terms of electrostatic ion waves produced by plasma instabilities associated with the electrojet. A *necessary* condition for the occurrence of these waves is a sufficiently large relative velocity between the Hall effect drifting electrons and the collision-dominated ions, which produces waves which travel nearly perpendicular to the ambient magnetic field. The latter property implies that the backscatter cross-section is maximum when the radar wave vector ("line-of-sight") impinges upon the irregularity very close to normal to the field line. Because of this aspect-angle sensitivity, auroral radars must be carefully sited to achieve normality of the wave vector and magnetic field lines in the E-region in the latitudinal area of interest.

Similar plasma instabilities and unstable electrostatic waves also occur in the equatorial E-region, and coherent backscatter has been observed and described by Farley et al. (1981), but we will concentrate on descriptions of OBS systems which are utilized to probe the *high-latitude* ionosphere irregular structure, since the effects of the magnetosphere and of energetic particle precipitation are most pronounced in these latitudinal regions.

Harang and Stoffregen (1938) were the first to observe and describe the scattering of VHF radio waves by auroral ionization, while Lovell et al. (1947) were the first to observe this ionization using radio techniques. Radio amateurs also noted that VHF signals were influenced by the aurora and even utilized auroral ionization poleward of their stations to communicate with each other over much greater than line-of-sight ranges (Moore 1951). Summaries of this early work have been published by Kaiser (1956) and by Little et al. (1956), and reviews of radio aurora research up until 1960 were presented by Booker (1960) and Chamberlain (1961). A more recent review has been published by Haldoupis (1989).

The main characteristics of auroral radar echoes based on the results summarized by Booker and by Chamberlain are:

Latitude: associated with discrete auroral arcs in the auroral oval, wherever the radar line-of-sight intersects the magnetic field line near normal incidents at low E-region heights

Azimuth: within a few degrees (depending on operating frequency of the radar) of magnetic north from the radar site

Height: at low-UHF, most of the echoes come from the 90–120 km height range. In the VHF region, echoes may come from a somewhat wider range of heights because of refraction effects

Range: at VHF/UHF most echoes seem to come from ranges ~ 400–1200 km magnetically north of the radar, while at lower VHF frequencies somewhat shorter ranges occur occasionally

Diurnal variation: whereas the visual aurora is only observable from ~ sunset-sunrise, the radio aurora is usually observed a few hours before and after sunset, depending on operating frequency and degree of magnetic activity

Seasonal variation: similar to visual aurora

Sunspot cycle variation: similar to visual aurora

Polarization: echo polarization seems to depend upon the polarization of the transmitted wave and upon the orientation of the irregularity, i.e., field-aligned.

Figure 4.15 summarizes some geophysical phenomena which may be studied using coherent radars of the AIS and OBS types, and Fig. 4.16 shows schematically how one approaches the investigation of properties of the ionospheric plasma.

The basic types of auroral radars recently or currently deployed may be characterized as:

1. Fixed antenna, range-time display.
   Example: NOAA Anchorage, Alaska radar system (Gray and Ecklund 1974).
2. Antenna moveable in azimuth, either PPI or sector-scan display.
   Examples: SRI-I Homer, Alaska radar (Presnell et al. 1972).

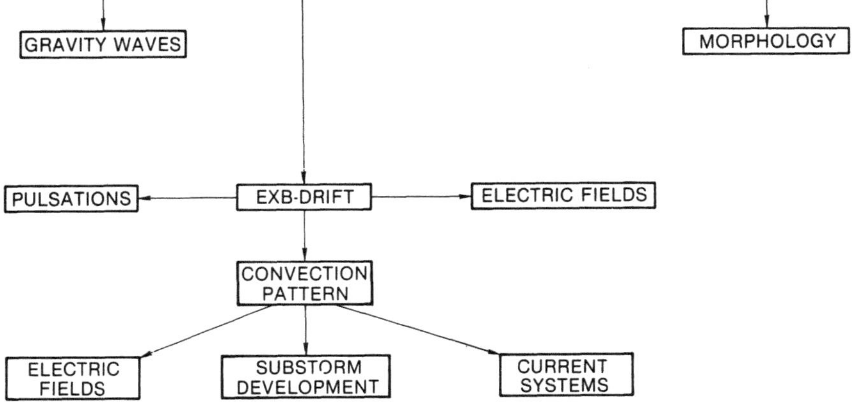

Fig. 4.15. Summary of coherent radar investigations made from a geophysical point of view

3. Electronically scanned (phased-array) antennas, sector-scan or range-time display.
   Examples: STARE, SABRE, etc. (Greenwald et al. 1978).
4. "Interferometer" systems.
   Example: CUPRI System (Riggin et al. 1986).

Another new OBS technique using the *direct* backscatter mode is called radar interferometry (Farley et al. 1981) and is based on a method developed by Woodman (1971), who used the north and south quarters of the large 50 MHz array at Jicamarca to accurately measure the location of F-region scatterers near the magnetic equator.

In its simplest configuration (a two-antenna interferometer) the technique depends on sampling the signals from each of the elements at a number of ranges, Fourier transforming them, computing the normalized cross spectra for each range, and averaging the results. The magnitude of the cross spectrum at each Doppler frequency and the phase information determine the echo location.

The vertical incidence radar interferometer technique used at Jicamarca was adapted to oblique incidence and operated from Ithaca, N.Y., to investigate auroral E-region irregularities associated with discrete auroral forms (Providakes et al. 1983). This system was also used to obtain detailed information on the temporal and spatial structure of plasma turbulence in a discrete auroral arc (Providakes et al. 1988). A detailed description of the Cornell University Portable Radar Interferometer (CUPRI) is given by Providakes (1985), and the radar parameters are given below:

Frequency	49.92 MHz
Typical peak pulse power	25 kW
Pulse width	
Range-time-intensity (RTI) mode	50 μs
Spectral mode	10–50 μs
Pulse repetition frequency	
RTI mode	100 Hz
Spectral mode	400–1000 Hz
Antenna azimuthal beamwidth	5.5°

The CUPRI system has also been operated at St. Croix Island in the Carribean to investigate mid-latitude sporadic-E ionization (Riggin et al. 1986) and has also been deployed in Scandinavia. A CW 50 MHz auroral radar interferometry system has been designed and tested recently by University of Saskatchewan scientists Haldoupis et al. 1988; Andre et al. 1988; Montbriand 1988).

4.3.3 Backscatter from the F-Region

At geomagnetic latitudes greater than $\approx 60°$, precipitating electrons and sometimes secondaries produce large field-aligned irregularities in the F-region

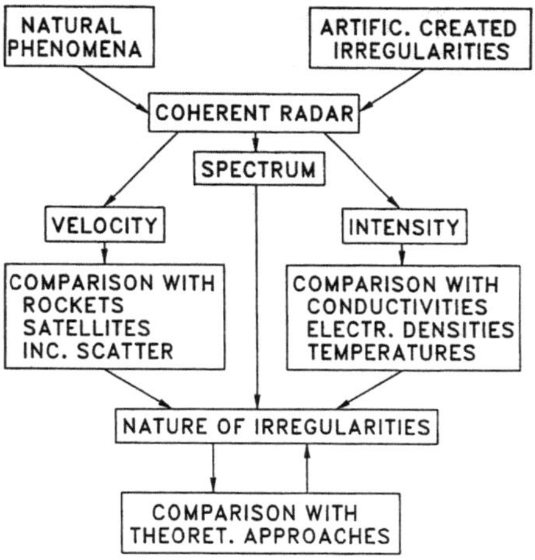

Fig. 4.16. Summary of coherent radar investigations made from a plasma physical point of view. (Schlegel 1984)

as well as in the E-region. The scale sizes of the irregularities in the F-region vary from tens to hundreds of meters along the magnetic field lines. Figures 4.17 and 4.18 (Schlegel 1984) are simplified sketches of the geometry for backscatter from the E- and F-layers.

One of the first to use an HF direct backscatter system to study the characteristics of high-latitude F-layer irregularities was Bates (1959, 1961), who utilized an NBS model C-4 ionosonde modified for oblique sounding to probe the auroral ionosphere poleward of College, Alaska. Figures 4.19 and 4.20 show

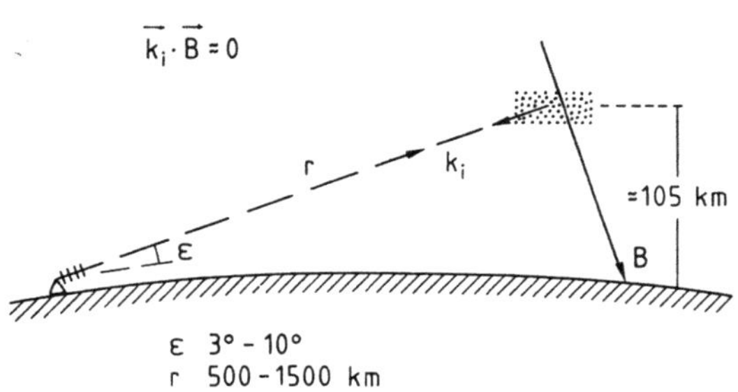

Fig. 4.17. Simplified sketch of direct backscatter from the E-layer. (Schlegel 1984)

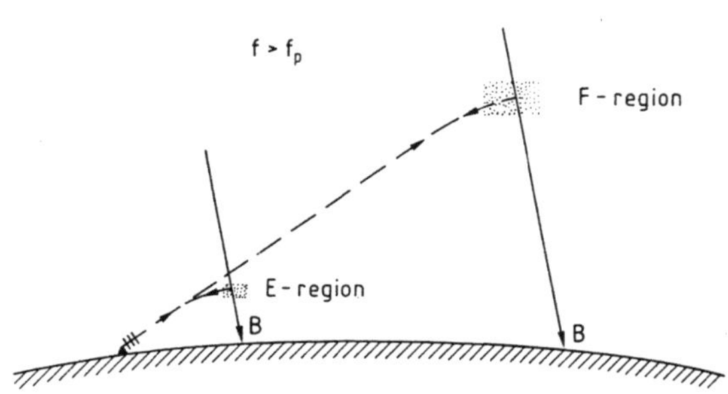

Fig. 4.18. Simplified geometry of direct backscatter from the F-layer. (Schlegel 1984)

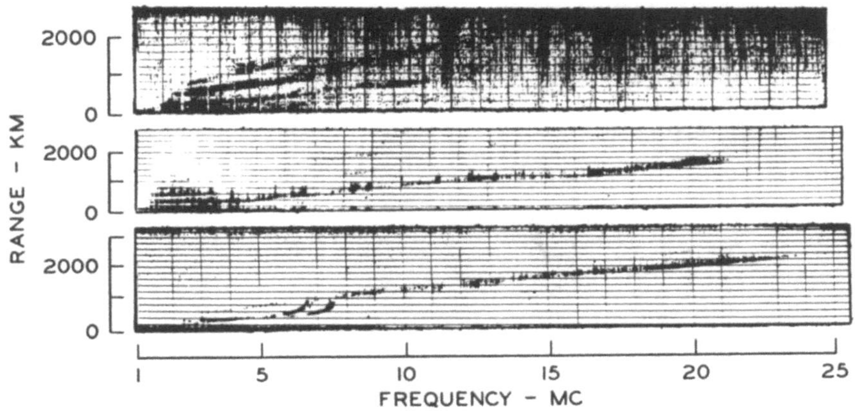

Fig. 4.19. Oblique backscatter ionogram. (Bates 1961)

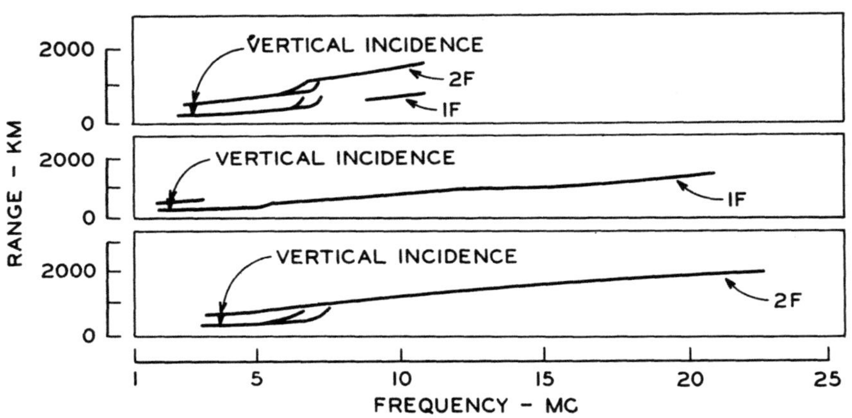

Fig. 4.20. Line drawing of the backscatter ionogram of Fig. 20. (Bates 1961)

oblique ionograms and line sketches, respectively, of typical records obtained with this system.

Bates devised a simplified model for direct backscatter from F-layer field-aligned irregularities, assuming that the wave achieved normal incidence upon a vertically aligned irregularity. He also used plane-earth and ionosphere geometry and ignored collisions. Figure 4.21 shows a direct F-layer backscatter echo observed with his sounder with the calculated direct backscatter trace shown below. A summary of the most important results obtained by the OBS system concerning high-latitude F-region irregularities is given by Bates and Albee (1970).

A rather sophisticated HF backscatter sounder is presently deployed at Goose Bay, Labrador, and has been described by Greenwald et al. (1985). This system was developed at the Applied Physics Laboratory of Johns Hopkins University

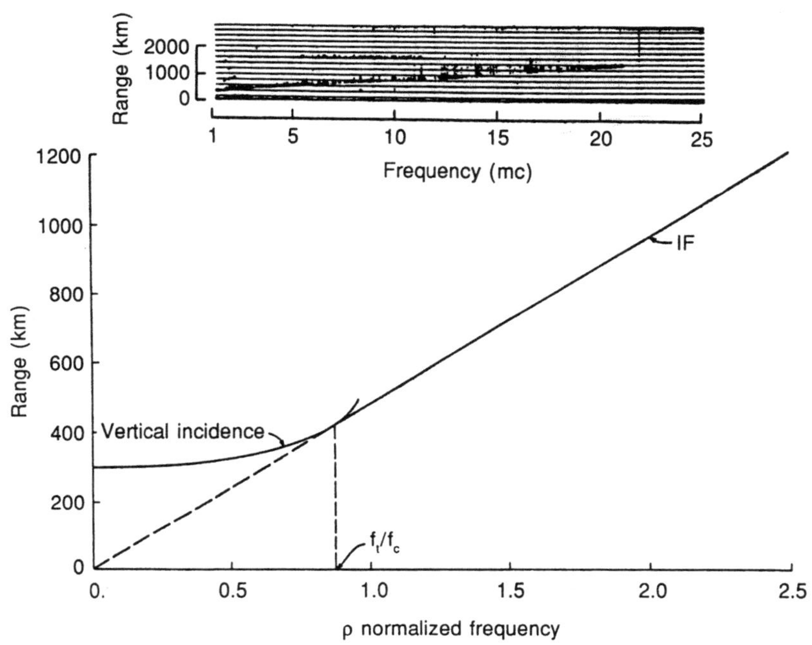

Fig. 4.21. *Above* Direct F-layer backscatter echo obtained from College, Alaska, and calculated trace shown *below*. (Bates 1961)

in the early 1980's and became operational in October 1983. Recent results have been published by Baker et al. (1986), Villain et al. (1990), and Sampson et al. (1990). The salient operating parameters are:

Location:	Goose Bay, Labrador – invariant latitude 65°
Frequencies:	8 to 20 MHz
Power output:	1000 W into each of 16 antennas
Pulse mode:	seven-pulse code
Antennas:	azimuth scan: linear spaced array of 16 log-periodics with electronic phasing and TR switches for transmission and reception. Bore-sited on 4° east of geographic north

Figure 4.22 shows a photograph of the installation and Fig. 4.23 is a block diagram of the system. The field of view of the Goose Bay radar is shown in Fig. 4.24; Fig. 4.25 illustrates the relative sensitivity of three contemporary coherent radars, and Fig. 4.26 shows the overlapping conjugate fields of view of the Goose Bay and (Antarctic) radar, which together comprise the PACE radar system (Baker et al. 1989).

Fig. 4.22. View of Goose Bay radar showing phasing-matrix shed and antennas of original transmit-receive array in foreground and antennas of receive-only array in background. (Courtesy of Ray Greenwald)

4.4 Capabilities and Limitations of OBS Systems

4.4.1 HF Groundscatter OBS Capabilities

1. Using narrow beam scanning antenna systems, deducing morphology of ionospheric irregularities at midlatitudes and at high latitudes (Croft 1972; Hunsucker 1971; Greenwald 1985).
2. Using sophisticated signal-processing techniques obtaining information on the character of the earth's surface (Blair et al. 1969).

Limitations

1. High cost and large sites for sophisticated HF systems with capabilities number (1) above.

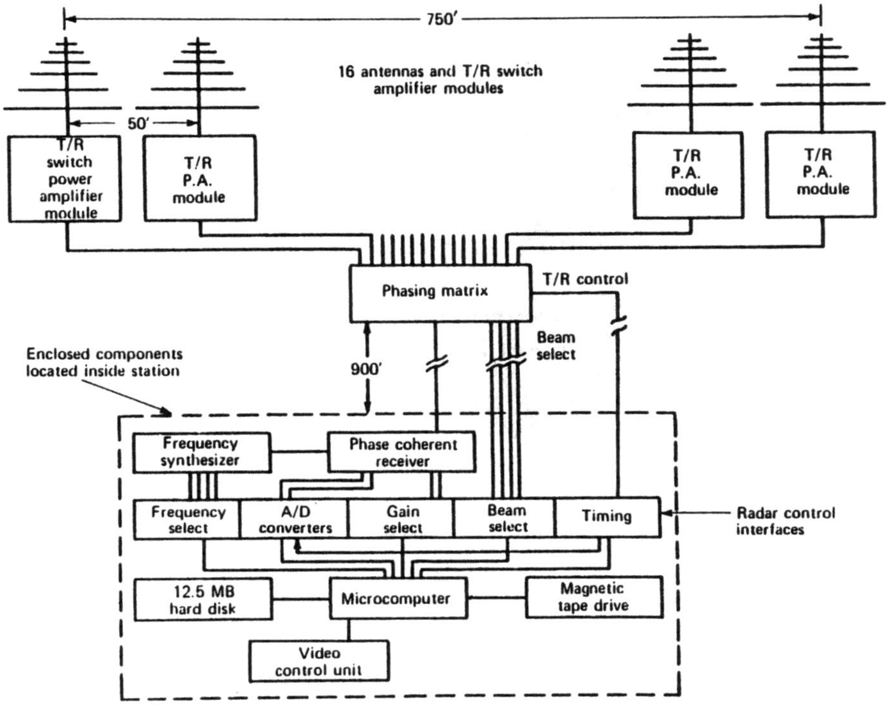

Fig. 4.23. Simplified block diagram of the Goose Bay HF radar system. (Courtesy of Ray Greenwald)

2. Differences in ionospheric "focusing" mechanisms for determination of communication coverage in number (1) (Hunsucker 1969).
3. Ionosphere "model-dependency" for quantitative information on numbers (1) and (2) (HF systems).

4.4.2 Direct HF Backscatter OBS Systems

Capabilities:

1. Determination of scale sizes, orientation, and anisotropy of electron density irregularities in the auroral and polar ionosphere.
2. Determination of irregularity motion.
3. Detection of plasma drift, waves, and tides in high-latitude ionosphere.
4. Determination of plasma convection characteristics over a large area.

Limitations include:

1. Relatively high cost (somewhere between cost of an "advanced ionospheric sounder" and an incoherent scatter radar system).
2. Relatively large antenna site required for HF systems.
3. Some "ionospheric model dependency" when interpreting results.

Fig. 4.24. Fields of view of several radar systems over eastern Canada and Greenland: Goose Bay HF radar, Canadian HF radar, Sondrestrom ISR, and Millstone Hill ISR. (Courtesy of Ray Greenwald)

4.4.3 Auroral Radar Systems

Capabilities:

In general, *auroral radar* data can, with proper interpretation, provide solutions to problems in geophysics and plasma physics such as:

1. Some details of auroral oval plasma convection pattern and electric field morphology.

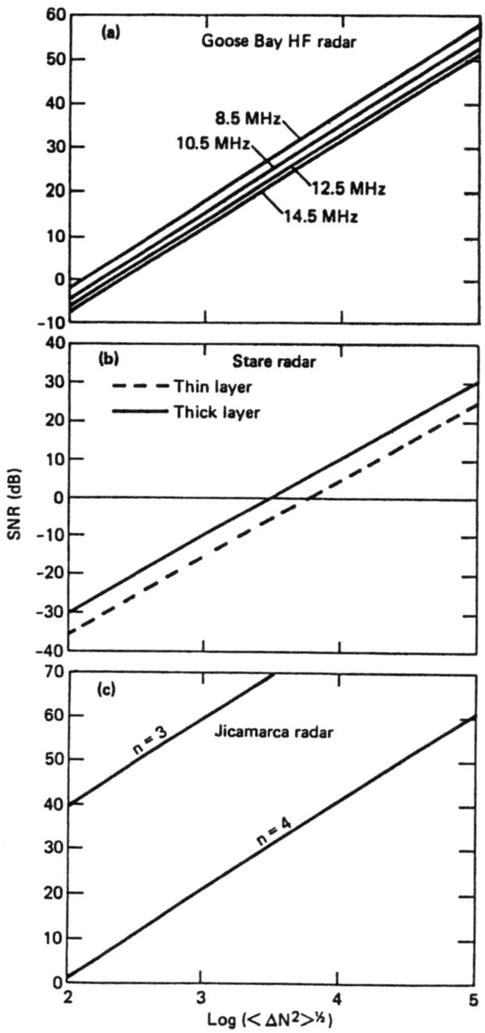

Fig. 4.25 a–c. Relative sensitivity of three contemporary coherent radar systems (**a, b, c**). (Walker et al. 1986)

2. Validity of auroral electrojet plasma instability theories, i.e., gradient-drift, two-stream, or other theories.
3. Detection of magnetic micropulsation-related radar signatures. Information on sources and propagation of MHD waves and other magnetospheric phenomena may also be inferred.

It should be emphasized, however, that the only quantities which are *directly* observable with ionospheric radars are the amplitude, time-delay and (in most cases) the power spectrum of the echoes.

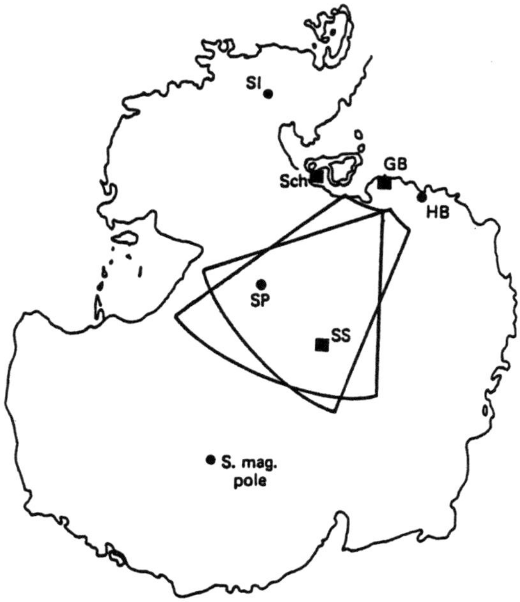

Fig. 4.26. Overlapping fields of view of the PACE HF radar system (Goose Bay and Halley Bay radars. The *dark circles* are: *HB* Halley Bay; *SI* Siple station; *SP* South Pole. The *dark squares* denote conjugate points: *GB* Goose Bay; *SS* Sondrestrom; *Sch* Schefferville. (Courtesy of Ray Greenwald)

Limitations:

1. Because of the aspect-angle restriction on auroral backscatter, the siting of auroral radars is quite restricted by geographic, topographic, and logistic considerations.
2. At low-VHF frequencies, one must make model-dependent corrections for E-layer refraction effects for proper echo-location in altitude.
3. In order to achieve a narrow azimuthal beamwidth, a moderately large antenna array must be utilized, typically ~ 10 acres (~ 4 hectares) for *HF* OBS systems. At VHF/UHF, much less land is needed.

In general, we can summarize this chapter on oblique backscatter sounder (OBS) systems by stating that this technique, using frequencies from HF through UHF, can provide HF communicators with approximate descriptions of geographical coverage areas as a function of frequency and time, and ionospheric physicists with rather quantitative information on the irregular structure of the global ionosphere.

Table 4.4 shows the *high-latitude* coherent radars operating at the time of writing of this book.

Table 4.4. Presently operating coherent radars at high latitudes

Radar	Location	Type	Frequency, MHz	Power (kW)	Antenna Arrays	Beam width	Special features
BARS	Canada	VHF (auroral)	48.5	60	Two 16 × 4 8-element yagi	3.6°	Two stations, phased array at Nipawin, Saskatchewan, and at Red Lake, Ontario
CUPRI	USA	VHF (Auroral)	49.92	25	Two 26-el.	3.5°	Interferometer
DAR	USSR	VHF (Auroral)	88	75	Four 10-el. Six 5-el.	17°, 9°	Two interferometers (lake)
SABRE	UK/Sweden	VHF (Auroral)	150	—	—	—	—
STARE	Scandinavia	VHF (Auroral)	140,143.8	60	Four 16 × 4 8-el. yagi	3.6°	Two stations, phased array
SLOPE-POINT	New Zealand	VHF (Auroral)	53.5	15	26-el. coax	5.6°	Interferometer (sea)
Syowa	Antarctica	VHF (Auroral)	50,112	15	Three 14-el. coax	3–4°	Meteor radar, double beam
Goose-Bay	Canada	HF	8–20	16	Two 16-el arrays of 10-el. log. periodics	3.3°	Interferometer, phased array
SOUSY	Scandinavia	VHF (MST)	53.5	200	576 4-el. yagi	3.0°	Steerable in four directions
SHERPA	Canada	HF	8–20	1	Eight 10-el. log. periodics	10°	Phased array
PACE	Antarctica	HF	8–20	16	One-16 el. of 10-el. log. arrays periodics	3.3°	Interferometer, phased array

Chapter 5 Incoherent Scatter Radar (ISR) Systems

An outline of the fundamentals of incoherent scatter theory and its historical development, a description of the milestones of the early development of the technique, and an example of how some ionospheric parameters are measured are given in Section 2.3.2 of this book. In this chapter we will describe: the ISR systems currently in operation, the most common modes of operation, specific data analysis techniques, how to obtain data from specific ISR's, and the capabilities and limitations of the ISR technique. A rather complete up-to-date presentation of incoherent scatter theory and special considerations in employing ISR's to probe the high latitude ionosphere is published in a book edited by Brekke (1977), and concise descriptions of the ISR technique are given by Hargreaves (1979) and Jursa (1985).

5.1 Introduction

The incoherent scatter radar technique, since its introduction in the early 1960's, has proven to be the *one* ground-based radio technique which can provide the *most* information about the terrestrial ionosphere. It is also the most expensive of these techniques! Another important feature of ISR's is their ability to make these measurements when the ionosphere is extremely perturbed, i.e., during a large geomagnetic storm. Like all experimental methods, it *does, however*, have its limitations – which we will discuss in Section 5.4 of this chapter.

5.2 Basic ISR System Considerations

The development of ISR systems from the early 1960's until the present has, as one would expect, been "driven" by technological advances. During the decade of the *1960's* increased sensitivity was achieved by increasing transmitter peak power to approximately the 5 MW level, increasing antenna aperatures up to $\approx 9 \times 10^4\,m^2$, and using uncooled parametric amplifiers to reduce antenna noise temperatures to ≈ 150 K at VHF and UHF. The decade of the *1970's* saw technological advances in VLSI, digital techniques, computers, pulse-code modulation, and antenna/preamplifier performance, all of which were employed advantageously at most of the ISR installations – perhaps best illustrated in the EISCAT system. Advances in the theory and practice of signal processing seem to be one of the strongest of the "technological driving functions" influencing

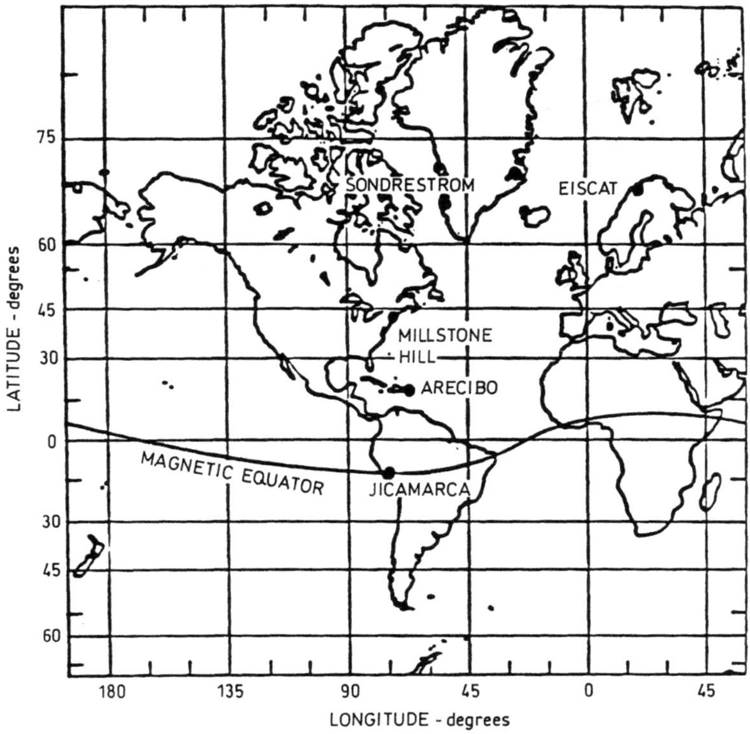

Fig. 5.1. Global distribution of incoherent scatter radar (ISR) systems which are presently operational (1991), excluding the MU Radar in Japan, which is primarily a "middle atmosphere" radar

improvements in sensitivity and resolution for *all radio techniques* (including ISR's) in the present decade.

In the later 1960's attempts were made to utilize a powerful radar located at Prince Albert, Saskatchewan, Canada for incoherent scatter studies. The radar was located at geographic coordinates of 105.9°W longitude and 53.2°N latitude at a magnetic latitude of 61.7°N. The transmitter operated at a frequency of 448 MHz with peak power output of 1.6 MW into a 84 ft (25.6 m) diameter (see Jaye et al. (1969) for more details of the system). Unfortunately, the radar system was destroyed by fire before it could become fully operational, otherwise it would have been the first high latitude ISR system.

5.3 Currently Operational ISR Systems

As of the time of publication of this book (1991), there are five ISR's deployed globally (all but one in the northern hemisphere) which are used primarily for investigations of the terrestrial ionosphere. Table 5.1 lists these ISR's and Fig. 5.1 shows their global distribution. Two more radars (Kwajalein and Japan) are occasionally used as ionospheric probes.

Table 5.1. Presently operable ISR systems

Facility	Geographic latitudes	Geographic longitude	Dip latitude (N)	L value	References and remarks
Jicamarca, Peru 1963-	11.9°S	16.00°W	1	1.1	Bowles (1963)
Arecibo, Puerto Rico 1963-	18.3°N	66.75°W	30	1.4	Gordon (1964)
Altair radar, Kwajalein	9.4°N	167.5°E			Tsunoda et al. (1979) f = 155.5 and 415 MHz
MU radar Shigaraki, Shiga, Japan 1985-	35.85°N	136.10°E			Kato et al. (1984)
Millstone Hill, MA, USA 1960-	42.6°N	71.5°W	57	3.2	Evans and Lowenthal (1964)
EISCAT, Tromso, Norway 2. 1981-	69.58°N	19.21°E	67	6.1	Baron (1984) 2. transmitter location
Sondrestrom, Greenland 1983-	67.0°N	51.0°W	71		Kelly (1983)

Table 5.2. Operational ISR systems 1970–1982

Facility	Geographic latitude	Geographic longitude	Dip latitude and L-value	Interval of operation	Organizations, references and remarks
St. Santin France	44.6°N	2.2°E	41°N/3.2	1965–1986	CRPE/CNET-CNRS Bauer et al. (1974) Bouvet and LaFeville (1982)
UK ISR Malvern, UK	52.1°N	2/3°W	50°N/2.6	1968–1975	Williams and Taylor (1974)
Chatanika, Alaska, USA	64.9°N	147.7°W	65°N/5.8	1971–1982	SRI-I/Menlo Park, CA, USA Leadabrand et al. (1972)

Other ISR systems were in operation during the period ≈ 1970–1982, and data for certain selected periods from seven of the ISR systems may be obtained by contacting the appropriate organization and/or persons. Table 5.2 lists the names of these ISR's the periods of operation and other pertinent information.

5.3.1 Descriptions of Individual Operational ISR's

Operational Systems

In this section we will list salient parameters, data analysis techniques, examples of data, and cognizant organizations for each of the ISR's currently employed in ionospheric research – starting with Jicamarca at 11.9°S latitude and working our way from these "sunny climes" toward the north pole.

Jicamarca Radar Observatory:

The Jicamarca ISR is unique in several respects:

1. Location: At 11.9°S, it is the only ISR in the southern hemisphere. $d \approx 20\,\text{km}$ from Lima, Peru; it probes the *equatorial* ionosphere essentially at right angles to the geomagnetic field.
2. Frequency: it has the lowest operating frequency ($f \approx 50\,\text{MHz}$) of all the ISR's *used primarily for ionospheric research.* This feature permits probing the upper regions of the ionosphere, and makes it very sensitive to turbulence fluctuations of scale size $\approx 3\,\text{m}$.
3. Antenna: it has the greatest antenna aperture of the operational ISR's ($A \approx 9 \times 10^4\,\text{m}^2$) but not the greatest *gain.*

This system became operational in 1961 and data were obtained on a fairly regular basis starting in 1964. Electron concentration data from May 1964 through November 1966 are available in a report by McClure et al. (1970). From 1970–1980 the sensitivity of the Jicamarca facility decreased due to lack of replacement for failing final amplifier tubes and antenna deterioration. Since 1984, the antenna has been rebuilt, the final amplifier has been refurbished, and

Fig. 5.2. Jicamarca ISR facility, taken in April 1985. (Courtesy of Wes Swartz)

new computer codes have been implemented for data analysis (Swartz et al. 1986, unpubl. report). Figure 5.2 is a view of Jicamarca Observatory, looking toward the north.

Jicamarca ISR parameters:

- Frequency: 50 MHz
- Peak output power: 5–6 MW
- Configuration: monostatic, pulsed
- Antenna: 300 × 300 m linear array of 9216 crossed dipoles, with manual beam steering over $\pm 3°$ off-vertical angles. Bandwidth = 1 MHz.
- On-line computer: Harris
- Transmitter: Pulse length – 1 to 500 μs; Duty cycle – 5%; Bandwidth – 50 kHz.
- Pulse coding scheme: double pulses at seven different spacings are used, the second pulse of each pulse pair is transmitted on the opposite circular polarization to that of the first pulse, to eliminate "clutter".
- Receiver system: four independent receivers are employed, sharing a common local oscillator.
  The signals are converted to 30 MHz first IF, then converted down to two quadrature signals for complex coherent detection at zero frequency.
  The noise figure is ~ 3 dB (actually determined by sky noise).

The US National Astronomy and Ionosphere Center (NAIC) at Arecibo

As mentioned in Section 2.3.2, the Arecibo and Jicamarca ISR's were the "first generation" incoherent scatter radars and may have been somewhat "overdesigned" in terms of transmitter power and antenna gain – based on our present understanding of the theory and practice of incoherent scatter from the terrestrial ionosphere. These features have, however, been put to very good use in investigations of the equatorial and low latitude ionosphere, as well as for radio and radar astronomy observations at Arecibo. An airborne photo of the Arecibo installation is shown in Figure 5.3.

Arecibo ISR parameters:

- Frequency: 430.0 MHz
- Peak output power: 2 MW
- Configuration: monostatic, pulsed
- Antenna: 300 m diameter spherical reflector with movable line feed. This is the only ISR with a spherical reflector.
- On-line computer: Harris
- Transmitter: Interpulse period – 12 ms; Pulse lengths – 52 μs; Duty cycle – 5%
- Pulse coding scheme: Several schemes have been used. An example of a recent scheme is described by Sulzer (1986) who developed a binary phase modulation system which results in either *increasing* the time resolution by a factor of 7, or

Fig. 5.3. Airborne photograph of Arecibo ISR Facility. (Courtesy of Craig Tepley)

decreasing the statistical errors by a factor of $\sqrt{7}$. This technique is most applicable to high S/N situations such as those observed at Arecibo.

- Receiver system: Includes a phase code with seven phase shifts, correlation function = 11 ms, Barker coded 108μs, with resolution of 600 μs (see Sulzer 1986).

The MU Radar

The most recent ISR to become operational is located at Shigaraki Shiga, Japan, associated with the Radio Atmospheric Science Center of Kyoto University. The MU (middle-and-upper atmosphere) radar is a pulsed monostatic Doppler radar which is used primarily as an MST (Mesosphere-Stratosphere-Troposphere) radar, but which also has the capability of operation as an ISR (Fukao et al. 1985; Kato et al. 1984; Sato et al. 1988) and Oliver et al. (1990).

A unique feature of the MU radar is the "active" antenna array. Each of the 475 Yagi crossed antennas which comprise the array has its own solid-state power amplifier, receiver preamplifier, and T/R switch – as shown in the block diagrams of Figs. 5.4 and 5.5.

This "active" array design philosophy permits very fast switching (during an interpulse period), almost continuous operation, and fast forming of independent

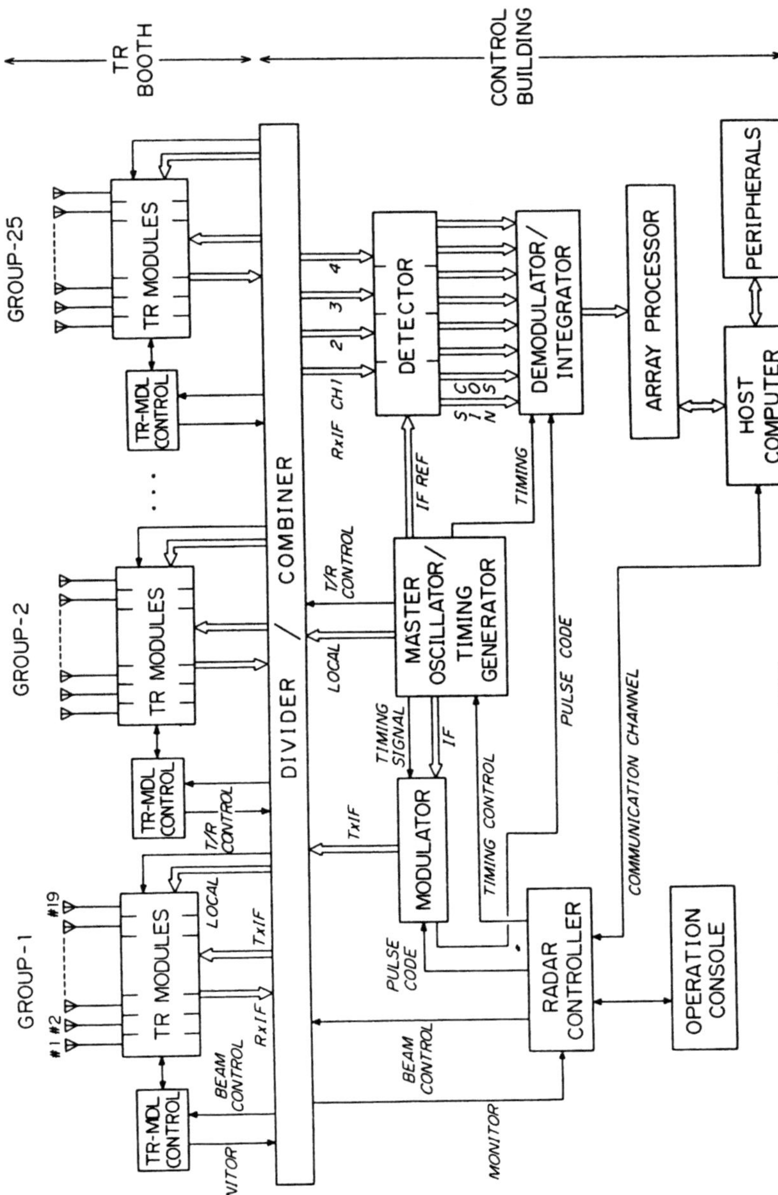

Fig. 5.4. Block diagram of MU Radar system. (Fukao, 1985)

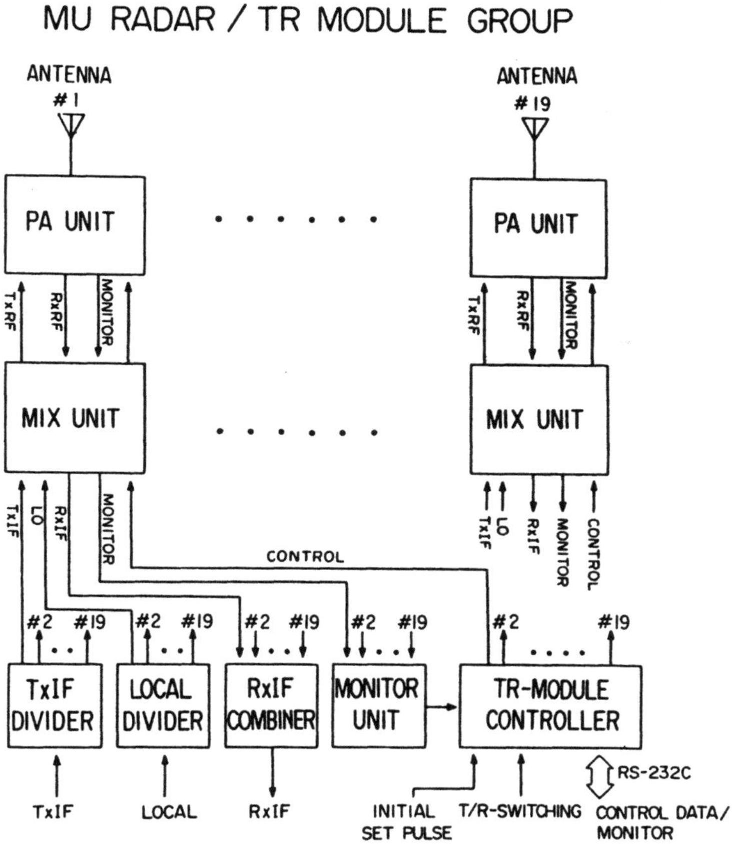

Fig. 5.5. Block diagram of MU Radar TR module. (Fukao 1985)

subarrays of the main antenna array. Three-dimensional probing of specific atmospheric regions should be possible with this type of system. Figure 5.6 is a photo of the MU system taken shortly after it became operational.

Pertinent parameters of MU radar:

- Frequency: 46.5 MHz
- Peak output power: 1 MW
- Configuration: monostatic, pulsed
- Antenna: circular array of 475 crossed Yagi's in circular array 103 m in diameter (aperture = $8.33 \times 10^3\,m^2$); beamwidth = 3.6° (one-way, half power); polarization-linear circular; steerability – 0 to 30° from zenith in interpulse period of 400 μs.
- Transmitter: duty cycle = 5%; bandwidth = 1.65 MHz for maximum pulse length; pulse length = 1–512 μs variable; PRF = 25 Hz.

Fig. 5.6. The MU Radar System – Shigaraki, Shiga, Japan. (Courtesy of S. Kato 1986)

- Receiver: bandwidth = 1.65 MHz; dynamic range = 70 dB; IF frequency = 5 MHz.
- Pulse coding scheme: Binary phase-coding of the 32-bit length of pulse (maximum). Barker and complementary codes are used.
- A/D converter: 12 bits × 4 channels
- On-line computers: VAX 11/750 and MAP-300 array processor.
- System control: HP9835A.

The Millstone Hill ISR

The incoherent scatter radar at the Haystack observatory of MIT is an example of how an *existing "coherent" radar system* was modified to permit probing the ionosphere using this powerful technique. The first reported ISR measurements using this facility were made in 1960 by V.C. Pineo, and in 1962 a regular observational program was started by J.V. Evans. Under Evans' leadership, for the next 20 years the incoherent scatter radar technique emerged as the most powerful tool available for probing the terrestrial ionosphere. Figure 5.7 is an aerial photo of the present antennas employed at the Millstone Hill ISR.

Fig. 5.7. The steerable and fixed UHF parabaloidal antennas at the Millstone Hill ISR facility. (Photo courtesy of John Foster)

Table 5.3. Characteristics of Millstone Hill ISR

Operating frequency	Pulse peak power output	Antenna type gain	Remarks
I. 440.0 MHz	5 MW	68 m Fixed parabolic reflector, G = 48 dB	Fixed position
II. 440 MHz	5 MW	46 m Parabolic reflector G = 44 dB	Fixed position Covers ±40° north and south of the Millstone latitude at F-region heights
III. 1290 MHz	2.5 MW	25 m Parabolic reflector G = 43 dB	Steerable in azimuth and elevation
Transmitters:	Monostatic pulsed nominal PRF 20–500 Hz	Pulse lengths ≈ 1 μs-2 ms	Duty cycle 3%

Computers used: Harris S-125 and FPS AP-120B array processor and CAMAC interface. All signal processing is performed under software control in the array processor.

Millstone Hill ISR parameters (Table 5.3)

Pulse coding schemes. At this time, single pulses are being used for incoherent scatter observations, but a phase coding (complimentary code) scheme is utilized for troposphere/stratosphere scatter studies and a microprocessor-controlled Barker coding scheme for ionospheric studies is now operational.

UHF receiver system: the pre-amp is a GaAsFET, resulting in system noise temperature of ≈ 95 K. A standard down-conversion system used.

The EISCAT Facility

The newest and most sophisticated of the incoherent scatter facilities now operational is the European Incoherent Scatter Facility (EISCAT) located in the Scandinavian countries. It is a dual frequency, multistatic ISR sited in the auroral zone. The EISCAT organization is founded as a Swedish corporation, and the scientific associates are Suomen Akatemia (SA) of Finland, Centre National de la Recherche Scientifique (CNRS) of France, Max-Planck Gesellschaft (MPG) of Germany, Norges Almenvitenskaplige Forskningrad (NAF) of Norway, Naturvetenskapliaga (NF) of Sweden, and the Science and the Engineering Research Council (SERC) of the United Kingdom. EISCAT headquarters are located at Kiruna, Sweden (see Fig. 5.8 and Table 5.4).

Fig. 5.8. **a** The VHF (*left*) and UHF (*right*) transmitting antennas at the Ramfjordmoen site near Tromso. The VHF antenna panels (each 30 × 40 m) are pointed in different directions to illustrate the construction details. **b** The UHF receive site near Kiruna, Sweden. **c** The UHF receiver site located near Sodankyla, Finland. (Photos courtesy of Murray Baron)

a

b

c

Table 5.4. EISCAT ISR characteristics

Station	Geographic location	L-value	Antenna	Frequency (MHz)	Transmitted power	
					Peak (MW)	Duty cycle (%)
Tromsø UHF	69.6°N 19.2°W	6.46	32 m Paraboloid (transmit/receive) (G = 48.4 dB)	933.5	1.5	12.5
Tromsø VHF	69.6°N 19.2°W	6.46	Parabolic Cylinder (transmit and receive) 120 m × 40 m (steerable in meridian plane from elevations of 30°N to 60°S). It can also be steered 21° in the east-west direction. Beamwidth = 1.7° in elevation and 0.6° in azimuth. Circular polarization is employed.	222.4 $225 + 0.2n$ $n = 0, \ldots, 15$ MHz	5	12.5
Kiruna UHF	67.9°N 20.4°E	5.63	32 m Paraboloid (receiver only)	933.5		
Sodankyla UHF	67.4°N 26.6°E	5.25	32 m Paraboloid (receiver only)	933.5		

The EISCAT facility includes two independent radar systems; a tristatic system operating at UHF and a monostatic system operating at VHF. The transmitters for both systems are located at Ramfjordmoen near Tromsø, Norway and receivers are located at the transmitter site, near Kiruna, Sweden, near Sodankyla, Finland, and a fourth receiver is planned for Murmansk, USSR. To give an idea of the magnitude of effort required to establish a "state-of-the-art" ISR, the EISCAT facility deeply involved the scientific resources of six countries for approximately 10 years for a total estimated cost of 121.6 MSEK (≈ $17 million US in 1984). The EISCAT system is described in detail by Folkestad et al. (1983), Baron (1984), and Rishbeth and Williams (1985).

Salient parameters of the EISCAT system: (*Table 5.4*)

Notes: (1) The EISCAT transmitter's frequency can be changed in 1 μs, and this frequency agility makes it possible to transmit several pulse schemes within one pulse repetition period (prf). Each transmitter has 16 different frequencies available to transmit short pulses, long pulses, coded pulses, and multipulse groups within the same prf. A complete documentation of the newest pulse coding scheme is given by Turunen (1985). (2) The antenna polarization is selectable for linear to circular, with circular being used for transmitting.

- Configuration: multistatic, pulsed.
- On-line computers: Norsk-Data.
- Receivers: Low noise parametric amplifiers are of course, installed at the antennas of all the receiving sites. As of 1987, helium-cooled parametric amplifiers are used at Kiruna and Sodankyla with noise temperatures of ≈40 K. An uncooled GaAsFET paramp is used at Tromsø to achieve system noise temperatures of ≈140 K, which is adequate in the RFI environment at the transmitter site.
  The first IF frequency is (120 ± 4) HMz, and the second IF = 30 MHz, the signals then going through a band pass filter with optional bandwidths of 1.2, 8, or 30 MHz, thence to hybrid detector.

Supporting instrumentation: several ionosondes, magnetometers, riometers, spectrophotometers, all-sky cameras, etc. are in operation in northern Scandinavia to complement the EISCAT observations. Other radio facilities can provide simultaneous information on ionospheric parameters. These facilities can provide simultaneous information on ionospheric parameters. These facilities include: a HILAT satellite earth station (F-region irregularity structure); STARE and SABRE VHF radars (E-region irregularity structure and dynamics); SAFARI auroral radars (E and F region irregularity behavior); rocket ranges at Andoya, Norway; and the ESRANGE near Kiruna to provide in-situ measurements of the ionospheric plasma; the MPAE HF ionospheric heating facility at Ramfjordmoen for modifying the ionospheric plasma; and the Viking satellite program. (Note: the acronyms are defined in Appendix A. 2 and elsewhere in this book.)

The Sondrestrom IRS

The ISR located near Sondre Strømfjord, Greenland is the only "peripatetic radar" of all the ISR's – having been located in California, Alaska, and Greenland! From 1965–1970 it was located near Menlo Park, California (geomagnetic latitude ≈43°N) operated by SRI-I personnel as an ISR (Bates and Baron 1969; Baron et al. 1970); from July 1971 through February 1982 it was located near Chatanika, Alaska (lat. ≈65°N) where it was used by SRI-I and several other institutions (Leadabrand et al. 1972).

The operation of this ISR at Chatanika is an excellent example of the usefulness of the ISR technique – during the 11 years that the radar was located at Chatanika, much new information was obtained on the morphology and physics of auroral ionospheric phenomena and ionosphere-magnetosphere interaction by some 20 organizations and over 100 individual scientists using the facility. Over 250 papers and reports of Chatanika ISR observations were published, and the data are still being analyzed, with new results still being reported.

Fig. 5.9. View from the top of a neighboring hill of the Sondrestrøm ISR in July 1985. (Photo courtesy of Don Rice)

Between March and October 1982 this ISR was moved to Sondrestrøm, Greenland (lat. ≈ 75°N) and has been in operation since early 1983 (Kelly 1983; Wickwar et al. 1984; Kelly and Vickrey 1984). See Figure 5.9 for a photograph of the Sondrestrøm ISR.

Based on the northward progression of this ISR from California to Alaska to Greenland, one is tempted to extrapolate the future location of this radar to a geomagnetic latitude of ≈ 85°N by ≈ 1994! (if funding is available!)

Sondrestrom ISR parameters:

- Operating frequency: 1290 MHz
- Peak pulse power output: 3.5 MW
- Duty cycle: 3%
- Configuration: monostatic, steerable antenna.
- Antenna: fully steerable 32 m diameter parabolic reflector with Cassegranian feed. gam ~ 50 dB; Beamwidth ≈ 0.5 between halfpower points; Efficiency-55%; Driver-Az-El.
- Receiver: thermoelectrically cooled GaAsFET, system noise temperature ≈ 90 K. Three receiver chains, double conversion using 30 MHz and 75 kHz as IF's. New radar signal compander (RSC) operational in 1989.

- On-line computer: Harris
- Pulse coding scheme: Barker coding system operational.
- Supporting instrumentation: a digital ionosonde is located near the site. The Danish Meteorological Institute also operates a riometer and magnetometer in the vicinity of the radar, and the University of Michigan maintains a spectrophotometric facility (including a Fabry-Perot interferometer) at the radar site. A D256 ionosonde and an "IRIS" riometer system were installed at the site in 1989.

It should be noted that two of the presently operating ISR's are located in the auroral or polar regions, and Millstone Hill ISR has the capability of probing the aurorally disturbed ionosphere using long integrating periods. The Chatanika ISR also obtained much information on the high latitude ionosphere over one sunspot cycle – some of which is still being analyzed today. This emphasizes the point that the ISR technique is the most powerful single technique for learning the structure and dynamics of the terrestrial ionosphere during very disturbed conditions, especially at high latitudes.

An interesting historical footnote is that the *first* ISR sounding of the high latitude ionosphere was obtained by W.L. Flock of the Geophysical Institute at the University of Alaska in 1962 during the "shakedown" operational phase of the U.S., Ballistic Missile Early Warning System (BMEWS) radar at Clear, Alaska. This radar operated in the 400 MHz region with very high power, using a narrow beam antenna directed toward the northwest at low elevation angles. Examples of profiles obtained in 1962 are shown in Fig. 5.10 for summer night periods (Flock 1962).

5.3.2 *How to Obtain Data and Information from Currently Operational ISR's*

If additional information is required concerning the description, operating schedule, or data obtained by any of the presently operating ISR's, the following addresses (at the time of publication of this book) should be of use:

Sondrestrøm	Coordinator Sondrestrøm Radar Geoscience & Engineering Center SRI International Menlo Park, CA 94025, USA	(415)859-4782
EISCAT	Director, EISCAT EISCAT Scientific Assoc. Box 705 S-981 27 Kiruna, Sweden	090-187-40 TELEX: 8778 EISSWE S
Millstone Hill	Director, Millstone Hill Field Station MIT Haystack Observatory Westford, MA 01886, USA	(617) 863-5500, X5621

Arecibo	Director Arecibo Observatory P.O. Box 995 Puerto Rico, 00612	(809) 878-2612
Jicamacra	Director, Jicamarca Observatory School of Electrical Engineering Phillips Hall Cornell University Ithaca, NY 14853, USA	(607) 256-6471

One may also obtain a catalog *The NCAR CEDAR Data Base Catalog* which lists data from several of the sounders by contacting:

HAO NCAR
Box 3000
Boulder, CO 80307 USA

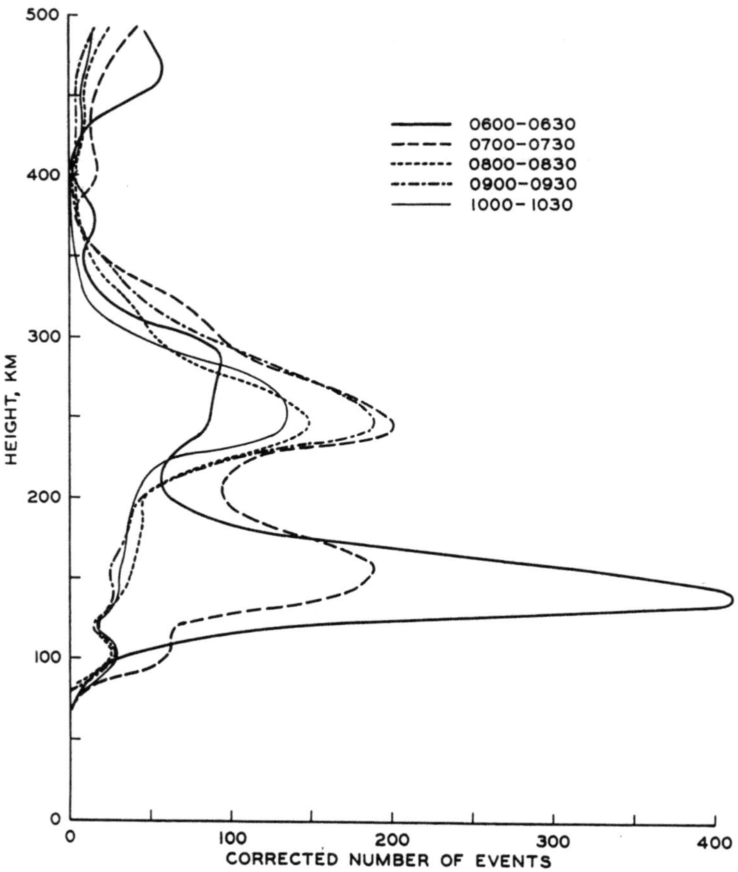

Fig. 5.10. Average incoherent scatter echo intensity (proportional to the electron density profile) obtained from BMEWS radar at Clear, Alaska, by Warren Flock in 1962)

5.4 Capabilities and Limitations

The *capabilities* of incoherent scatter radars can be summarized as follows:

Directly measured quantities:

- Signal-to-noise ratio (S/N) of echoes as a function of radar range along the radar line-of-sight.

Derivable quantities:

Power spectral density (spectra) of individual or averaged returned pulses (after autocorrelation and fast-Fourier transform)

From which one can obtain:

- frequency shift of spectrum from center frequency, hence, line-of-sight Doppler shift of echo
- amplitude of spectrum
- width of spectrum
- shape of spectrum

From the application of incoherent scatter theory, one can then deduce directly:

- T_e/T_i as a function of height
- electron density as a function of height
- vector ion velocity (from several antenna positions) as a function of height
- electric fields in the F-region (from plasma velocities and a geomagnetic field model)
- ion composition as a function of height
- estimates of thermospheric composition as a function of height (from shape of spectrum).

By assuming model neutral densities, one can also measure:

- Hall and Pedersen conductivities
- meridional neutral wind
- current perpendicular to B
- joule heating
- energy deposition by auroral electrons
- neutral temperature
- heat flux at high altitudes
- Birkeland current
- energy spectrum of auroral electrons
- other strange and wondrous quantities.......

Most of the above quantities may be obtained during periods when the ionosphere is extremely agitated, as during a major geomagnetic storm.

Limitations:

An obvious *disadvantage* of this system is cost. Estimates for construction of a new state-of-the-art facility are ≈ \$15–25 million-US in 1991. There are also problems in siting such a facility to minimize RFI and "auroral clutter" problems, and in obtaining the required "radio horizon" terrain elevations as a function of azimuth. The logistics of the facility supply and maintenance are also quite expensive at some of the rigorous locations required for exploration of certain latitudinal regions of the ionosphere. If a *monostatic* ISR system is used, there is the additional problem of the convolution of spatial and temporal measurements. For instance, if one is making azimuth scans or a sequence of three-position measurements, the F-region range gates might be typically separated by ≈ 200 km horizontally, and in order to resolve velocities one must assume that the temporal and spatial variations of the ion velocities are small over the distances and times required for the scan mode.

There are also certain limitations in the proper application of incoherent scatter, signal detection, and pulse coding theory to the analysis of ISR data, and some inherent errors in ISR systems. Some specific examples are:

- In order to obtain the frequency of the *plasma line* as a function of height, the plasma line S/N must be enhanced by suprathermal electrons, photoelectrons, or precipitating auroral electrons. An example of the complexity of this problem can be found in discussions on the effects of electron-neutral collisions on the plasma frequency (see Swift 1974).
- Systematic errors in the electron density must be corrected by use of critical frequency values from an ionosonde, integrated electron content to a satellite, Faraday rotation measurements, or plasma line observations.
- In the bistatic mode, the phenomenon of "self-noise" is introduced – which adds to the system noise budget. It can be ameliorated by using an orthogonally polarized double pulse scheme or other methods.
- The technique for obtaining *error estimates* for such parameters as: T_e, T_i, $([H^+] + [He^+])/N_e$, line-of-sight velocity, etc. seems to increase in complexity with the degree of sophistication of the ISR system, to the point where one seldom sees error bars on ISR data. (Or, if they are present, one wonders how realistic they are!).
- In a typical program to determine parameters such as electric field, Hall and Pedersen conductivities, current densities, neutral winds, and joule heating, one makes assumptions such as:
  - the ion-neutral collisions are negligible above ≈ 150 km;
  - the electron collision frequencies are negligible above 85 km;
  - the field lines are equipotentials down to 85 km;
  - an empirical equation is used to describe the "smearing" of velocities over the width of the spectral range gate.
- In more esoteric computer codes used to estimate ionospheric composition, precipitating particle energy spectra, etc. one must resort to many iterations of atmospheric models from a "library", raising the question of the "uniqueness" of the particular model used, among other concerns.

– Also uncertainties $(O^+)/(NO^+)$ and electric field effects in the F-region make it difficult to unambiguously interpret some ISR results.

In the applications of *any theory to a particular experiment* one *must*, of course, make assumptions and approximations, but in the case of ISR data analysis, the developers of the theory must be particularly careful to list these assumptions and carefully outline them to users of the data. Good examples of detailed documentation of data analysis software for the Chatanika/Sondrestrom and EISCAT ISR users are found in reports by Turunen (1985), Beaujardiere et al. (1984), Brekke (ed., 1977), and Rishbeth and Williams (1985).

In our examination of the ISR technique for probing the terrestrial ionosphere, it seems rather obvious that, in spite of its expense and limitations, it is the most powerful ground-based radio technique available at this time. Indeed, sophisticated ISR systems such as EISCAT, when used to the limit of their capabilities, are, without much doubt, the best single technique of *all* techniques available. However, because of high electrical power requirements and relatively high maintenance costs, most ISR facilities are only operated 100–200 h/month. Thus, they are employed in the "campaign" mode and do not provide continuous data. Further improvement of the ISR technique most likely will result from advances in theory and implementation of advanced signal processing and modulation. As this book goes to Press, tentative plans are being made to build 4 more ISR's in northeastern Canada, in Alaska, in Taiwan, and on Spitzbergen (Svalbard) island.

Chapter 6 Ionospheric Modification by High Power Radio Waves

Introduction

Ionospheric Cross Modulation

The first published observation of the phenomenon of modification of the terrestrial ionosphere by high power radio waves was made by Butt (1933), who observed that the transmission from Radio Paris was superimposed on the signal transmitted by Radio Luxembourg. The most often quoted reference in the literature, however, is that reported by Tellegin (1933), who reported the reception at Eindoven, Holland, of the program from Radio Luxembourg (the "unwanted signal," p = 200 kW, f = 250 kHz) on top of the "wanted" signal – Radio Beromünster – at 650 kHz. Tellegin correctly interpreted this observation as the interaction of the two signals in the ionosphere.

"The phenomenon described above was first known as the Luxembourg effect (Soviet scientists refer to it as the Luxembourg-Gorkii effect to include the simultaneous, independent work of Gorkii), then as radio wave interaction, and later as ionospheric cross modulation. Huxley and Ratcliffe (1949) recommended that the latter term be used and that the interacting radio waves be labeled the "wanted wave" and the "disturbing wave." Figure 6.1 illustrates the geometry and nomenclature for an oblique incidence, pulsed, cross-modulation experiment.

Ionospheric Heating

In addition to *passive* observations of the natural ionosphere, scientists have also used *active* experiments to investigate the dynamics of the ionosphere. Experiments can monitor the response of the ionosphere to controlled external heating, as well as its recovery to ambient conditions after the perturbing force is removed. This allows measurement of ionospheric parameters such as heating and cooling rates as a function of altitude that are important elements of ionospheric computer models. These studies also aid in the understanding of disturbed ionospheric effects by producing nonequilibrium conditions in a relatively controlled experimental environment.

Bailey (1937a) was apparently the first to suggest that the ionosphere could be "heated" by a sufficiently powerful HF transmitter using a high gain antenna. The state of the art in HF transmitters and antennas was not sufficiently advanced at that time to carry out this experiment and by the time the technology became

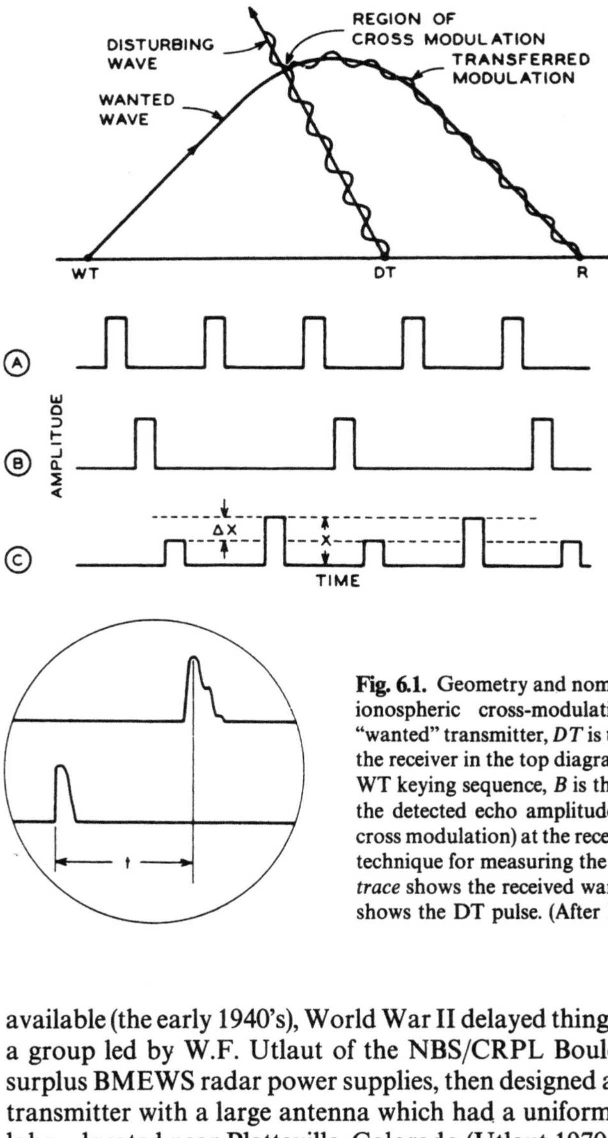

Fig. 6.1. Geometry and nomenclature describing a "generic" ionospheric cross-modulation experiment, *WT* is the "wanted" transmitter, *DT* is the "disturbing" transmitter, *R* is the receiver in the top diagram. *In the middle section, A* is the WT keying sequence, *B* is the DT keying sequence, and *C* is the detected echo amplitude of the wanted wave (for 50% cross modulation) at the receiver. The *bottom panel* shows the technique for measuring the height of interaction. The *upper trace* shows the received wanted echo, and the bottom trace shows the DT pulse. (After Benson 1963)

available (the early 1940's), World War II delayed things further. In the late 1960's a group led by W.F. Utlaut of the NBS/CRPL Boulder laboratories acquired surplus BMEWS radar power supplies, then designed and constructed a CW HF transmitter with a large antenna which had a uniform vertically directed major lobe – located near Platteville, Colorado (Utlaut 1970; Utlaut and Cohen 1971). This system was then used to modify the ionosphere, yielding many interesting results (see the special issue of *Radio Science*, edited by Utlaut, of November 1974).

The idealized sketch in Fig. 6.2 illustrates some of the effects produced by ground-based HF transmitter systems with power-aperture products of the order of 10^4 MW/m^2. The subject of ionospheric modification by high power HF transmissions has been reviewed by Perkins (1974), Carlson and Duncan (1977), Fejer (1979), Stubbe and Kopka (1980), and Gordon and Duncan (1980). A thorough theoretical treatment of ionospheric modification by powerful radio

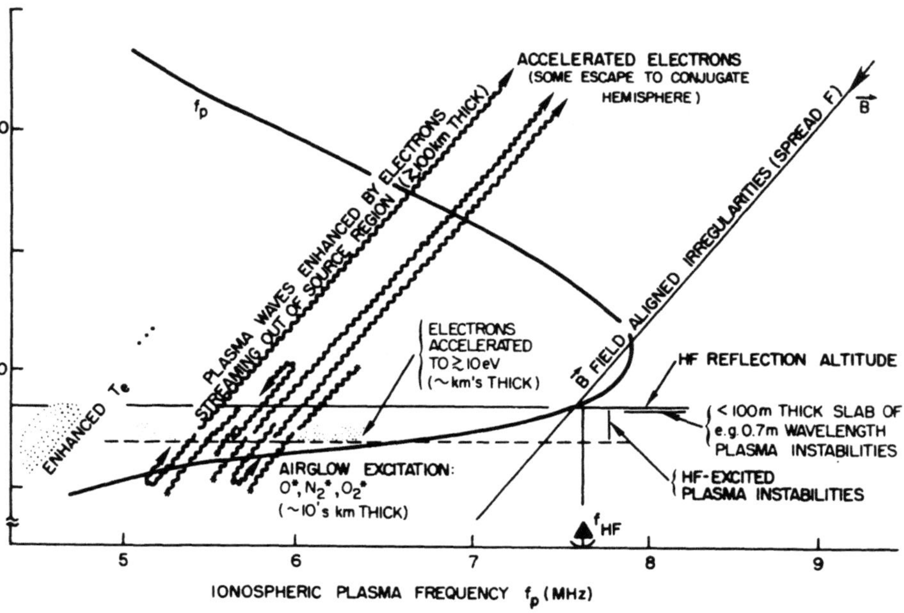

Fig. 6.2. Some of the effects produced by high-power HF heating facilities. (Carlson and Duncan 1988)

waves (including *both* heating and cross modulation phenomena) was presented by Gurevich (1978) in Volume 10 in the same series as this book.

6.1 Ionospheric Cross Modulation Techniques – Some Representative Results

From the middle 1940's through the 1970's several research groups utilized the ionospheric cross modulation method in carefully controlled experiments to obtain collision frequencies in the lower ionosphere. The wave interaction and the partial reflection techniques have been the major methods for studying the D-region for at least the last two decades. However, the incoherent scatter radar (ISR) technique (i.e., Arecibo) etc. has been used quite successfully to measure D-region parameters during this same period. Recent advances in pulse-coding and signal processing hardware and software indicate that incoherent scatter radars will soon be capable of very high-resolution probing of the D-region (Devlin et al. 1986).

The following sections describe variants of the wave interaction technique used by different investigators.

6.1.1 Modulated Disturbing Wave, Unmodulated Wanted Wave at Oblique Incidence, and Gyro Interaction

Ionospheric cross modulation experiments using a modulated disturbing wave and an unmodulated wanted wave have been performed by Ratcliffe and Shaw

(1948), Huxley (1950) and Shaw (1951). These investigators compared the phase and amplitude of the transferred modulation received on the ionospherically reflected wanted wave with the phase and amplitude of the modulation on the disturbing wave via the direct path DTR in Fig. 6.1, and obtained the electron collision frequency ν at a given height. They used power levels from 50 to 150 kW and frequencies from 200 kHz to 1 MHz.

The concept of *gyro-interaction* was developed by Bailey (1937a) in a revision of his earlier theory of cross modulation. The 1937 theory stated that a resonance effect should be observed as the frequency of the disturbing transmitter is varied about the gyrofrequency, f_H (see Section 2.2.1). He further stated that a disturbing transmitter radiating only a few kilowatts could, under certain conditions, produce observable interaction when the operating frequency was within a few percent of f_H, and he called this disturbing wave the *gyro-wave*.

Bailey (1937b) gave some observational evidence for the existence of the gyro-wave, and this was confirmed by other observations reported by him later that year (Bailey 1937c). The subject of gyro-interaction has, however, been the subject of considerable controversy (Huxley and Ratcliffe 1949, 1950), mainly concerning the magnitude of the effect predicted by theory. Bailey et al. (1952) observed the gyro-interaction resonance effect during experiments conducted in Australia and concluded that this technique can be used to measure the magnetic field in the E-region with a high degree of precision. Considerable work on radio wave interaction using "gyro-waves" was done by investigators at the University of New England in NSW, Australia as reported by Smith et al. (1965). This group used a 500 kW transmitter feeding a circularly polarized array of 40 dipoles to study the mid-latitude lower ionosphere at night and obtained accurate values of the energy loss coefficient, electron relaxation time, and electron collision frequency at heights of 69 to 85 km.

6.1.2 Self-Interaction and Gyro Self-Interaction

Cutolo (1951) observed that the modulation of the disturbing wave itself was modified during its passage through the ionosphere – this is called self-interaction or self-distortion. When the disturbing signal frequency is close to the gyrofrequency, the effect is called gyro self-interaction. A theory describing self- and gyro-interaction has been developed by Hibberd (1955, 1957) which predicts that the maximum reduction in modulation depth for a gyro-wave with 100 kW radiated power would be 1–2% under "favorable" ionospheric conditions.

6.1.3 Pulse Modulated Disturbing and Wanted Waves – Vertical Propagation

In 1954 Fejer performed experiments in South Africa using pulsed wanted and disturbing waves, obtaining profiles of electron density and collision frequency in the 65–90 km height region of the ionosphere (Fejer 1955). In this version of the cross modulation technique, the disturbing transmitter radiates one pulse for every other pulse radiated by the wanted transmitter, and the pulse repetition period of the wanted pulses is chosen to be much longer than the decay time of the elevated electron temperature caused by the disturbing pulse. This results in cross modulation producing a fractional change in the amplitude of every other pulse of the wanted wave. Using Fejer's technique, then, one can vary the height of interaction by varying the time t between the transmission of the disturbing pulse and the reception of the wanted echo pulse (see Fig. 6.2 and 6.3). Variations of the Fejer cross modulation technique have been used since 1957 by Barrington and Thrane (1962) , Rumi (1962a, b), and others.

6.1.4 Pulse Modulated Disturbing Wave – Cosmic Noise Wanted Wave, Using Vertical Propagation

At auroral latitudes it was found that the cross modulation technique was limited because of anomalously high D-region absorption much of the time. While it was necessary that absorption be high enough to absorb energy from the disturbing wave (to produce a heating effect), it also had to be low enough to permit reception of the lower frequency (thus more greatly absorbed) wanted wave. G.C. Rumi suggested that cosmic radio noise could be used as the wanted wave at College, Alaska. The cosmic noise method has several advantages over the standard cross modulation method: (1) it permits obtaining D-region parameters in the auroral regions over extended time periods, (2) several heights may be explored simultaneously by using multiple receiving channels, (3) continuous operation is possible without an operator (no adjustments are necessary to compensate for variations in the height of reflection of the wanted wave). Results of these studies are given by Benson (1962, 1963).

6.1.5 Outline of Cross Modulation Theory

Bailey and Martin (1934) originally developed the basic theory of cross modulation which was then revised by Bailey (1937a) and recast into nomenclature more familiar to ionospheric physicists by Shaw (1951), then further refined by Huxley (1953). Another approach to the theory of nonlinear effects in the lower ionosphere, based on Boltzmann's kinetic theory, was proposed by Ginzburg (1957). Huxley's treatment is outlined in the following section.

Consider the absorption of the wanted wave as it passes through an ionospheric layer,

$$E = E_0 e^{-\int_0^s K_w \, ds}, \qquad (6.1)$$

where:

E_0 = amplitude of the electric field of *incident* wave
E = amplitude of electric field of *emerging* wave
K_w = absorption coefficient for the wanted wave = $K_w(N, v)$
ds = element of path length

N = electron density
ν = electron collision frequency,

then let $\nu = \nu_0 + \Delta\nu$ and $N = N_0 + \Delta N$, where ν and N are functions of time t, ν_0, and N_0 are values measured in the absence of a disturbing wave, and Δν and ΔN are changes produced by the disturbing wave.

A Taylor expansion can then be used to obtain an expression for the absorption coefficient (K_w) as,

$$K_w(N,\nu) = K_w(N_0,\nu_0) + \frac{1}{1!}\left[\left(\frac{\partial K_w(N,\nu)}{\partial N}\right)_{N_0,\nu_0}\Delta N + \left(\frac{\partial K_w(N,\nu)}{\partial \nu}\right)_{N_0,\nu_0}\Delta\nu\right] + \frac{1}{2!}\left[\left(\frac{\partial^2 K_w(N,\nu)}{\partial N^2}\right)_{N_0,\nu_0}(\Delta N)^2 + 2\left(\frac{\partial^2 K_w(N,\nu)}{\partial N\partial\nu}\right)_{N_0,\nu_0}(\Delta N)(\Delta\nu) + \left(\frac{\partial^2 K_w(N,\nu)}{\partial \nu^2}\right)_{N_0,\nu_0}(\Delta\nu)^2\right] + \cdots. \tag{6.2}$$

If we assume that changes in ν dominate in this relation, then (6.2) can be written as

$$K_w(N,\nu) = K_w(N_0,\nu_0) + \frac{1}{1!}\left(\frac{\partial K_w(N,\nu)}{\partial \nu}\right)_{\nu_0}\Delta\nu + \frac{1}{2!}\left(\frac{\partial^2 K_w(N,\nu)}{\partial \nu^2}\right)_{\nu_0}(\Delta\nu)^2 + \cdots. \tag{6.3}$$

We can now write Eq. (6.1) as

$$E = E_0\left[\exp\left(-\int_0^s K_W(N_0,\nu_0)ds\right)\right] \cdot \left\{\exp\left(-\int_0^s\left[\frac{1}{1!}\left(\frac{\partial K_W}{\partial\nu}\right)_{\nu_0}\Delta\nu + \frac{1}{2!}\left(\frac{\partial^2 K_W}{\partial\nu^2}\right)_{\nu_0}(\Delta\nu)^2 + \cdots\right]ds\right)\right\}. \tag{6.4}$$

If we further assume that $e^{-x} \approx 1 - x$ for $x \ll 1$, then

$$E = E_s\left\{1 - \int_0^s\left[\left(\frac{\partial K_W}{\partial\nu}\right)_{\nu_0}\Delta\nu + \frac{1}{2!}\left(\frac{\partial^2 K_W}{\partial\nu^2}\right)_{\nu_0}(\Delta\nu)^2\right]ds\right\}, \tag{6.5}$$

where

$$E_s = E_0\exp\left(-\int_0^s(N_0,\nu_0)ds\right)$$

is the electric field amplitude of the emergent wanted wave. We can define the cross modulation T as

$$T = \frac{(E - E_s)}{E_s} = -\int_0^s\left[\left(\frac{\partial K_W}{\partial\nu}\right)_{\nu_0}\Delta\nu + \frac{1}{2!}\left(\frac{\partial^2 K_W}{\partial\nu^2}\right)_{\nu_0}(\Delta\nu)^2 + \cdots\right]ds. \tag{6.6}$$

Equation (6.6) can be simplified when the frequency of the wanted wave is such that $\nu \ll \omega$ in the region of cross modulation. For frequencies ≈ 1 MHz, this condition is satisfied for heights > 80 km. In this region $K \propto \nu/\omega$ and $\partial K_W/\partial \nu$ is independent of ν, the higher-order derivatives of K are all $= 0$, and 6.6 then reduces to

$$T = -\int_0^s \left(\frac{\partial K_W}{\partial \nu}\right)_{\nu_0} \Delta\nu \, ds. \tag{6.7}$$

In order to apply cross modulation theory to experimental observations, the term $\Delta\nu$ must be related to the energy supplied by the disturbing transmitter.

Consider now the case of pulse modulated radio waves vertically incident upon the ionosphere. The absorption of a wanted radio wave traveling downward through a thin homogeneous layer dh can be described by

$$E = E_0 \exp\left(-\int_{h+dh}^{h} K_W \, dh\right), \tag{6.8}$$

where the wanted parameters are:

E = amplitude of the electric field of the emergent wave.
E_0 = amplitude of the electric field of the incident wave.
K_W = absorption coefficient.

For infinitesimal layer thickness we can assume that K_W = constant, and

$$E = E_0 e^{-K_W dh}. \tag{6.9}$$

Further development of the theory using the results obtained by Bailey and Martyn (1934) and Phelps (1960) gives the following formula describing the observed cross modulation

$$T = \frac{Pg}{2\pi a k} \int_0^{h_0} \left(\frac{\partial K_W}{\partial \nu}\right)_{h'} \frac{(Kd)_{h'} D}{N\theta_e (h')^2} \exp\left(-\int_0^{h'} 2K_d dh\right) dh, \tag{6.10}$$

where:

D = the decay term

$$D = 1 - \exp\left(-2(h_0 - h')\frac{G\nu_a}{c}\right),$$

when

$$h_0 \geq h' \geq \left(h_0 - \frac{c\tau}{2}\right)$$

and

$$D = \left\{1 - \exp\left[-2\left(\frac{c\tau}{2}\right)\frac{G\nu_a}{c}\right]\right\} \exp\left[-2\left(h_0 - \frac{c\tau}{2} - h'\right)\frac{G\nu_b}{c}\right],$$

where $h' < h - c\tau/2$

P = transmitter peak power in watts
G = a constant to be determined
θ_e = electron temperature
a = a constant = 1, 4/π, 3/2 depending on whether the electron thermal energy is considered to be associated with the most probable electron velocity v_m, the mean electron energy, v, or v_{rms}
g = antenna gain in vertical direction
K = Boltzmann's constant = 1.38×10^{-23} joules/K.

The decay term D describes the exponential rise and decay of Δν as a function of height below h_0 and is illustrated for a few values of h_0 for $\tau = 50\,\mu s$ in Fig. 6.3.

Fejer (1955) considered the disturbing pulse to be a delta function, resulting in an instantaneous increase in Δν, while Rumi (1962b) did not include the exponential rise and decay of Δν in his theoretical development. Equation (6.10) and Fig. 6.3 describe the analysis of the response to a rectangular disturbing pulse. Other investigators have used different pulse shapes in their investigations, producing somewhat different results. Rumi (1962b) and Benson (1963) extended Fejer's (1955) theoretical treatment for wave interaction, resulting in the equation

$$T = -\int_0^{h_0} \left\{ \left(\frac{\partial K_w}{\partial \mu} \right)_{h'} - \frac{N}{\nu} \left(\frac{\partial K}{\partial N} \right)_{h'} \left[\theta_E \frac{\partial (\ln K)}{\partial \theta_e} \right] \right\} \Delta\nu \, dh, \tag{6.11}$$

where: h′ = virtual height.

Benson (1963) summarized the expression for total cross modulation for four cases,

Case I – the classical theory
Case II – the classical theory approximation to the generalized theory for low ionospheric heights such that $\nu \gg \omega \pm \omega L$
Case III – the generalized theory (consideration $\partial K_w/\partial \nu$ only)
Case IV – the generalized theory (considering $\partial K_w/\partial \nu$ and $\partial K_w/\partial N$,

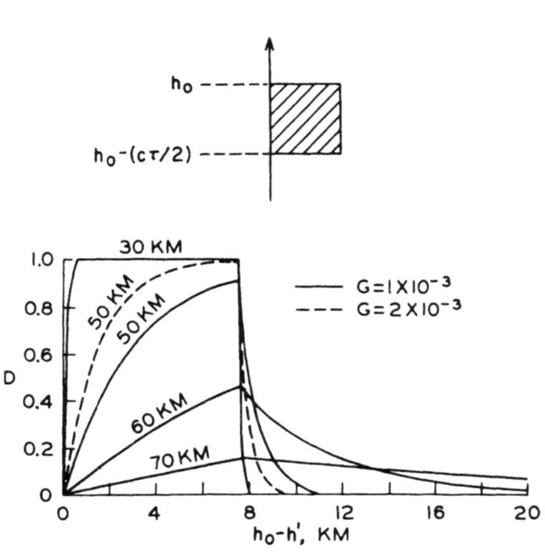

Fig. 6.3. Some effects of the ionospheric time constant (*G*). *Top* the upward traveling disturbing pulse when it has reached a height h_0. *Bottom* The decay term, *D*, parametric in height (30, 50, 60, 70 km) for two different values of *G*. (Benson 1963)

and concluded that the equations derived above *all* indicated that the major contribution to cross modulation results from variations in the electron collision frequency at all elevations except ≈ 30 km, where variations in electron *density* becomes important.

The basic experiment devised by Fejer (1955), which measures the *amplitude* of the electric field of the wanted wave as modified by the disturbing wave, has been employed by several groups of investigators to obtain electron density and collision frequency profiles in the 60–90 km height range (Landmark and Lied 1961; Barrington and Thrane 1962). An improved form of Fejer's experiment was suggested independently by Vilenskii (1962) and by Ferraro et al. (1963), which also measures *phase changes* in the waves, and the first results using this technique were reported by Weisbrod et al. (1964a, b). The path geometry of the basic Fejer experiment and a simplified block diagram of the system used to obtain these first results are shown in Fig. 6.4 and 6.5, respectively.

The expression for the "modulation transfer coefficient," T, including phase changes is derived by Weisbrod et al. (1964a) as

$$T = \int_0^{h_i'} T_h \left(\frac{1 - e^{-G\nu\tau}}{G\nu\tau} \right) \cdot \exp\left[\frac{-2G\nu(h_i h)}{c} \right] \left[-\frac{\partial F}{\partial h} \right] dh \tag{6.12}$$

$$+ \int_{h_i'}^{h_i} \frac{T_h}{G\nu\tau} \cdot 1 - \exp\left[\frac{-2G\nu(h_i - h)}{c} \right] \left[-\frac{\partial F}{\partial h} \right] dh,$$

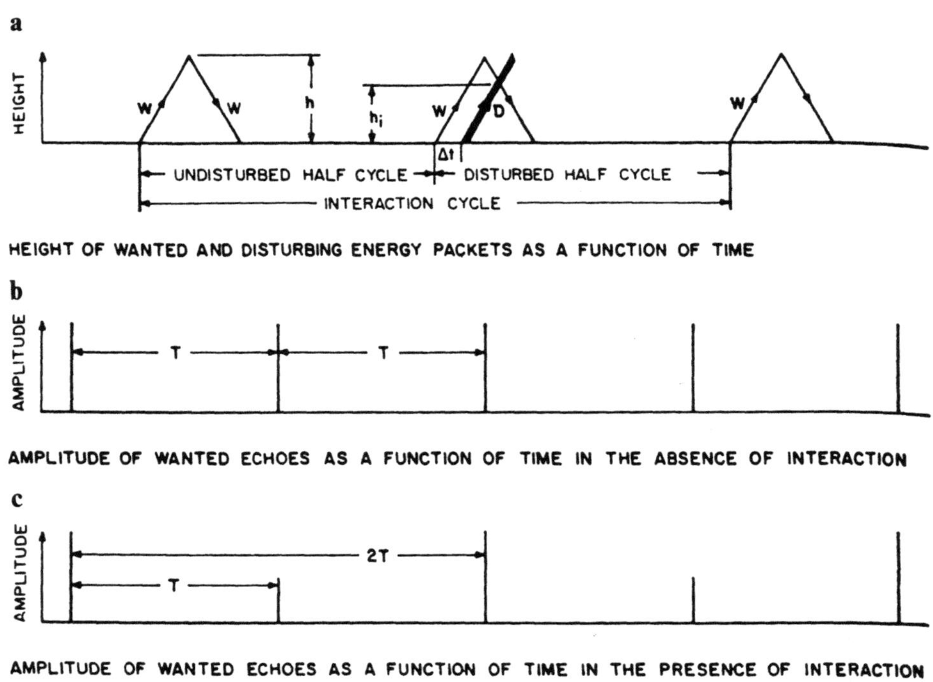

Fig. 6.4 a–c. Path geometry for the basic Fejer experiment using a finite disturbing pulse. (After Weisbrod et al. 1964)

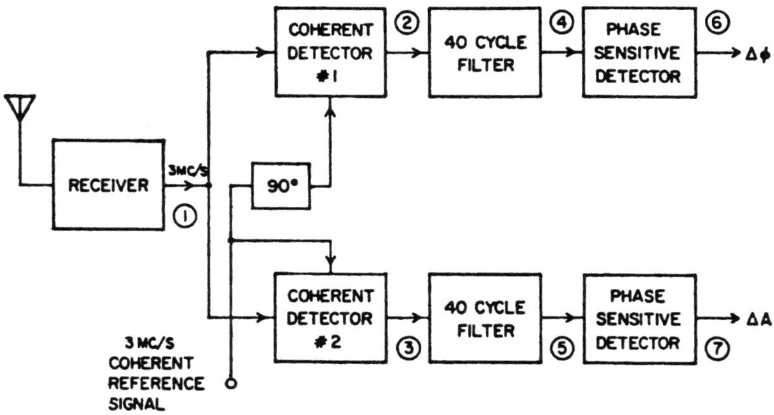

Fig. 6.5. Simplified block diagram of the Penn State University phase interaction equipment. (Weisbrod et al. 1964b)

where:

$$h_i' = h_i - \frac{c\tau}{2}$$

$$T_h = \frac{A}{h^2}\left[\frac{Z(h)}{(1 \pm Y_L)} - jZ(h)\right]^2$$

$$A = \frac{e^2 P_D G_D \tau B}{12\pi k\,\theta\, m\varepsilon\Omega^2}$$

P_D = disturbing transmitter power
G_D = disturbing antenna gain
G = energy loss factor, experimentally determined to be 1.7×10^{-3}
τ = duration of disturbing pulse
K = Boltzmann's constant
θ = electron temperature
k_w = free space propagation constant for the wanted wave
$Z = v/w$
Y_L = longitudinal component of Y.

All other quantities are shown in Fig. 6.4.

A mathematical technique for obtaining D-region electron density profiles using the experimental technique described above was reported by Hellrich and Ferraro (1965), and many D- region profiles were obtained using $P_D \approx 100\,\text{kW}$ at a frequency near 300 kHz, with f = 2.4 MHz as the wanted frequency (Ferraro and Lee 1966). An analysis by Manson (1967) indicated that a power level of $\approx 1\,\text{MW}$ and a receiver time constant of $\approx 5\,\text{s}$ should greatly improve the observational efficiency of pulse interaction experiments. This prompted the

Penn State University group to initiate construction of an improved facility in 1966 with the following parameters:

Disturbing system	Wanted system
P = 500 kW	P = 100 kW
f = 4.5 MHz	f = 2.2 MHz
$\tau = 10\,\mu s$–$1000\,\mu s$	
Antenna gain = 14.5 dB circularly polarized.	

A description of this system and results obtained were described by Ferraro and Lee (1968), and additional D-region measurements using the Penn State University system were reported by Lee and Ferraro (1969), Rowe et al. (1970), Cole et al. (1969), Hagenbuch (1973), and Rowe et al. (1974); also Lee et al. (1977) presented a detailed intercomparison between the wave interaction and partial reflection methods of measuring electron density.

Other investigators who utilized the wave interaction technique at different locations include Belrose et al. (1968) in Canada and Thrane (1974) in Norway.

Another version of the wave interaction technique called "complementary cross modulation" was proposed by Sulzer et al. (1976) and tested at the Arecibo HF heater facility. The results were somewhat ambiguous because of the limitations of the heater facility at that time.

Since the late 1970's no regular investigations of the D-region using the wave interaction technique have been carried out (to the author's knowledge). This is principally because of the expense of building, maintaining, and operating the required facilities when compared with the scientific results obtained from a relatively restricted height range of the ionosphere.

6.2 Ionospheric Heating Using High Power HF/CW Transmitters

A fairly comprehensive treatment of the theory of ionospheric heating by high power HF/CW transmitters is presented in a monograph by Gurevich (1978), and some of his most important results are summarized in Section 6.2.2. Readers are referred to Gurevich's book and the references cited therein for derivations of these results. At the time of writing of the present book, high power HF ionospheric heating facilities are operational in Norway, Alaska, Puerto Rico, and the Soviet Union. Descriptions of the theoretical basis for various phenomena associated with ionospheric heating are given by Perkins (1974), Fejer (1979), Stubbe and Kopka (1980), Tomko et al. (1980), Lee (1981), Lee and Kuo (1983), Stubbe et al. (1984), and Shoucri et al. (1987).

6.2.1 Basic Theory of Ionospheric Heating

Gurevich's elementary treatment for the heating of electrons and ions assumes an unmagnetized three-component isotropic plasma, for which the change in electron and ion temperature is given by

$$\frac{dT_e}{dt} = \frac{-2}{3} ev_eE - \delta_{ei}\nu_{ei}(T_e - T_i) - \delta_{em}\nu_{em}(T_e - T) \qquad (6.13)$$

and

$$\frac{dT_i}{dt} = \delta_{ei}\nu_{ei}(T_e - T_i) - \nu_{im}(T_i - T), \tag{6.14}$$

where:

T_e = electron temperature
T_i = ion temperature
e = electronic charge
v_e = average directional velocity of electrons
E = electric field intensity
ν_{ei} = collision frequency of electrons with ions
ν_{em} = collision frequency of electrons with neutrals
ν_{im} = collision frequency of ions with neutrals
δ_{ei} = average fraction of energy lost by the electron in one collision with an ion
δ_{em} = average fraction of energy lost by the electron in one collision with a molecule
ν_{em} = collision frequency of electrons with molecules
T = temperature of neutrals (assumed to be constant and independent of time).

For a strongly ionized plasma, when $\nu_{ee} \gg \delta_e$, the electron temperature increase is given by

$$\frac{dT_e}{dt} = \frac{2Q}{3N} - \frac{2e}{3N} Ej(T_e) - \delta(T_e)v_e(T_e)(T_e - T), \tag{6.15}$$

where:

Q = energy released by external source in the ionosphere in 1 cm/s.
N = electron density in a quasi-neutral plasma ($N_e \approx N_i \approx N$).
j = electric current density.

In the *D- and E-regions* of the ionosphere, where the heating of the ions is of little importance (since $\nu_{im} \gg \delta_{ei}\nu_{ei}$), the electron gas heating is expressed as

$$\frac{dT_e}{dt} = \left[\frac{e^2E_0^2\nu_{em}\phi_p}{3m(\omega^2 + \nu_{em}^2)}\right]K_\sigma\left(\frac{\omega}{\nu_{em}}\right) - \delta_{em}\nu_{em}(T_e - T) + \frac{2Q}{3N}, \tag{6.16}$$

where:

ϕ_p = the polarization factor

$$= \left[\frac{\omega^2 + \nu_{em}^2}{(\omega \pm \omega_H)^2 + \nu_{em}^2}\right] \cdot \left[\frac{K_\sigma\left(\frac{|\omega \pm \omega_H|}{\nu_{em}}\right)}{K_\sigma\left(\frac{\omega}{\nu_{em}}\right)}\right].$$

In the ionospheric F-region we must, of course, take the heating of the ions into account and Gurevich describes the electron and ion heating effects as

$$\frac{dT_e}{dt} = \left[\frac{e^2E_0^2\nu_e K_\sigma\phi_p}{3m(\omega^2 + \nu_e^2)}\right] - \delta_{em}\nu_{em}(T_e - T) - \delta_{ei}\nu_{ei}(T_e - T) + \frac{2Q}{3N} \tag{6.17}$$

and

$$\frac{dT_i}{dt} = \delta_{ei}\nu_{ei}(T_e - T_i) - \nu_{im}(T_i - T). \tag{6.18}$$

Gurevich further shows that calculations assuming that for $z = 300\,km$, $Q = 1.7 \times 10^4\,eV/cm^3/s$, $T_{e0} = 2000\,K$, $T_{i0} = 1350\,K$ indicate that appreciable heating of the electrons at $z = 300\,km$ is possible in fields $E_0 \approx E_p$. The temperature of the ions increases weakly up to $z \approx 600\,km$ then above that height it increases to approach T_e.

In a weak electric field the stationary perturbations of electron temperature are determined by the relation

$$\frac{\Delta T_e}{T_{e0}} = \left(\frac{E_0}{E_p}\right)^2 \phi; \quad \phi = \frac{\phi_p}{\phi_T}, \tag{6.19}$$

where ϕ_p is the polarization factor [defined in Eq. (6.16)] and ϕ_T is the non isothermal factor defined as

$$\phi_T = 1 + \frac{\left(T_{e0} - T\right)}{(\delta\nu_e)T_{e0}}\left[\frac{d}{dT_e}(\delta\nu_e)\right]_{T_{e0}}; \quad \delta\nu_e = \delta_{ei}\nu_{ei} + \delta_{em}\nu_{em}. \tag{6.20}$$

This assumes that $\nu_{im} \gg \delta_{ei}\nu_{ei}$ (which is satisfied at $z \geq 600\,km$), and using the stationary distributions of electron temperature given by Gurevich (1978). In the F-region an important part is also played by transport processes.

The *simplest* description of ionospheric ohmic heating is obtained by considering the absorption of an HF wave in the ionospheric height region where the impinging HF wave frequency is near the local plasma frequency. This absorption occurs when the electrons, which are excited by the incident heating wave, collide with heavy particles, and energy is extracted from the wave and added to the random thermal energy of the ambient electrons. The rate at which energy is absorbed (Q) is (neglecting nonlinear process).

$$Q = \tfrac{1}{2}\sigma E_0^2 = \frac{1}{2}\left[\frac{Ne^2\nu E_0^2}{m(\omega^2 + \nu^2)}\right], \tag{6.21}$$

where:

E_0 = maximum value of the heater waves' electric field
σ = electrical conductivity of the plasma
ω = angular frequency of the radio wave
ν = collision frequency (electron-ion)
m = mass of electron
e = electronic charge
N = electron number density.

In the F-region $\omega \gg \nu$, and (6.21) can be simplified to

$$Q = \frac{1}{2}\left(\frac{Ne^2\nu}{m\omega^2}\right)E_0^2.$$

The above development is strictly for *ohmic* heating. Nonlinear effects associated with parametric instabilities are quite substantial, even dominant under many circumstances.

6.2.2 Descriptions of Facilities and Some Results from Ionospheric Heating Investigations at Mid-Latitudes

US Department of Commerce/Boulder Labs Facility at Platteville, Colorado

As mentioned in the Introduction to this chapter, one of the first HF ionospheric heating facilities was established in Colorado by the US Department of Commerce Boulder research labs under the leadership of W.F. Utlaut. Descriptions of the facility plus a summary of results of the first 4 years' observations are given by Utlaut (1970, 1972, 1975); Utlaut and Cohen 1971 and in a special issue of *Radio Science* edited by Utlaut (Vol. 9. No. 11, November 1974). Simultaneously with the development of the heating facilities at Platteville, other heating systems became operational at Arecibo, Puerto Rico, and at Penn State University. Recent results of airglow diagnostics using the Arecibo heater are given by Bernhardt et al. (1988).

The most important parameters of this HF heating facility are shown in Table 6.1.

Figures 6.6, 6.7, and 6.8 illustrate features of the Platteville HF heater facility. The facility was utilized by many investigators using several different diagnostic techniques to probe the "modified" ionosphere from April 1970 until the early 1980's. The transmitters were then dismantled and shipped to Fairbanks, Alaska, in 1983 where they were reassembled and utilized at the HIPAS facility (Wong et al. 1983) – see Section 6.2.4 in this chapter.

Among the results obtained during the period 1970–1974 by investigators utilizing the Platteville HF heater facility are: the first observation of the generation of Spread-F and other signatures on ionograms caused by HF heating (Utlaut et al. 1970; Utlaut and Violette 1974); determination of the spatial extent,

Table 6.1. Platteville, Colorado, HF heater characteristics

Location	Platteville, Colorado (40.18°N, 104.73°W) Geomagnetic Lat. = 49°N, Long. = 317°E
Transmitters	Ten 160 kW cw units using vapor-phase-cooled tetrode tubes in the final amplifiers Frequency tuning range 2.7–25 MHz Power adjustable in 3 dB steps down to – 18 dB from full power Modulation scheme – full power CW ("continuous wave") operation or wide variety of pulsed modes available
Antennas	Two ring array systems, each consisting of a ten-element ring of crossed dipoles, nine elements in a ring of diameter 110 m and one antenna in the center First ring array covers 4.5 to 10 MHz, second covers 2.7–3.5 MHz Half-power beam width = 16° at 7.5 MHz Power-aperture product $\approx 10^4$ MW m^2

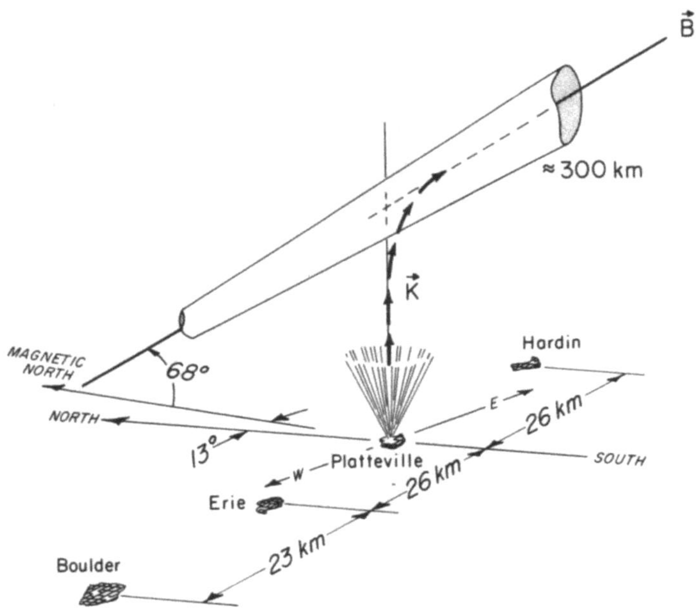

Fig. 6.6. A sketch of the Platteville HF heater geometry. (Courtesy of W.F. Utlaut)

Table 6.2. Salient parameters of Soviet MF and HF ionospheric modification facilities. (Gurevich and Migulin 1982)

Location	Operating frequency (MHz)	Started operation	Transmitter power (kW)
Near Moscow	1.35	1960–1967	10^6
	4.5	1979	8×10^4
Near Gorkii	5–6	1972	10^4
	4–6	1973	1.6×10^4
	5–10	1980	4×10^5
Near Murmansk	3.3	1976	10^4

structure, and time history of HF heater-produced ionospheric irregularities by use of an HF phased array backscatter sounder (Allen et al. 1974); Thome and Blood 1974; Fialer 1974; and other papers in the November 1974 issue of *Radio Science*). Field-aligned irregularities created by the heater were also demonstrated to be useful for HF-VHF communications on path lengths up to 2000 km (Stathacopolous and Barry 1974).

Facilities in the USSR

Other HF heating facilities at mid-latitudes have been used by Soviet scientists to obtain a wide variety of results, as described by Migulin and Gurevich (1985). The

Fig. 6.7. Elements of the Platteville antenna system. (Courtesy of W.F. Utlaut)

Fig. 6.8. One of the Platteville HF heater final amplifier units, with Mr Dal Whitaker in attendance. (Courtesy of W.F. Utlaut)

parameters of these facilities have been described by Gurevich and Migulin (1982) and are summarized in Table 6.2.

Some HF ionospheric modification results obtained by Soviet scientists using facilities in the Soviet Union have been summarized by Migulin and Gurevich (1985) and they include changes in ionospheric "ionizing effectiveness," discovery of an artificial defocusing lens, relatively long-lived scattering regions, striction-parametric-induced anomalous absorption, effects of accelerated electrons and ions due to heating from a satellite-borne high power pulsed transmitter, and enhanced sodium layer emissions in the lower ionosphere. Soviet scientists also operate a low-latitude HF heater facility (GISSAR) near Dushanbe (Yerukhimov et al. 1988).

6.2.3 The Low-Latitude HF Heating Facility at the National Aeronomy and Ionosphere Center (NAIC), Arecibo, Puerto Rico

The first HF heating experiments conducted at Arecibo in 1970 were reported by Gordon et al. (1971) using a 160-kW HF transmitter feeding a pair of log periodic curtain antennas mounted in the focus of the NAIC 1000 ft diameter antenna. The Arecibo ISR was used for diagnostics of the heated ionosphere and results obtained from 1971–1974 were reported by Gordon and Carlson (1974). Based on these early results, a separate, large HF heating facility was built in the vicinity of the Arecibo ISR and became operational in 1980. The parameters of this facility are listed in Table 6.3 and a photograph of the array is displayed in Fig. 6.9.

Some results obtained using the Arecibo HF heater facility and the ISR and other diagnostics were reported by Carlson and Duncan (1977). Since then, this facility has been used by many investigators and some examples of results include: HF enhanced plasma lines (HFPL) observed in the E- and F-regions (Wong and Taylor 1971; Carlson et al. 1972; Kantor 1974; Carlson and Duncan 1977; Muldrew and Showen 1977; Muldrew 1978; Showen and Kim 1978; Djuth 1984,

Table 6.3. Arecibo HF facility parameters

Geographic location	East of Arecibo, Puerto Rico 18°N, 67°W
Magnetic latitude	32°N
Magnetic dip	48.6°
Frequency	3–15 MHz
Power output	≈ 800 kw
Antenna:	32 log periodic radiating elements in a 4 × 8 array (2160 × 1080 ft)
Zenith gain	25 dBi
Half-power beam width	5–10°
Polarization	Either sense of circular or linear. Aligned ≈ NW or NE
Effective radiated power	160–320 MW
Diagnostics	The Arecibo ISR, ionosonde, spectro-photometric instruments, HF radar

Fig. 6.9. View of the Arecibo, P.R, HF heater antenna array. (Courtesy of Craig Tepley)

and Djuth et al. 1986). ELF radiation from the ionospheric current system modified by the HF heater was also demonstrated at Arecibo by Ferraro et al. (1984).

6.2.4 HF Heating Facilities at High Latitudes

Tromsø, Norway

One of the best documented and most successful of the HF ionospheric modification facilities is located at Ramfjordmoen – adjacent to the EISCAT ISR facility near Tromsø, Norway. A brief description of this installation and a succinct summary of the theories describing F-region modification are given by Stubbe and Kopka (1980) and another description is given by Stubbe et al. (1982), while the complete description may be found in a report by Stubbe and Kopka (1979). This facility was designed, constructed, and operated by personnel of the Max Planck Institut für Aeronomie of Lindau/Katlenburg, Germany. It became fully operational in September 1980. A brief description of this facility is given in Table 6.4 and an aerial view of the antenna array is shown in Fig. 6.10.

The MPAE heater at Tromsø is capable of producing an energy flux of $3\,mW/m^2$ at $h = 100\,km$ and $0.3\,mW/m^2$ at $h = 300\,km$ and electric field strengths of $1\,V/m$ at $h = 100\,km$ and $\approx 3\,V/m$ in the F-region. Summaries of the results obtained using the Tromsø heater to modify the D-, E-, and F-regions of the

Table 6.4. Description of MPAE HF heater facility near Tromsø, Norway

Location	Ramfjordmoen – near Tromsø, Norway 69.6°N, 19.2°E
Geomagnetic latitude	67.1°N
Geomagnetic dip	78°
	L = 6.2
Transmitter	12 Units of 125 kW "continuous wave" (CW) output each
Frequency range	2.5–8 MHz
Antenna system	Three separate arrays, covering 2.5–4, 3.85–5.65, and 5.5–8 MHz, respectively. Each array consists of six rows of six crossed full wavelength dipoles which are in the rhombic configuration to broaden the horizontal plane pattern. Each row of six dipoles is driven by one transmitter and the other six crossed dipoles of the array are driven by another transmitter. The rows of antennas are oriented east-west to permit beam tilting in the meridian plane
Antenna system gain	24 dBi ± 1 dB
Beam width	14.5°
Maximum ERP	360 MW
Nominal ERP	300 MW
Diagnostics	The EISCAT ISR facility, riometers, ionosonde, HF backscatter radars, rocket launch facility, VHF radar systems, HILAT/POLAR BEAR satellite ground station

Fig. 6.10. View of the MPAe Tromsø HF heater antenna array. (Courtesy of Peter Stubbe)

high latitude ionosphere have been reported by Stubbe et al. (1982, 1985), Jones et al. (1984), Barr et al. (1988) and Ishame et al. (1990). At the time this book went to Press, the MPAE heater at Tromsø had been upgraded to provide an ERP exceeding 1 GW.

Fairbanks, Alaska – The HIPAS Facility

In the late 1970's a high power HF ionospheric modification facility was built at the Chena Valley Field Site (40 km east of Fairbanks, Alaska, under the direction of A.Y. Wong of UCLA and J.G. Roederer of the Geophysical Institute of the University of Alaska). The facility had a 2-MW pulsed transmitter capable of transmitting pulses of ≈ 1 ms with a repetition rate of ≈ 1 Hz into a circular array antenna of nine crossed dipoles operating at f = 4 MHz. Results obtained in a laboratory model, at Arecibo and at the HIPAS facility, were presented by Wong et al. (1983). (The acronym HIPAS stands for "high power active stimulation" of the auroral plasma – with the object of creating nonlinear reactions which may, in turn, produce plasma irregularities such as "cavitons" e.g., Wong 1977 and Wong 1990).

The HF pulse transmitter was destroyed by fire in 1985, and the Platteville CW HF transmitters were shipped to Fairbanks and became operational in 1986 with a modified version of the circular antenna array. Table 6.5 summarizes the pertinent parameters of the HIPAS facility.

Table 6.5 HIPAS facility parameters

Location	Chena Valley Field Site – Central Alaska 64°54′22″, 146°50′06″ W
Geomagnetic latitude	64°52′ N
Geomagnetic dip	76°30′, 28°02′ E declination L = 5.6°
Transmitter	The Platteville, Colorado system – eight transmitters each capable of ≈ 100 kW output, CW ("continuous wave") f = 2.85 MHz, 3.349 MHz, 4.530 MHz, 4.905 MHz, P_0 = 800 kW
Antenna system:	Circular array of eight crossed dipoles. (Seven crossed dipoles equally spaced around a circle of radius 105.46 m, plus one crossed dipole in the center of the circular array)
Estimated gain	18.4 dB*i* at 2.85 MHz
Estimated half-power beam width	21° at 2.85 MHz
Estimated ERP	~ 50 MW at 2.85 MHz
Diagnostics	Two digital ionosondes, HF receiver for pulsed operation, HF "phased array" system, sensitive VLF/FLF receivers, photometers 50 MHz steerable beam radar system, EXOS-D, EXOS-C satellite receivers.

As this book goes to Press, plans are underway to build another HF Heater in the Fairbanks area, under the name of HAARP (HF Active Auroral Research Program). This facility will have an ERP exceeding 1 GW, over a frequency range of ≈ 1.7–10 MHz.

Monchegorsk (Kola Peninsula — USSR)

Investigations from the Polar Geophysical Institute at Apatity have used a commercial transmitter (Radio Moscow) in Monchegorsk to conduct ionosphere stimulation experiments at night. The transmitter operates in the CW mode with 100 kW output into an antenna with gain = 23 dB (Gorokhov 1988, pers. commun.).

6.3 Capabilities and Limitations of Radio-Wave Interaction and HF Heating Experiments

6.3.1 Capabilities

Wave Interaction (Cross-Modulation) Experiments

By measuring the amplitude and phase of the "wanted wave" as a function of height in the "disturbed" region, profiles of electron density and collision frequency over the height range of ≈ 50–90 km can be obtained.

HF Heating Experiments

Capabilities and results obtained by HF heater facilities at middle and low latitudes are given in Section 6.2.2, describing results obtained in the 1970's and early 1980's. More recently, Stubbe et al. (1985) have given a comprehensive report on the results from several heating experiments using the MPAE heater facility at Ramfjordmoen, and from other HF heater facilities:

In the D-region at *high latitudes*, partial reflection diagnostics of HF heating reveal information on the physics of electron temperature enhancements and on the effects of electron temperature reaction rates upon ion chemistry changes. VLF propagation multimode theories have been tested by monitoring OMEGA VLF propagation through a volume of the D-region which is subjected to controlled heating.

In the high latitude E-region, the auroral electrojet has been subjected to amplitude-modulated heating and acts as an antenna radiating at the heater modulating frequency. Because of the combined effects of the time constants of the electron temperature and the electron density on the total ionospheric conductivity, however, strongest modulation effects are observed at ULF, although the electrojet also radiates at VLF and ELF as shown by Rietveld et al. (1989).

E-region-irregularities observed by the STARE radar have also been created by HF heating at high latitudes. Analysis of these results sheds new light on E-region mechanisms for irregularity generation.

High latitude *F-region* HF heating has been shown to produce both small- and large-scale irregularities. HF backscatter sounders located at distances up to 1100 km from the heated volume have detected these striations which have a narrow spectral width. A 140 MHz radar has also detected "plasma line scatter." Whereas HF heaters at mid-latitudes can "create the spread-F signature" observed on ionosondes, this signature is not observed at high latitudes, but radio star scintillation at 224 and 933 MHz has indicated that heating results in *decreased irregularity scale size* when the heater power is reduced. These results require some new theoretical development.

Anomalous F-region absorption has also been observed at both high and mid-latitudes. New anomalous absorption results have been obtained at high latitudes, including "hysteresis" effects, a dependence on ionospheric variability, heater wave polarization, and output power, and on the difference between heater and diagnostic frequencies. In addition to the previously mentioned long-term F-region absorption effects (scale of seconds), millisecond-scale effects have also been observed when O-mode HF heating is employed at high latitudes.

O-mode HF heating has also produced excited plasma lines and ion line modification, as observed at Arecibo (Gordon and Carlson 1974) and by Fejer et al. 1985), and at auroral latitudes as observed by the EISCAT system. These effects are thought to be the result of the parametric decay instability and the oscillating two-stream instability and gross behavior of these enhanced lines depends upon the wavelength of the ISR, the power density achieved at the proper height interval and the angle the radar k-vector makes with the local geomagnetic field lines. Some new insights into plasma instability mechanism have been obtained by ISR observations of HF heating under a variety of conditions.

O-mode heating can also produce stimulated electromagnetic emission in a frequency band of ± 100 kHz of the primary wave. Analysis of these data has revealed new information on the scattering process involving Langmuir waves and cascade modes as they relate to the creation and structure of irregularities.

6.3.2 Limitations

Both the wave-interaction and the HF heating experiments are expensive to build, require large areas for the necessary high gain HF antenna arrays, and consume relatively large amounts of primary electrical power.

Recent advances in modulation coding techniques as applied to ISR systems make it possible to obtain D-region electron densities and other parameters (Devlin et al. 1986) in a more cost-effective manner than by using wave-interaction techniques.

Most of the HF heating and plasma stimulation experiments require that the heater be operated near the ionospheric critical frequency (f_0F_2). Since this parameter changes diurnally, seasonally, and with sunspot number, the operating frequency of the heater, hence the antenna dimensions required for optimum gain, must also be changed. This is an added expense, especially when the heater is operated over periods approaching half a sunspot cycle in length.

Particular care must be taken in the design and construction of the HF antenna arrays used in the heating experiments in order to concentrate most of the power into the main antenna lobe, and the avoidance of a multi-lobe pattern which could radiate large amounts of power in undesired directions. It is sometimes necessary to make airborne measurements of the actual antenna pattern in order to verify the model calculations. Such "calibrations" are very expensive to carry out properly.

Chapter 7 Methods of Measuring Ionospheric Absorption and Other D-Region Parameters

The essential elements of the theory of absorption of electromagnetic waves in the ionosphere, including both deviative and nondeviative absorption, have been presented in Section 2.2.2, and in considerable detail by Davies (1969, 1990). The absorption coefficient (per unit length) for EM waves propagating in an *unmagnetized plasma* is

$$k = \left[\frac{e^2}{2}\varepsilon_o mc\mu \cdot \frac{N\nu}{\omega^2 + \nu^2}\right]. \qquad \text{[Eq. (2.46)]}$$

For a QL wave, nondeviative absorption, and $\mu^2 \gg X^2$, Eq. (2.53) describes the absorption coefficient in a *magnetoionic medium*,

$$k = \left[\frac{e^2}{2\varepsilon_o mc} \cdot \frac{N\nu}{(\omega \pm \omega_L)^2 + \nu^2}\right], \qquad \text{[Eq. (2.53)]}$$

where all the quantities are defined in Section 2.2.2.

We will use the URSI nomenclature to describe the basic techniques which are employed to measure absorption, namely: *method A1* (vertical incidence pulse reflection); *method A2* (cosmic noise absorption – the "riometer"); *method A3a* (oblique field strength observation on frequencies above 2 MHz); and *method A3b* (oblique field strength measurements on frequencies in and below the MF broadcasting band). The partial reflection (sometimes referred to as the MF radar) technique will also be covered. Geisweid et al. (1968) have described the A1, A2, and A3 techniques and the relationships between the data obtained by these methods in the context of data obtained during the International Year of the Quiet Sun (IQSY), while Rawer (1976) has presented a more detailed description of these techniques in the form of an *URSI Manual on Ionospheric Absorption Techniques*, which is available from World Data Center A (WDC-A) in Boulder, Colorado. A very brief treatment (two pages) of the A1, A2, and A3 methods is also given by Giraud and Petit (1978), and early results were reported by Appleton and Piggott (1954).

It should be pointed out that the A1, A2, and A3 methods do not yield comparable results, mainly because the variation of absorption with height is a strong function of frequency. In general, those techniques using frequencies below ≈ 1 MHz or above ≈ 10 MHz are most influenced by absorption *below* ≈ 70 km, whereas methods using frequencies from 1–10 MHz essentially measure absorption *above* 70 km.

We will follow Rawer's (1976) outline describing absorption techniques and update it with descriptions of newer techniques and/or modifications of existing techniques in use at the present time.

7.1 A1 – The Pulse Reflection Method

7.1.1 Vertical Incidence

If a pulse of RF energy of frequency $f < f_o$ is transmitted vertically, it will pass through the highly absorbing D-region, be reflected by the E- or F-layers, then traverse the D-region again as it returns to the earth's surface. The A1 technique employs an accurate, stable frequency for transmitting and receiving, a "constant gain" receiver, a "constant output power" transmitter, an antenna system producing a uniform lobe directed vertically, plus careful and frequent calibration. The technique actually measures relative variations of field strength of a particular echo reflected from the ionosphere. The attenuation of the received signal consists of the spatial attenuation, focusing effects, scattering by small irregularities and ionospheric absorption – but by using a proper calibration system, the ionospheric absorption can be separated out. The reflection coefficient of the reflecting layer must also be accounted for (total or partial reflection). Rawer (1976) describes each of the elements of the technique in considerable detail.

Based on the short- and long-term variation of the ionosphere, a "standard frequency" for A1 measurements of (2.2 ± 0.2) MHz has been found to work well, except at high latitudes, where anomalous absorption strongly attenuates signals at these frequencies. Gregory et al. (1970) have used the A1 technique at *subauroral* (60° invariant) latitudes. A second frequency at least 0.5 MHz above the "standard frequency" is also usually employed in this technique. When possible, this second frequency is changed to adapt the system to ionospheric variations, but the standard frequency is kept constant. Modern versions of the A1 technique usually employ six or more frequencies reasonably spaced between 1.6 and 6 MHz. The multifrequency system makes it possible to discriminate between deviative and nondeviative contributions to absorption.

For vertical incidence reflection of an RF pulse, the amplitude E_1 of the once reflected wave is

$$E_1 = \frac{E_0}{2h'(\rho)}, \tag{7.1}$$

where:

E_0 = the amplitude of the transmitted pulse at a distance of 1 km
h' = the vertical height of reflection
ρ = the apparent reflection coefficient of the ionospheric reflecting layer.

The amplitude of this reflection coefficient is given by Schultz and Gallet (1970) as

$$\rho = \exp\left[-2\int_0^h k(z)\,dz\right], \tag{7.2}$$

where k is the absorption per unit distance.

In the A1 technique, the reflection coefficient (ρ) is measured each day at noon on several frequencies between 2 and 5 MHz, as discussed by Piggott (1953) and

Piggott et al. (1957). Piggott (1953) also developed a graphical analysis technique involving plotting log(ρ) versus $(\omega \pm \omega_H)^2$ in order to eliminate the part of (ρ) thought to be due to the deviative absorption. This graphical analysis results in an index, A, which is independent of frequency, and which gives a useful index of nondeviative absorption in the lower ionosphere. A is related to (ρ) by the

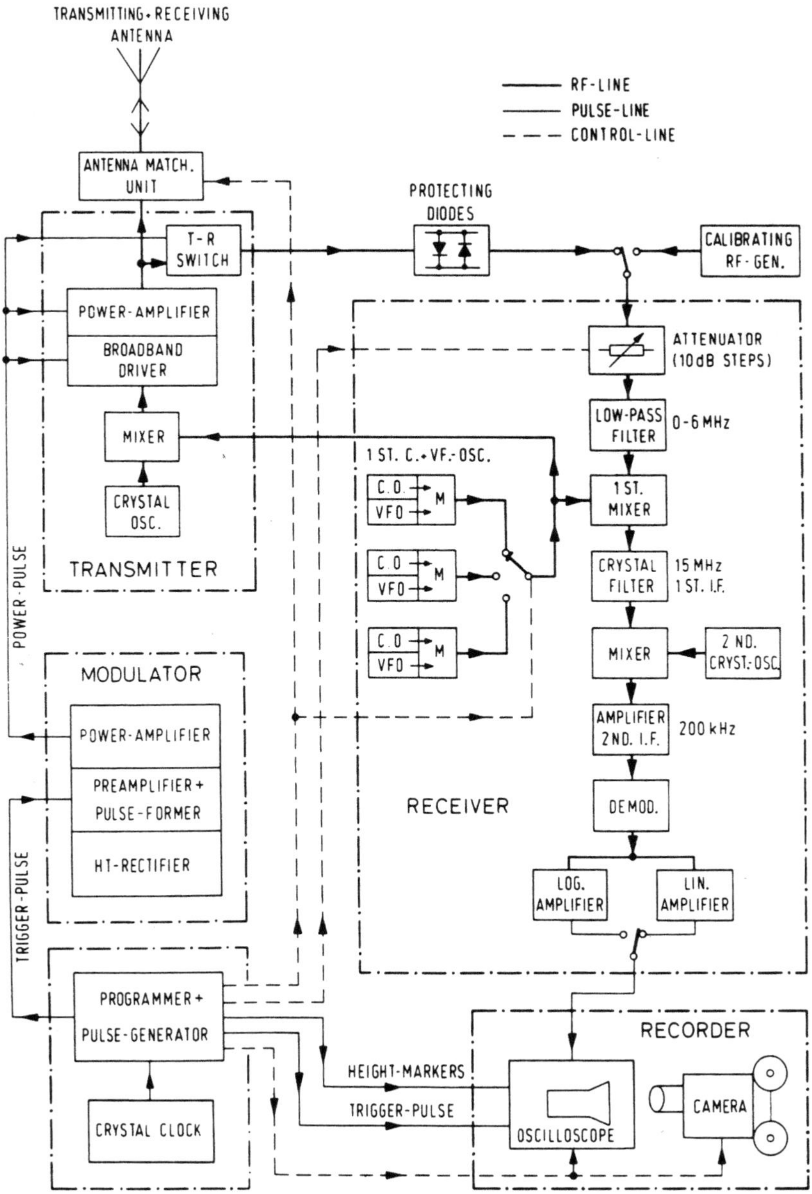

Fig. 7.1. Block diagram of A1 system used aboard MV Hanau. (Barke et al. 1974)

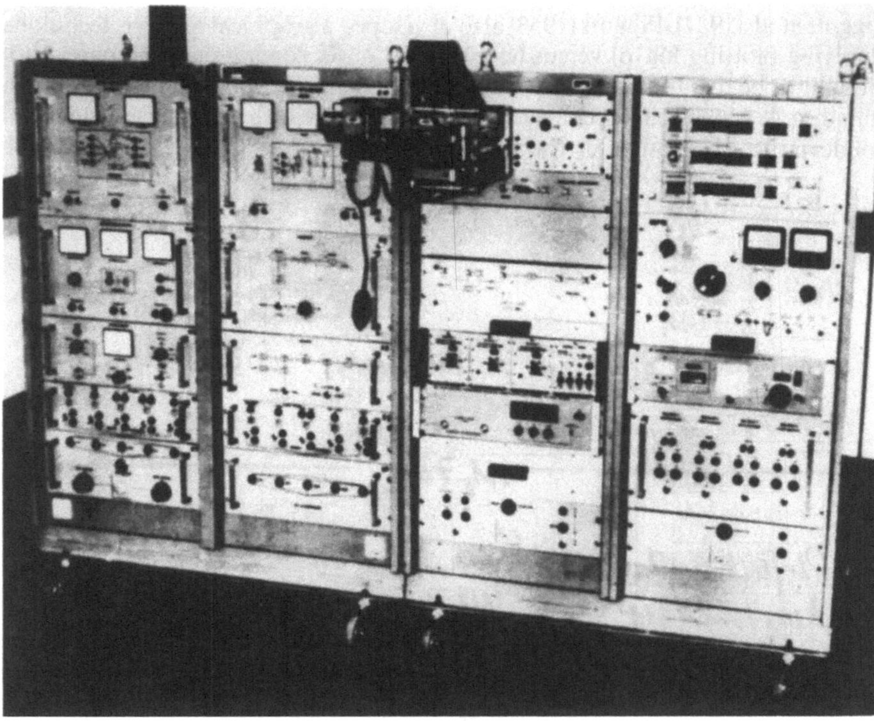

Fig. 7.2. A1 mobile equipment used by Barke. (Barke et al. 1974)

relation (George 1971)

$$A(\mathrm{dB\,MHz^2}) = 8.7(f + f_H)^2 \log_e(\rho) \tag{7.3}$$

the absorption index A can be used to obtain the absorption at other frequencies (not near a critical frequency, f_o – that is, for frequencies at which deviative absorption is not significant). A quantity called the vertical incidence transmission loss L_v in dB, is given by

$$L_v(\mathrm{dB}) = \frac{A}{(f + f_H)^2} \tag{7.4}$$

(this relation is not valid near the dip equator.)

Figures 7.1 and 7.2 show the block diagram and a photograph, respectively, of circa 1970 version of an A1 sounder. This particular system was deployed on the ship Hanau traversing routes from Hamburg-Guaira, Venezuela, in an absorption measuring campaign (Barke et al. 1974).

7.1.2 A1 Technique – Oblique Incidence

For oblique incidence at short distance we can multiply Eq. (7.4) by Sec Φ where Φ is the angle of incidence at the midpoint of the absorbing region for a single hop mode. We can modify Eq. (7.4) to apply to times other than local noon by multiplying it by a suitable factor which also includes effects of sunrise/sunset time at h = 100 km, and is given by Schultz and Gallet (1970) as

$$L_v(\mathrm{dB}) = A(\mathrm{dB\,MHz}^2)\left[\frac{\cos 0.893x}{\cos 0.893x_{12}}\right]^d \cdot d\left\{\frac{\operatorname{Sec}\Phi}{(f+f_H)^2}\right\}, \tag{7.5}$$

where:

X = solar zenith angle
X_{12} = solar zenith angle at noon
d = experimental constant = (0.6–1.3)
A = from Eq. (7.3), modified empirically by geometry of ray path as outlined by Schultz and Gallet (1970).

7.2 A2 – The Cosmic Noise Technique

The basic theory, instrumentation advances, and results for the A2 method up to 1966 were covered in a review by Hultqvist (1966). We will attempt to outline the basic technique and discuss significant advances in instrumentation since that time in this section.

Radio noise from extraterrestrial sources covering an extremely wide range of frequencies is continuously incident on the top of the terrestrial ionosphere. Our ionosphere acts as a "high pass filter," passing those frequencies which exceed the critical frequency of the F-region (or in some instances at auroral latitudes, the E-layer). The major contribution of extraterrestrial noise from HF through low VHF comes from our own galaxy and reaches maximum strength in the plane of the galaxy (towards the Milky Way). Discrete sources such as those near Cygnus and Cassiopeia and our Sun contribute lesser noise power densities at the surface of the earth – except for the Sun when it is very active. So, except for the active Sun, the galactic radio noise received by a typical A2 method antenna varies smoothly with sidereal time.

If we consider an antenna (whose main lobe is directed vertically) located at a particular point on the earth's surface at a given sidereal time, the incident cosmic noise power should be constant – if there is no radio wave absorption in the ionosphere. Therefore, we should be able to deduce the amount of absorption (integrated over the ray path in the ionosphere) at a particular frequency (greater than the ionosphere critical frequency) by making relatively simple measurements of signal voltage or power.

Shain (1951) was the first to realize that the cosmic noise technique should provide relatively simple measurements of ionospheric absorption, then Mitra and Shain (1953), Blum et al. (1954), and Little (1954) were among the first to use this technique to measure cosmic noise absorption.

When one attempts to measure absorption by the cosmic noise technique, the total power output of the receiver can be written as:

$$P_0 = G(P_s + P_i + P_a + P_c + P_r + I), \tag{7.6}$$

where:

P_s = effective noise power from the sky at the receiver input
P_i = noise power from the ionosphere
P_a = noise power generated by the ambient temperature at the antenna
P_r = noise power of receiver
P_c = noise power from the cable
I = interference
G = power gain of the receiver.

If the receiver is operated in the low-VHF range, and if good engineering practices are followed in its design, the dominant factor in Eq. (7.6) will be the cosmic noise power. Among the serious problems, however, are interference from radio transmitters and instabilities in receiver gain – the modern riometer being designed to eliminate these and other less serious problems.

The first cosmic-noise absorption devices were invented by Machin et al. (1952) and refined by Little and Leinbach (1959) at the Geophysical Institute of the University of Alaska – the latter investigators coining the name RIOMETER. The IGY riometer, which utilized vacuum tubes, can be considered to be the "first generation" riometer. It was followed by the first transistorized model (Schiffmacher 1966), and the "second generation," which in various forms was commercially produced and used in the 1960's. Further advances in VLSI technology resulted in a very compact, low-power-consumption version developed by HJA Chivers, which may be considered to be the "third generation" model. At this time there are ≈ 150 of the Chivers riometers deployed in the northern and southern high latitude regions.

The Little/Leinbach model was deployed during the IGY-IGC and IQSY periods with great success (Leinbach and Basler 1963; Basler 1963; Hook 1968; Hargreaves 1966; Berkey and Parthasarathy 1964; Ansari 1965, and Parthasarathy and Hessler 1964). The word riometer is an acronym standing for Relative-Ionospheric-Opacity-Meter-Extra-Terrestrial-Electromagnetic-Radiation. Figure 7.3 is a simplified block diagram of the basic riometer system, which can be described as a "self-balancing receiving system in which a noise source is continuously equal to the noise power from the antenna" (Little and Leinbach 1959). The following description of operation of a basic riometer is extracted from the *URSI Absorption Manual* (edited by Rawer), Chapter 5.

In the riometer illustrated (in Fig. 7.3), the outputs from the antenna and the local noise diode are alternately switched into the receiver input through a radio frequency switch controlled by a reference audio oscillator. The error signal which results from the inequality between the two inputs is detected in the receiver, amplified, and applied to a phase detector driven by the same reference oscillator. The phase detector produces a DC signal whose polarity is determined by the phase of the error signal, which is then applied to the servo noise diode in such a way as to reduce the error signal to zero, thereby maintaining the noise output equal to that of the antenna. The noise power generated by the diode is proportional to its plate current, and this quantity is normally recorded as the riometer output. The chart recording is calibrated in terms of the test diode plate

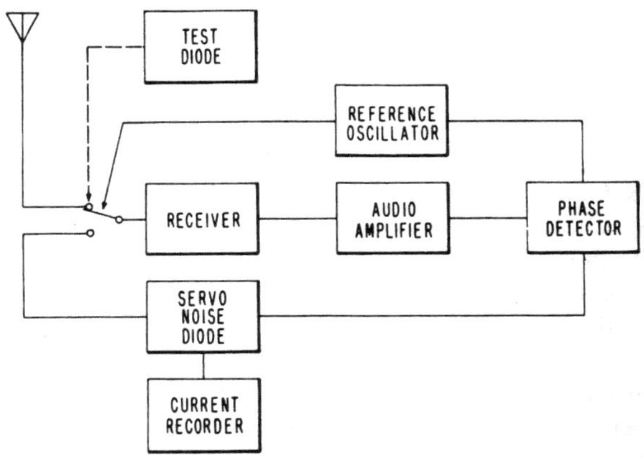

Fig. 7.3. Simplified block diagram of basic (IGY) riometer system. (Rawer et al. 1976)

current, and the absorption is calculated by the simple relation

$$A(dB) = 10\log_{10}\left(\frac{P_0}{P}\right), \tag{7.7}$$

where:

$A(dB)$ = cosmic noise absorption in decibels
P_0 = power output in the absence of the ionosphere
P = signal power output of riometer.

The basic IGY riometer provided our first information on polar cap and auroral absorption during the IGY and IGC (Leinbach 1962, and Basler 1963). A review of riometer results at high latitudes was presented by Hargreaves (1969), and Foppiano and Bradley (1985) have produced a model of auroral absorption as a function of latitude, longitude, time of day, season, and solar epoch.

Despite the success of "first, second, and third generation" riometers in measuring cosmic noise absorption, the technique has some serious problems, such as: finite efficiency antennas, noise originating in the ionosphere, RFI from man-made sources, noise in the receiving system, and receiver power gain instabilities. Another important problem is accurate determination of the quiet day curve.

The quiet-day curve (QDC) for a given riometer system represents the variation of cosmic noise power at a given location as a function of sidereal time in the absence of ionospheric absorption. It was derived during the first 25 years of riometry by obtaining a least-square fit of the mass data plot for a given period (i.e., one month) of the selected quiet-days cosmic noise values. Using this QDC technique, for a typical system the accuracy of riometer absorption data scaling was ≈ 0.25 dB.

Another method of obtaining a QDC was to use a percentage criterion, in which case the QDC was defined as the curve which separates the highest 10% of the data points in the distribution from the remaining 90%. The "mass plot" and "percentage" criterion methods are described by Basler (1963), Hargreaves (1969), Ranta and Ranta (1977), and Armstrong et al. (1977) and the "maximum density" method is described by Drevin et al. (1987).

A considerably improved technique has been described by Krishnaswamy et al. (1985) called the inflection-point method for calculating a QDC. In the inflection point method, the QDC value for a given sidereal time interval is taken to be the signal level corresponding to the inflection point on the high-signal side of the peak of the distribution of the measured cosmic noise power. Krishnaswamy et al. (1985) show that the uncertainty in the determination of the QDC for a riometer operating at Siple, Antarctica on 30 MHz for 1975 and 1980 was only ≈ 0.1 dB, even in the presence of strong RFI.

In order to obtain a truly quantitative value of absorption, one must account for the finite antenna beam width. The apparent absorption, which is the actual value measured, is greater than the zenithal absorption of a plane wave passing through the ionosphere because of the contributions of the oblique rays in the finite antenna pattern. Hargreaves (1969, Fig. 3) has assembled a plot showing this relation (assuming a horizontal uniform absorbing layer). More recently, Hargreaves and Jarvis (1986) have calculated the relation between true and apparent absorption at Halley Bay, Antarctica, as:

$$B = 1.4\,A^{0.980}, \tag{7.8}$$

where:

B = measured absorption
A = true zenithal absorption

As may be seen, this is a rather small correction. Figure 7.4 and Table 7.1 show results obtained at Halley Bay.

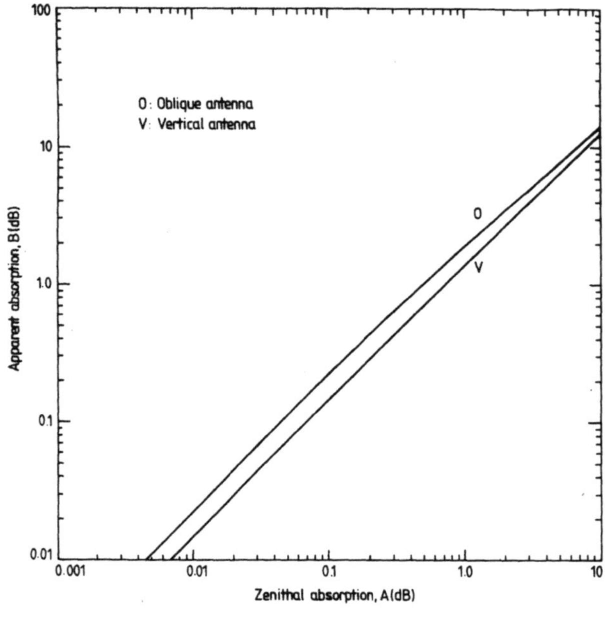

Fig. 7.4. Corrections relating zenithal and apparent absorption. *O* is curve for oblique antenna pointed at 45 to the zenith, and *V* is for antenna main lobe directed toward the zenith. (Hargreaves and Jarvis, 1986)

Table 7.1. Equations relating apparent (B) and zenithal absorption (A) assuming a uniform slab ionospheric model [see Eq. (7.8)]

Antenna	Equation	Standard error
Oblique	$E = 1.87\,A^{0.929}$	4%
	$A = 0.50\,B^{1.075}$	4%
Vertical	$B = 1.40\,A^{0.980}$	1%
	$A = 0.709\,B^{1.020}$	1%

Lerfald et al. (1964) has computed the effect of a nonuniform ionosphere (with small-scale irregularities) on the above relation. More recently, Stoker (1987) has shown that the riometer absorption measured using a wide angle antenna is ~ 1.4 times the zenithal absorption, and has derived a relationship between this absorption and precipitating election flux.

Most riometers operate on frequencies near 30 MHz, this being a compromise between the ionospheric layer critical frequency, and avoidance of interference from the crowded shortwave bands. Quite a bit more information about the D-region can be obtained if one utilizes a *multiple frequency* riometer instead of a single frequency system, as shown by Parthasarathy et al. (1963) and Little et al. (1964). The early riometers usually used crossed three- or four-elements Yagi antennas directed vertically on a single mast. An example of a typical antenna system used on a modern riometer is shown in Fig. 7.5 below.

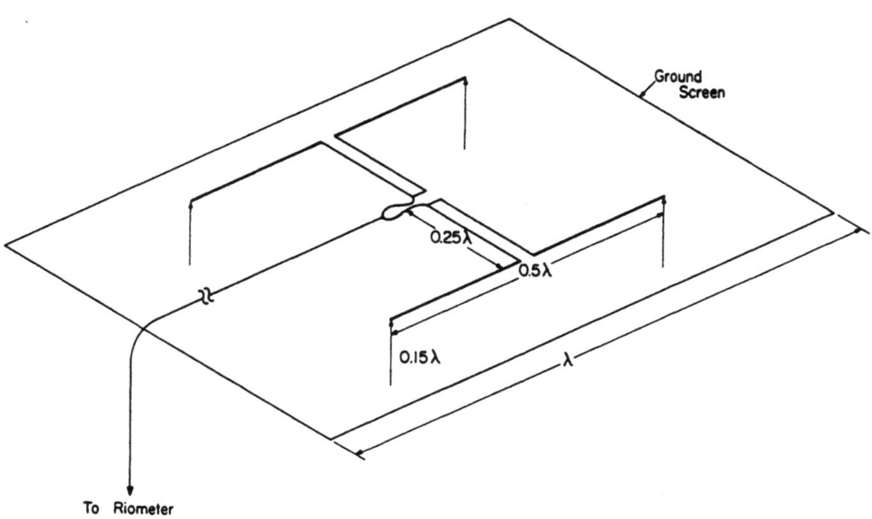

Fig. 7.5. A simple antenna system for a single frequency antenna. (Rawer et al. 1976)

In addition to utilizing several frequencies simultaneously, some investigators have employed narrow antenna beams in an attempt to delineate the spatial variation of D-region absorption. One of the first successful uses of a narrow-beam switchable lobe riometer was reported by Ansari (1963). He used a 4 × 4 linear array of three-element Yagi antennas spaced 1.1 λ apart. The array had a half-power beam width of 12° which was switchable from 12° north of the zenith to 12° south of the zenith and operated on a frequency of 36 MHz. Ansari used this system to determine some of the "fine structure" of auroral absorption (Ansari 1964, 1965).

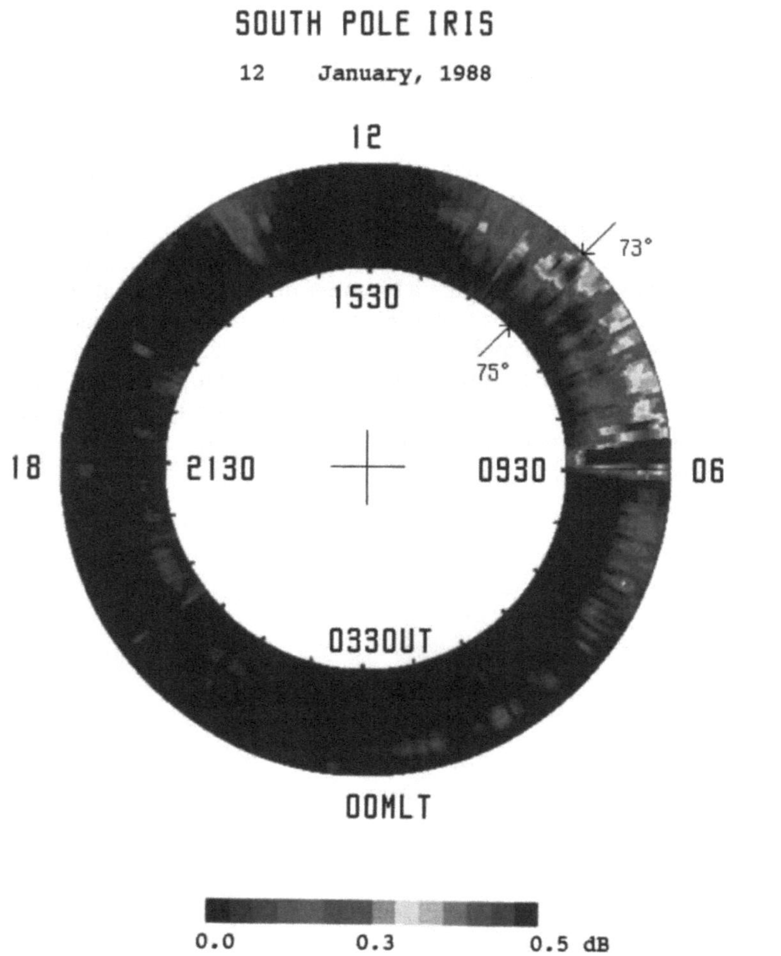

Fig. 7.6. Clock-dial latitude-time representation of ionospheric radiowave absorption in the vicinity of South Pole station during 1 day. Absorption is shown color-coded in dB for the seven beams of IRIS that measure along the magnetic meridian through the station. The latitudinal extent of the coverage is from 73° to 75°. The data display is incremented every 4 min. Universal time is indicated along the *inner circle*; magnetic local time along the *outer circle*. (Rosenberg pers. commun. 1989)

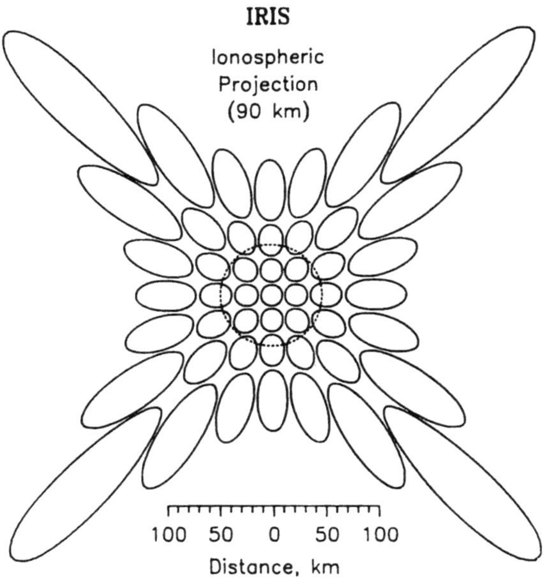

Fig. 7.7. IRIS antenna beam pattern. The projection of each IRIS beam (−3 dB contours) onto the ionosphere at a height of 90 km. The *dashed circle* in the center represents the viewing area of a conventional broad-beam riometer. (Rosenberg pers. commun. 1989)

A modern version of a very sophisticated riometer is the Imaging Riometer-Ionospheric Studies (IRIS) developed at the University of Maryland and presently operating in Antarctica and at Sondrestrøm, Greenland. This system consists of seven solid-state riometers operating on 38.2 MHz. The antenna system is a $4\lambda \times 4\lambda$ square array of 64 Turnstile antennas employing circular polarization, which produces 49 antenna-independent beams which are sampled in 1 s. The half-power beam width of the zenithal beam is $\approx 12°$ (Detrick 1984; Rosenberg, 1990; see Figs. 7.6, 7.7, and 7.8).

Attempts have also been made to relate absorption measured with a riometer using an oblique antenna lobe with the lowest – observed – frequency (LOF) on a transpolar HF synchronized step-sounding circuit (Hunsucker 1964a). Since the riometer only measured the absorption on the upward leg of the first hop of a multi-hop transpolar path, excellent agreement between absorption and LOF would not be expected, but as Fig. 7.9 shows, there is a good correlation – at least for the period 0000–1400 UT. From 1400–2400 UT there were very little step-frequency data because of high interference levels in the shortwave (SW) bands.

In a related study, Hunsucker (1964b) found that absorption measured with an oblique antenna 27.6 MHz riometer at College, Alaska, had a very low correlation ($R = 0.1$–0.2) with 18 MHz signal attenuation on the College-Thule circuit. This low correlation was attributed to the patchy nature of auroral absorption and possibly to non-great-circle (NGC) propagation (Bates et al. 1966).

A high correlation between 30 MHz riometer absorption and electron density at ≈ 90 km height observed by the Chatanika incoherent scatter radar was reported by Hunsucker et al. (1972) and Hunsucker (1974).

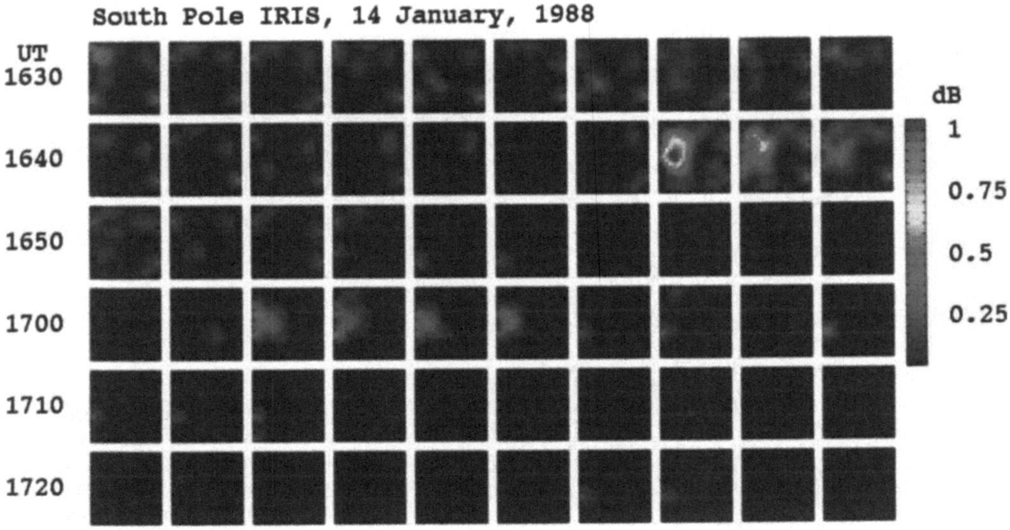

Fig. 7.8. The *square images* in this figure are IRIS images showing two localized regions of cosmic noise absorption caused by short duration pulses of auroral ionization. Each image is oriented such that the magnetic poleward (equatorward) is *up* (*down*) and west (east) is to the *left* (*right*) as looking at the ionosphere from below. (Rosenberg pers. commun. 1989)

It should be noted that the A2 method of measuring absorption is less sensitive than the A1 and A3 methods because of the higher A2 operating frequencies. The A2 method is, therefore, most useful in measuring auroral and polar cap absorption, while the A1 and A3 methods are most useful for measuring "normal solar controlled" D-region absorption.

Another effect which must be considered in the A2 method is the F-region iris effect, which occurs when the riometer frequency is close to f_oF2, as outlined in Chapter 5 of the *URSI Absorption Handbook*. Since most riometers currently in operation are on frequencies $> f_oF2$, this effect is not of great importance.

Results of riometer investigations up to 1979 are summarized in Chapter 6 of the *Proceedings of the International Solar-Terrestrial Symposium* held in Boulder, Colorado, in April 1979 (Donnelly 1979), and results since that time have been reported by Hargreaves (1985).

7.3 Oblique Field Strength Measurements (the A3 Method)

This technique applies to short one-hop ionospheric paths using transmissions in the LF, MF, and HF bands at mid-latitudes. Basic descriptions of this method may be found in articles by Gockel (1948), Schwentek (1958), Triska (1962, 1967), Hense (1969), Best et al. (1936), Lauter and Sprenger (1966), and in the *URSI Manual on Absorption Techniques* (Rawer 1976).

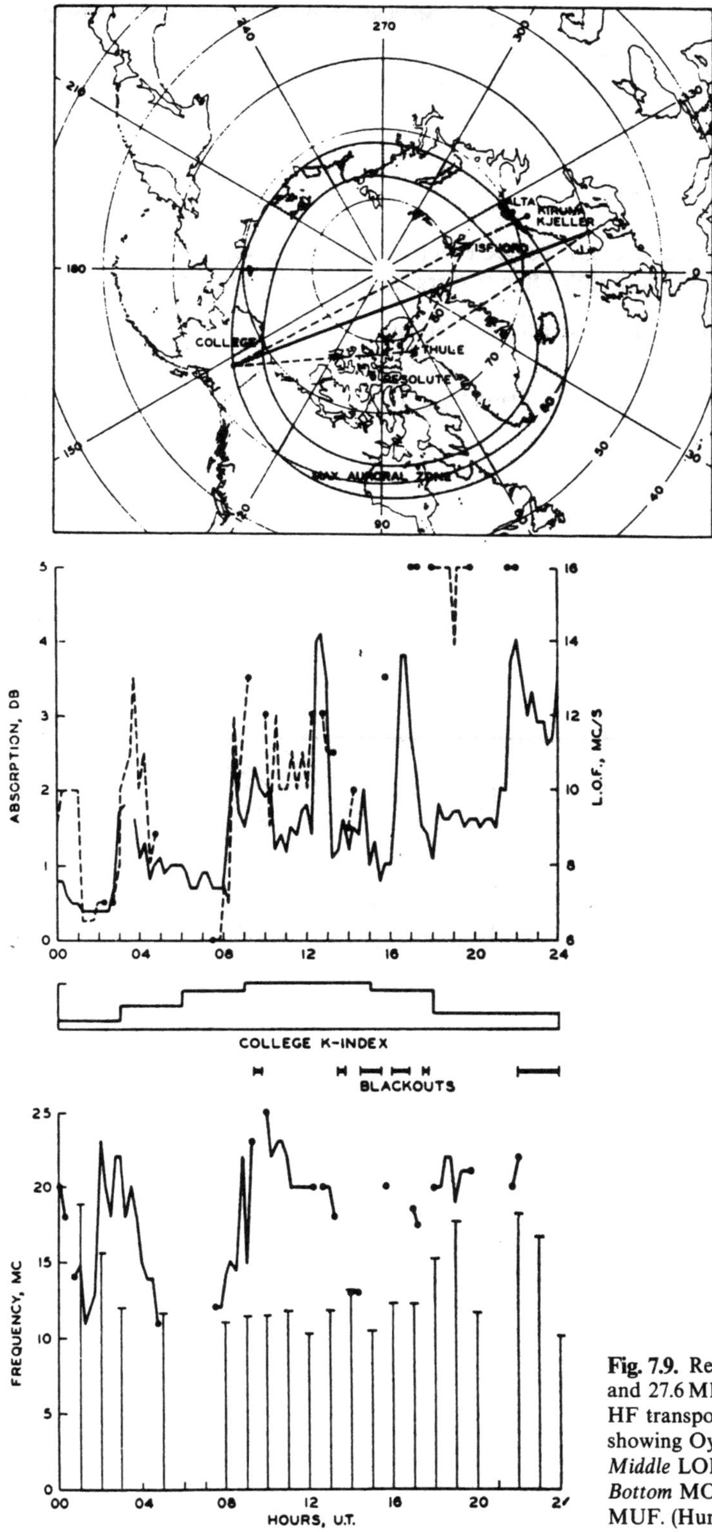

Fig. 7.9. Relation between LOF and 27.6 MHz absorption on an HF transpolar circuit. *Top* map showing Oya College HF path. *Middle* LOF and absorption. *Bottom* MOF and Resolute Bay MUF. (Hunsucker 1964)

We start by defining the absorption coefficient k by the relation

$$\int k\,ds = -\ln\left(\frac{E}{E^*}\right), \tag{7.9}$$

where:

E^* = electric field strength before entering absorbing region (volts/m)
E = electric field strength after traversing absorbing region,

then we can express the absorption in dB as

$$A(dB) = -10\log\left(\frac{P}{P^*}\right) = -20\log\left(\frac{E}{E^*}\right), \tag{7.10}$$

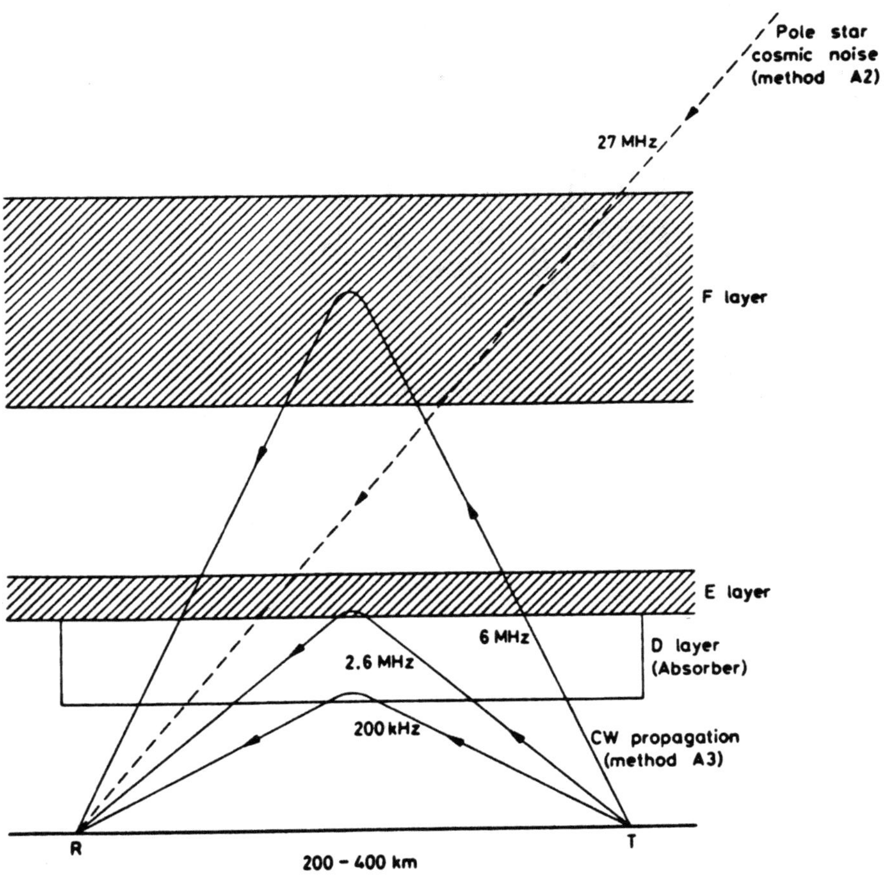

Fig. 7.10. Geometry for A3 absorption measurement method. The *dashed line* from R to the "pole star" is an idealized ray path for A2 (riometer) method. (Rawer et al. 1976)

where:

P* = power in wave before entering the absorber
p = power in wave after leaving the absorber.

7.3.1 The A3a Method – Oblique Incidence Field Strength Observation on Frequencies Above 2 MHz

Frequencies most often used for this technique are between 2 and 3 MHz and around 6 MHz, the usual transmission mode is CW, and typical ground distance between transmitter and receiver is 200–400 km as shown in Fig. 7.10.

In this experimental procedure, the vertical plane polar diagrams of the transmitting and receiving antennas are configured in such a way that small changes in height of the reflecting layer are relatively unimportant, provided that the wave is reflected from the same layer. It is most desirable to use E-layer reflection throughout the observing period, if possible, to calibrate the system using sporadic-E reflection at night when the absorption can be neglected.

Some of the technical requirements for this method include:

1. Stable transmitter power output.
2. Stable receiver sensitivity.
3. Well-designed transmitting and receiving antennas.
4. No appreciable ground wave received.
5. The first E-layer or F-layer reflection stronger than other modes.

Usually the main purpose of the A3 technique is to measure the diurnal, seasonal, and sunspot variations in D-region absorption. Because of the relatively short path length, the plane-earth approximation is valid. The theory and experimental procedure for this method are presented in exquisite detail in the *URSI Manual on Ionospheric Absorption Measurements* (Rawer 1976).

7.3.2 Method A3b – Oblique Incidence Field Strength Observation on Frequencies in and Below the MF Broadcasting Band

The main advantage of this technique is its ability to follow D-region changes using a relatively simple, inexpensive system. The method provides a full-time "patrol" of the 60–90 km height region of the ionosphere. Effects which are studied using the A3b method include:

1. Daily and seasonal variations.
2. Sunrise effects.
3. Sunspot cycle variations.
4. Solar flare effects.
5. Storm-time and after-storm effects.
6. Eclipse effects.

Most A3a measurements are made at frequencies of 2.5 MHz and higher, where the ground wave is strongly attenuated. This is not the case for the A3b method, so one must incorporate a method for rejecting this signal. This separation of sky and ground wave is usually accomplished by using a vertical loop antenna with its plane perpendicular to the direction of arrival of the ground wave. An auxiliary antenna and a special input circuit are usually also employed. Additional details on theory, experimental technique, and observational programs may be found in Rawer's 1976 URSI absorption manual.

Because of the low frequencies used, the A3 Method is not suitable for measuring absorption at auroral latitudes.

7.4 Partial Reflection Experiments (PRE)

This technique was pioneered by Gardner and Pawsey (1953), and usually uses frequencies near 2–3 MHz. Basically it consists of a high powered transmitter, a sensitive receiver and a receiving antenna array with lobes directed vertically, which distinguishes between the downcoming o and x wave polarizations. By measuring the amplitudes of both magnetoionic components it is possible to obtain information on electron density, collision frequency, and absorption in the D-region.

7.4.1 Theory

The basic theory for this technique was developed by Gardner and Pawsey (1953) and refined by Belrose and Burke (1964) who used Sen and Wyller's (1960) magnetoionic equation (which includes absorption effects). The QL approximation formulation is then

$$n^2 = \left[\mu_{x,o} \frac{-jc_o k_{x,o}}{\omega}\right]^2 = 1 - \left(\frac{\omega_N^2}{\hat{v}\bar{\omega}}\right) y_{x,0} C_{3/2} y_{x,0} - j\frac{5\omega_N^2}{2\hat{v}\bar{\omega}} C_{5/2}(y_{x,o}), \qquad (7.11)$$

where:

$\mu_{x,0}$ = real part of refractive index
$k_{x,0}$ = absorption coefficients (x = extraordinary, o = ordinary component)
$\omega = 2\pi f$ = angular operating frequency
$\omega N = \sqrt{Nq^2/m_e \varepsilon_o}$ (plasma frequency)
N = electron density
$\hat{v}$ = electron-neutral collision frequency
$y_{0,x} = (\omega \pm \omega L)/v$
$\omega L = 2\pi f_H \cos \Theta$ = longitudinal component of the gyro-frequency
C = C-script integrals (Burke and Hara 1963).

The operating frequency must not be close to the plasma frequency.

Assuming a Fresnel reflection from a sharp boundary of infinite horizontal dimension in the D-region, the reflection of the upgoing wave is

$$R = \frac{n_2 - n_1}{n_2 + n_1} = \frac{\Delta n}{2n}, \qquad (7.12)$$

where n_2 and n_1 are the refractive indices below and above the boundary, respectively.

Belrose and Burke (1964) showed that (7.12) was valid for "blobs" of enhanced electron density (volume scattering) as well as for an infinite slab, and Flood (1968, 1969) has developed a more rigorous treatment of volume scattering.

The ratio of the amplitudes of the x and o components can be expressed as

$$\frac{A_x}{A_o} = \left|\frac{R_x}{R_o}\right| \exp\left[-2\int_0^h (k_x - k_o)\,dh\right]. \tag{7.13}$$

Since A_x and A_o can be measured, it is necessary to express the other parameters in terms of known quantities. We can use Eq. (7.13), and ignoring changes in ν at the boundary, we can write

$$\frac{R_x}{R_o} = \frac{\dfrac{dn_x^2}{dN}}{\dfrac{dn_o^2}{d^2N}}. \tag{7.14}$$

Equation (7.14) assumes that the influence of irregularities is several orders of magnitude less than that of electron density blobs. Manson (1966) showed that one must, however, take into account irregularities in collision frequency under certain circumstances. Rawer (1976) obtains the final result

$$\frac{A_x}{A_o} = \frac{R_x}{R_o} \exp\left[-2\int_0^h NF\,dh\right] \tag{7.15}$$

or,

$$\ln\left\{\frac{A_x}{A_o}\right\} = \ln\left\{\frac{R_x}{R_o}\right\} - 2\int_0^h NF\,dh, \tag{7.16}$$

where F = logarithmic measure of E, usually dB over 1 μV/m.

7.4.2 Experimental Technique

If we measure A_x and A_o at two adjacent heights h_1 and h_2, where Δh is sufficiently small so that N and F in the integral will remain nearly constant with height, then the electron density can be written as

$$N = \left(\frac{1}{2F\Delta h}\right)\left[\ln\left\{\frac{A_x}{A_o}\right\}_1 - \ln\left\{\frac{A_x}{A_o}\right\}_2 - \ln\left\{\frac{|R_x|}{|R_o|}\right\}_1 + \ln\left\{\frac{|R_x|}{|R_o|}\right\}_2\right]. \tag{7.17}$$

The only implicit unknown or the right side of (7.17) is the collision frequency, $\hat{\nu}$. We assume that at the lower "boundary" of the D-region the electron density is close to zero and the absorption coefficients will also be zero, then from (7.15) we have

$$\left\{\frac{A_x}{A_o}\right\}_{ho} = \left\{\frac{|R_x|}{|R_o|}\right\} = f(\omega, \omega_L[h_o], \nu[h_o]). \tag{7.18}$$

Now since the left-hand quantity is known, we can calculate the collision frequency at h_0. This value is used to construct an exponential N profile, taking the gradient of the collision frequency profile from a standard atmosphere or from rocket data.

A much more satisfactory method for estimating the effects of collision frequency on PRE data consists of measuring both the *phase* and amplitude of the echoes. This technique is usually referred to as Differential Phase Measurements. The phase difference between two waves can be expressed as

$$\Delta\Phi = \Phi_x - \Phi_o = \frac{2\pi}{c_o}\int (n_x - n_o)dh + (\eta_o - \eta_x), \tag{7.19}$$

where:

$\eta_{o,x}$ = the phase change at the reflecting point.
c_o = velocity of EM wave in free space.

When differential phase is measured and combined with amplitude measurements, it is possible to obtain independently and simultaneously both the collision frequency and electron density profiles.

There are various methods for measuring the differential phase (Austin 1971; Belrose 1970), and Meek and Manson (1987) have presented in some detail how one uses the full polarization measurement in PRE work. Most investigators now use the term MF radars to describe PRE systems, and recently Manson and Meek (1985) have shown how MF radars may be used in the interferometric mode to obtain information on the nature of radar scattering in the middle atmosphere.

Details of the choice of probing pulse parameters, probing frequency, as well as design considerations for the transmitter, receiver, antenna, and data recording systems for the PRE method are given in Sections 8.5.1 through 8.6 of Rawer's *URSI Absorption Handbook* (1976). The FM-CW technique has also been successfully applied to PRE (Kramm et al. 1972).

7.5 Other Techniques

Several other radio techniques have been utilized to measure ionospheric absorption – without much success. In principle, monitoring HF transmissions from an orbiting satellite should provide absorption values, but uncertainties in transmitted wave characteristics, power, and polarization have proved too difficult to handle. Various investigators had used differential height (Kazantsev 1959), differential amplitude for oblique and vertical propagation (Mass 1963), and differential polarization (Amayenc 1968) with less than spectacular results.

Since absorption $\approx 1/f^2$, the variation of lowest-observed-frequency (LOF) on an ionogram gives some idea of how absorption varies over the ionosonde site, but uncertainties in transmitter power, receiver sensitivity, and antenna pattern precludes quantitative measurements. A thorough investigations of the use of the lowest-observed frequency (LOF) on a 2000-km oblique incidence sounder path was presented by Rose (1965) with promising results. Bradley (1976) has also used ionosonde absorption data, as have Wakai and Ishizawa (1962).

Jones et al. (1972) used a LORAN-C transmitter as an LF sounder along with D-region modeling and propagation theory in an attempt to obtain electron density profiles – with reasonably good results.

7.6 Capabilities and Limitations

Table 7.2 summarizes the salient characteristics of the techniques described in this chapter.

Table 7.2. Summary of capabilities and limitations

URSI designation or other name	Capabilities	Limitations	Remarks
A1 method	Quite sensitive	Interference, cannot measure high values	For mid or low latitudes
A2 method	Large dynamic range	Not as sensitive as some other methods	Passive low cost, can be used to measure polar cap and auroral absorption
A3a method	Quite sensitive	Interference, more complex than A1	Mid or low latitude
A3b method	Sensitive	Interference	Mid or low latitude
PRE method	Can obtain electron and collision frequency profiles	Interference, complex system, more expensive than others	MF radar can be used to probe the middle atmosphere
f_{min} method	Can give a qualitative indication of variation of abs.	Very sensitive to equipment parameter variation	Really not too useful
LOF	Gives a value that can be applied fairly directly to HF circuits	Interference, difficult to interpret	Not used very much
Satellite HF beacon	Global coverage	Too many variables	Not too useful

Part III Satellite, Rocket, and Other Techniques

Whereas Part II of this monograph dealt solely with earth-based radio techniques for probing the ionosphere, Part III will cover passive and active radio experiments aboard satellites and rockets, as well as various other ground-based radio techniques which are variants of previously discussed methods. A thorough review of ionospheric effects upon earth-space radio propagation was published by Lawrence et al. (1964) and most of the essentials of the theory of the interaction of radio waves with a magnetoionic medium which are relevant to this subject have been presented in Section 2.2 of this book.

Chapter 8 Earth Satellite Radio Studies of the Ionosphere

The first instance of the use of extraterrestrial radio sources to study the ionosphere occurred shortly after World War II, when Hey et al. (1946) noticed short period fluctuations in 64 MHz radiation from Cygnus. At first, these signal variations were attributed to fluctuations of the source, but subsequent correlation studies at spaced receiver sites suggested that their origin was in the earth's atmosphere (Smith et al. 1950; Little 1952). Hewish (1952) pointed out that these "radio star" emanations could, in fact, be used to study the irregular nature of the ionosphere. Representative results from some of these early investigations of the ionosphere using radio stars are given in papers by: Parthasarathy et al. (1959), Little et al. (1962), Owren et al. (1964), and Fremouw (1966).

The use of the earth's *natural* satellite – the moon – as a reflector of radar pulses[3] also made possible the measurement of ionospheric *total electron content* (*TEC*) by the Faraday rotation technique. Murray and Hargreaves in 1954 announced the discovery of the ionospheric Faraday effect as it applied to moon-earth signals, followed by other early reports by Browne et al. (1956) and by Evans (1956, 1957) (the Faraday rotation method will be covered in detail in Section 8.1.3).

8.1 Satellite Radio Experiments

Man first started to use *artificial* earth-orbiting satellites to investigate his environment with the launch of Sputnik I in October 1957 and the measurement of fading characteristics of the Sputnik I Majak beacon transmission (Aitchison et al. 1958; Al'pert 1958; Kent 1959). In the 30 years since Sputnik I, radio beacons aboard earth-orbiting satellites have provided much information on the global morphology and magnetic storm-time behavior of the terrestrial ionospheric TEC. [See the paper by Al'pert (1976) for a thorough review of results of the first two decades of satellite beacon studies.] Deep space probes employing transmitters at various frequencies using large earth station antennas have even provided information on planetary ionospheres by occultation and other types of experiments (see Bauer 1973, and Mahajan and Kar 1988).

Beacon satellites can also be used to study ionospheric irregularities. The amplitude, phase, and angle of arrival of a radio wave traversing the ionosphere

[3] The use of the moon as a "calibration" target for sophisticated radar systems has been recently re-emphasized by Mathews et al. (1988).

will be distorted when the signal encounters drifting or stationary irregularities. These distortions, or "*scintillations*", when measured with ground-based receivers, will provide information on the structure and motion of the irregularities (see Davies 1969).

A third generic use of artificial satellites is a "topside sounder" by installing a suitably miniaturized ionosonde in the satellite, which then permits one to obtain the global morphology of the *topside* (above the F-layer peak) ionosphere. Until the advent of the high altitude rockets, satellites, and incoherent-scatter sounders, the topside ionosphere was inaccessible by standard ground-based radio techniques. Lawrence et al. (1964) and Millman (1967) have published reviews on propagation effects on trans-ionospheric signals, and *reviews* on the three "generic" uses of earth satellites to measure ionospheric characteristics (TEC and scintillations of beacon signals, and topside sounding) are given by; Chapman and Warren (1968), Evans (1977), Davies (1980), Jackson et al. (1980), Aarons (1982), Yeh and Liu (1982). Other pertinent papers are in the June 1969 special issue of the IEEE Proceedings (*Topside Sounding and the Ionosphere*), (Schmerling and Langille 1969) two special issues of *Radio Science* on *Ionospheric Studies by Means of Beacon Satellites* (1970 and 1984), and in Sections 10.7, 10.8, and 10.9 of the *Handbook of Geophysics and the Space Environment* published by the US Air Force Geophysics Laboratory (edited by Jursa 1985). The technique using rf probes to make in-situ measurements of the ionospheric plasma from rockets and satellites will be covered in Chapter 9.

8.1.1 A Short History of Radio Beacon Experiments (RBE)

Some information on ionospheric spatial structure and electron content was obtained from the early Sputniks (Al'pert 1976, and Garriott and Little 1960). By 1965 the variation of TEC with sunspot number and time of day was reasonably well established in terms of the equivalent ionospheric slab thickness (Bhonsle et al. 1965). Some other early results were reported by Bowhill (1958), who measured the differential Doppler between two signals (see bibliography in Al'pert's review) and by investigators using the Beacon Explorer satellite (Kersley et al. 1988).

The next new results were obtained using radio beacons on U.S. Navy navigational satellites (specifically, Transit 4A; see Yeh and Flaherty 1966 and Bertin and Papet-Lepin 1970).

The first results using *geostationary* satellites were reported by Garriott et al. (1965). The usefulness of geostationary beacon measurements of TEC was further illustrated by Klobuchar and Allen (1970) using Faraday rotation measurements from the VHF beacon from ATS-3. Hargreaves developed a proposal for a geosynchronous satellite beacon designed specifically for ionospheric studies – resulting in the ATS-6 Radio Beacon Experiment (RBE) (Davies et al. 1975, 1976). This multi-frequency beacon was designed to measure the absolute total columnar electron content N_T between the transmitter and the receiver, and the Faraday (ionospheric) columnar content N_F along the ray path between the receiver and a height of ≈ 2000 km. Davies was designated the Principal

Table 8.1. Locations and intervals for the various ATS-6 phases

Phase	Time intervals	Location	Orbital inclination	Average height (km)	Other information
1	June 1974–May 1975	94°W	$\approx 1.5°$	35791	Frequencies used were 40,140 and 360 MHz
2	Aug. 1975–Aug. 1976	35°E	$\approx 0.6°$	35789	Modulation side-band frequencies were ± 0.1 and 1.0 MHz
3	Sept. 1976–June 1979	140°W	$\approx 1.6°$	35790	ERP = 250 to 2000 W

See Davies (1980) for more specific details.

Investigator of the ATS-6 RBE, and the transmitter/receiver systems were designed by Grubb (1972), and Grubb et al. (1976).

The most important feature of the ATS-6 beacon was its combined use of the Faraday effect and of the modulation phase technique; the former was weighted toward the lower ionospheric levels by the geomagnetic field, and the latter was sensitive to electrons *anywhere* along the raypath. Locations and intervals for the various ATS-6 phases and other information are shown in Table 8.1.

More recent beacon satellites include the geostationary ETS-1 and ETS-2; and the six orbiting US Navy NNSS satellites which are planned to be phased out in 1996–1997; Wideband (Fremouw et al. 1978), HILAT (Fremouw 1985) and Polar Bear (Basu et al. 1988). Both the HILAT and Polar Bear beacons have ceased operations as of the completion of this book.

8.1.2 Methods of Determining Electron Content

Faraday Rotation

One physical principal which makes possible the determination of ionospheric columnar electron content is *Faraday rotation.* This effect was discovered by Michael Faraday (1845), when he subjected a block of glass to a strong magnetic field. He observed that a plane-polarized monochromatic beam of light passing through the glass in a direction parallel to the imposed magnetic field had its plane of polarization rotated. The amount of rotation was given by the expression.

$$\Omega = \mathrm{KHl}, \tag{8.1}$$

where:

Ω = the angle of rotation

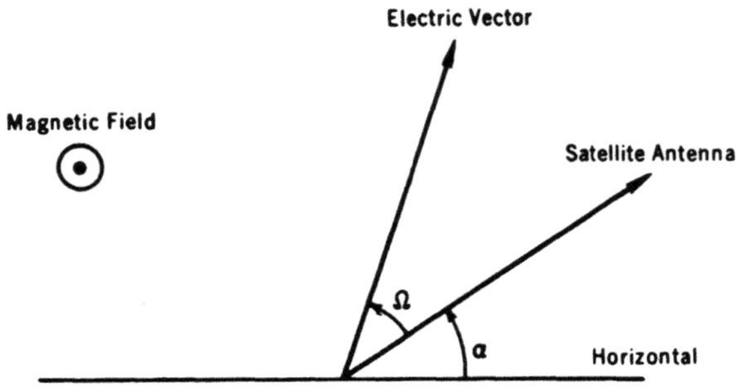

Fig. 8.1. Simplified plane geometry of satellite-earth propagation path in the northern hemisphere to explain Faraday rotation effects as defined in Eq. (8.2). (After Davies 1980)

K = a constant associated with each substance
l = length of the path of light through the substance (m)
H = magnetic field intensity in amps/m.

This phenomenon has been observed in many different solids, liquids, and gases – including the ionosphere.

The Faraday rotation of the electric vector of a radio wave propagating from a satellite radio beacon in a direction parallel to the earth's magnetic field (as seen by an observer looking up, in the northern hemisphere) is *anticlockwise*, as shown in Fig. 8.1. An observer looking up in the southern hemisphere would observe *clockwise* rotation.

Ignoring refraction, the Faraday rotation of the electric vector is given by (Davies 1980 – in his notation) as

$$\Omega = \pi f/2c \int_R^S \frac{X\sqrt{Y_T^4 + 4(1-X)^2 Y_L^2}}{(1-X)(1-Y_L^2) - Y_T^2}\, ds, \tag{8.2}$$

where:

Ω = Faraday rotation in radian
f = wave frequency (Hz)
c = 2.998×10^8 m/s
X = kN/f^2
k = 80.61
N = electron density (el/m^3)
$Y_L = (eBo \cos \Theta)/2\pi mf$
$Y_T = (eBo \sin \theta)/2\pi mf$
e, m = charge and mass of electron
θ = angle between geomagnetic field B_o and ray path s
α = aspect angle to satellite.

The integration is between the receiver R, and the satellite S. For VHF frequencies the quasi-longitudinal approximation holds (see Section 2.2), and we can express Eq. (8.2) as

$$\Omega \approx \left(\frac{\pi K}{cf^2}\right) \int_R^S f_L N\, ds \tag{8.3}$$

evaluating the constants, yields the relation

$$\Omega \approx 8.447 \times 10^{-7} f^{-2} \int_R^S f_L N \, ds \text{ rad}, \tag{8.4}$$

where $f_L = 2.80 \times 10^{10} B_L$, the electron gyrofrequency corresponding to the longitudinal component of the geomagnetic field along the ray path.

For some cases, we can obtain sufficient accuracy by replacing f_L in Eq. (8.4) with an average value, thus taking it outside the integral and inverting the integral equation to obtain two mean "shape parameters" as defined below

$$\bar{M} = \frac{\int_O^S NB \cos\theta \operatorname{Sec} i \, dh}{\int_R^S N \, dh} \text{(tesla)} \tag{8.5}$$

and

$$\bar{F} = \frac{N f_H \cos\theta \, ds \delta}{\int_R^S N \, ds} \text{(MHz)}, \tag{8.6}$$

where:

f_H = the electron gyrofrequency
$ds = dh \operatorname{Sec} i$
i = the angle between the ray path and the vertical.

Titheridge (1972) found that M gives a first approximation of the electron content up to a height $h_F \approx 2000$ km. A technique for obtaining F and S_F is given by Davies (1980) which involves plotting the quantity

$$\frac{\int_O^S f_L N \, ds}{\int_O^h N_{Sec} i \, dh} \text{ versus } h$$

for a variety of electron density profiles. If the curves pass through a single point of convergence, then that point defines appropriate values of F and h_F. Different techniques for determining these values are compared in Davies (1980). Most investigators take a height of ≈ 400 km to obtain the value to F, since this is about two scale heights above h_m F2.

In interpreting Faraday rotation results it is necessary to take into account the variation of f_L with height in the vicinity of the ray path, since it will affect M, F, etc. It is also necessary to calibrate this technique by computing the polarization angle in the absence of the ionosphere (Davies 1969).

Modulation Phase or Group Delay Technique

This method is essentially independent of the geomagnetic field and yields the total columnar electron content between the beacon and the earth station receiver. It utilizes the variation of phase or group velocity with frequency, and the version we shall describe applies most closely to the ATS-6 RBE. It is based on the fundamental relationship that the time-of-flight, t_g, of a wave group is the frequency derivative of the phase,

$$t_g = \partial\Phi/\partial f \approx \Delta\Phi/\Delta f \,(\text{sec}), \tag{8.7}$$

where: $\Delta\Phi$ = phase difference in cycles between waves with frequency difference Δf Hz

Since the RBE frequencies are much higher than the ionospheric critical frequencies, refraction of the ray path is negligible, and the phase difference between the transmitter and receiver is

$$\delta\Phi = \frac{2\pi f}{C} \int \mu \, ds, \tag{8.8}$$

where μ = phase refractive index (see Davies 1969, Chapter 4). We can further simplify the calculations by using the nondeviative and quasi-longitudinal approximation for the refractive index

$$\mu = \frac{1 - 80.6}{2(f \pm f_L)}. \tag{8.9}$$

All the terms in this equation have been previously defined, and the + and − refer to the ordinary and extraordinary waves, respectively (discussed in Chapter 2). Since $f > 40$ MHz and $f_L \approx 1$ MHz, the effects of the geomagnetic field are rather small. Substituting Eq. (8.7) into Eq. (8.6), we find that there is a term which depends upon the total path length (the line-of-sight motion of the satellite). In the modulation phase method, this term is removed by using two sets of frequencies spaced by Δf. If $\bar{f}_2$ and $\bar{f}_1$ are the arithmetic averages of these two frequency sets, then the difference $\Delta\Phi$ between the two ordinary wave phase differences is

$$\Delta\Phi = -\frac{k\Delta f}{2c}[1/(f_2 + \bar{f}_L)^2 - 1/f_1 + \bar{f}_L)^2] \int_R^S N \, ds (\text{cycles}). \tag{8.10}$$

Therefore, from the measured $\Delta\Phi$ and the wave frequencies, the $N_T = \int_R^S N\,ds$ term can be determined (see Davies et al. 1975); e.g., for the ATS-6 RBE, 1 MHz of modulation at 140 MHz is compared with the coherent 1 MHz modulation at 360 MHz.

The Carrier Phase Method

Another method for measuring TEC is to reduce two widely spaced carrier frequencies f_2 and f_1 to a common reference frequency, f_R and measure the phase difference Φ^c in cycles as defined by

$$\Phi^c = f_R(\Phi_2/f_2 - \Phi_1/f_1) = \frac{kf_R}{2c\left[\frac{1}{f_2(f_2+\bar{f}_L)} - \frac{1}{f_1(f_1+\bar{f}_L)}\right]} N_T. \tag{8.11}$$

When we compare Eq. (8.10) and (8.11), we see that we can write

$$\Phi^c \approx (f_R/\Delta f)\Delta\Phi. \tag{8.12}$$

Dropping the f_L term from Eq. (8.9) results in an error of only 2% at carrier frequencies > 40 MHz at midlatitudes, and a change of 1% in the geomagnetic field for a given profile results in a change of 0.03% in the "time-of-flight." Thus the Carrier Phase method is essentially independent of geomagnetic field effects; however, the "ambiguity problem" (cycle identification) is greater than for other methods of measuring TEC. The cycle ambiguity in Faraday rotation measurements can be resolved for a two-or-more frequency beacon by utilizing the differential rotation (Ω_f) on closely spaced frequencies (Δf) (Garriott et al. 1970) using the relation,

$$\frac{\Omega_f}{\Omega} = \frac{2\Delta f}{f} \tag{8.13}$$

where Ω is the Faraday rotation.

The absolute accuracy of the Faraday rotation and Group Delay techniques depend upon many factors, such as: antenna construction and siting, "aspect angle," the angle between the raypath and the geomagnetic field, delay times, etc. Evans (1977) and Davies (1980) discuss most of these problems and methods of calibration in their review papers.

8.1.3 Global Variation of TEC

There have been hundreds of papers and reports published on the morphology of electron content using RBE satellite transmissions (see papers by Onwukwe 1974; Davies et al. 1975; Bertel et al. 1976; Evans 1977; Essex 1977; Klobuchar et al. 1977; Huang 1978; Davies 1980; Hunsucker et al. 1984; the *Special Edition* of *Radio Science* for May–June 1984. Other investigators have used RBE data to investigate travelling-ionospheric disturbances (TID's) produced by acoustic-gravity waves (AGWs), Hines (1960) (see Hunsucker 1982; Evans and Wand 1983, and Hunsucker and Hargreaves 1988).

Most investigators using the RBE method for obtaining TEC use some standard parameters and nomenclature, as outlined in this section (see Davies 1980, p. 376, and Fritz 1976). The total columnar electron content N_T can be obtained from group phase delay measurements, while the Faraday rotation, Ω, gives the integral $\int_R^S f_L N\, ds$ from the receiver, R, to the satellite, S, (up to a height of ≈ 2000 km). The latter term is called the *Faraday content*, N_F. The foregoing quantities are related by

$$N_P = N_T - N_F = N_T(1 - F/\bar{F}), \tag{8.14}$$

where:

N_P = the residual, protonospheric or plasmaspheric content
$\bar{F}$ = a shape factor depending on the distribution of electrons with respect to the geomagnetic field,

$$F = \int_R^S f_L N\, ds/N_T. \tag{8.15}$$

Alternatively, one can use the following "shape parameters" to analyze RBE data. The *slant total slab thickness* is given by

$$\tau_{TS} = N_T/N_m. \tag{8.16}$$

The *slant Faraday slab thickness* is measured by most RBE techniques, and is given by

$$\tau_{FS} = N_F/N_m, \tag{8.17}$$

where N_m = the peak electron density. The *vertical* slab thickness are defined in a similar manner. Also, by definition,

$$\int_R^S f_L N\, ds = \bar{F}\, N_F \tag{8.18}$$

combining Eqs. (8.16) and (8.18), we obtain

$$F\, N_T = \bar{F}\, N_F. \tag{8.19}$$

So, in terms of measured quantities, we can write

$$F = \{R_f \Omega_f / \Phi_{f+\Delta f}\}\, \Delta f\ (\mathrm{MHz}), \tag{8.20}$$

where:

Ω_f = Faraday rotation on a beacon frequency f, and
$\Phi_{f+\Delta f}$ = modulation phase difference on frequencies f and Δf
R_f = the "frequency coefficient" (varies as shown below).

From Eq. (8.17) we can write

$$F = \tau_{TS} = \bar{F}\tau_{FS}. \tag{8.21}$$

Table 8.2. Frequency coefficients for different latitude zones

	f_L (MHz)	R_{40}	R_{140} (MHz)
Middle latitudes	1.0	0.917	0.83
Low latitudes	0.1	0.96	0.84
High latitudes[a]	0.7	—	—

[a] From Hargreaves et al. (1980).

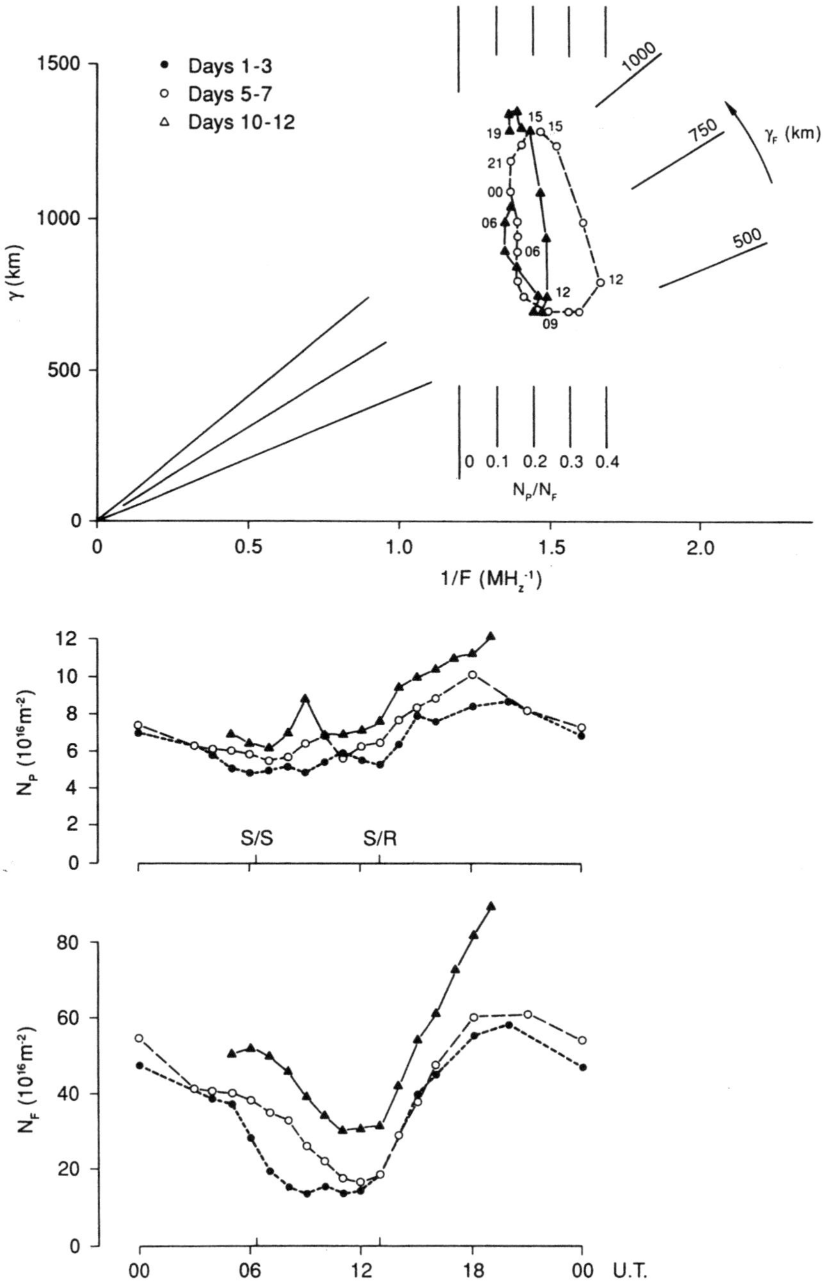

Fig. 8.2. Slab thickness (τ) versus 1/F plots (*upper plot*); and Faraday electron content (*NF*), Protonospheric content (*Np*) as a function of time for Fairbanks, Alaska. (After Hargreaves et al. 1980)

As a first interpretation, Hargreaves (1978) has used τ vs. 1/F plots to investigate the effect of the variation of profile parameters, as shown in Fig. 8.2.

Analyses of RBE data using the τ vs. 1/F plots provided a valuable input to ionosphere/protonosphere modeling because they include a valid protono spheric contribution (Poulter et al. 1981).

8.1.4 RBE Studies of Ionospheric Scintillation

Ionospheric scintillations are fluctuations of amplitude, phase, and angle-of-arrival of a VHF/UHF signal passing through irregularities located mainly in the F-region. Either extra-galactic sources such as radio stars, or satellite beacon transmitters may be used as the signal sources for earth-observed ionospheric scintillations, and both geostationary and orbiting satellite beacons have been used.

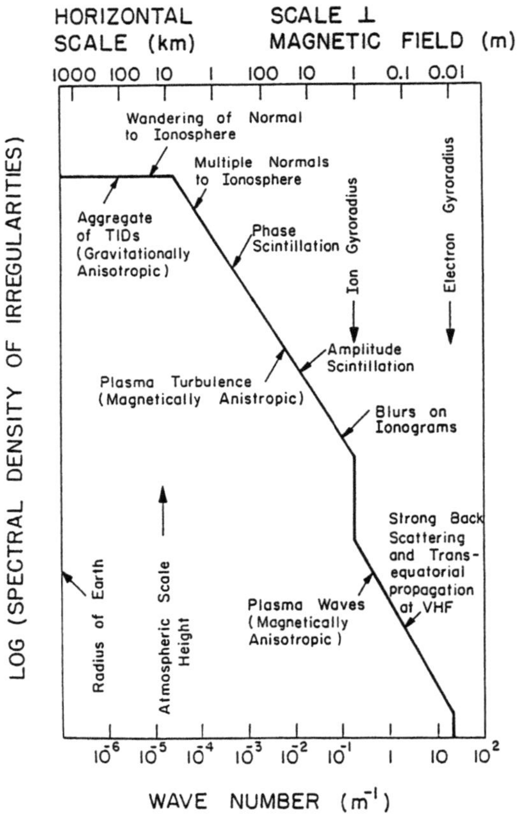

Fig. 8.3. A composite spectrum summarizing the intensity of ionospheric irregularities as a function of wavenumber over a large spatial scale. (Debye length = 10^{-3} m). (After Yeh and Liu 1982)

Fig. 8.4 a, b. Phase (**a**) and amplitude (**b**) spectra for weak scintillation observed from ATS-6 satellite at Goose Bay, Labrador. (AFGL Handbook 1985)

Some of the first reports on the global *morphology* of ionospheric scintillation were published in a report edited by Aarons (1970), and in papers by Aarons and Allen (1971), and Fremouw and Bates (1971); also a worldwide model of F-layer scintillation was published by Fremouw and Rino (1971). Some scintillation *results* using the ATS-6 and Wideband satellite beacons were given by Davies et al. (1975) and Fremouw et al. (1978). More recent *reviews* of the morphology of ionospheric scintillation were presented by Davies (1980, Section 9), and by Aarons (1982), and Basu et al. (1988) have presented some recent results on polar cap scintillation modeling using the DE-2 satellite. The basic elements of the *experimental technique and theory* will be outlined in this section.

The experimental technique is quite straightforward for the RBE's designed for scintillation studies (ATS-6, Wideband, HILAT, Polar Bear, ETS-2). One must, of course, pay careful attention to the sensitivity, frequency/gain stability, and bandpass characteristics of the receiver; as well as to the antenna gain, beamwidth, sidelobe level, front-to-back ratio, and siting criteria (see the papers by Grubb et al. 1976; Fremouw 1985, and *Antenna* books by Johnson and Jasik 1987 and by Kraus 1988). Amplitude and phase data from satellite beacons are digitized, recorded on magnetic tape, then analyzed to produce power spectra of amplitude and phase.

The scintillation of trans-ionospheric radio signals in the VHF/UHF part of the radio spectrum is (qualitatively speaking) due to multiple scatter (see Section 2.3 of this book) from ionospheric irregularities of meter scale size normal to the geomagnetic field. A very wide range of irregularity scale size exists in the ionosphere, as shown by the plot in Fig. 8.3 developed by Booker (1979) and modified by Yeh and Liu (1982).

"Typical" scintillation data obtained from ATS-6 RBE measurements made at high latitudes are shown in Fig. 8.4.

Theoretical explanations of ionospheric scintillation are based on the effects of one or of multiple "phase-changing screens" on VHF/UHF propagation (see Davies 1969, Section 13.6.5). A plane wave front impinging on a "lossless" screen (whose thickness varies in the transverse direction in some manner), after emerging from the screen, will have its wavefront distorted by the screen. At any point some distance under the screen there will be constructive and destructive interference between parts of the wavefront traveling in different directions. The amplitude variations along the wavefront will increase in magnitude with increasing distance from the screen. Hewish (1952) showed that at a distance D, below the screen the amplitude variation across the wave becomes "fully

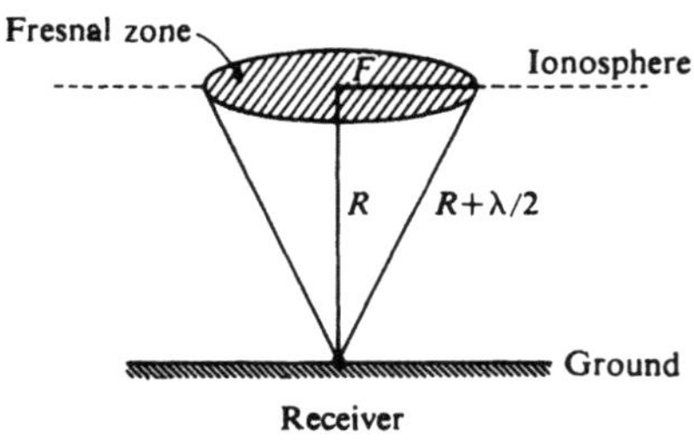

Fig. 8.5. Simplified geometry of Fresnel zone for scattering by an ionospheric irregularity. (Davies 1980)

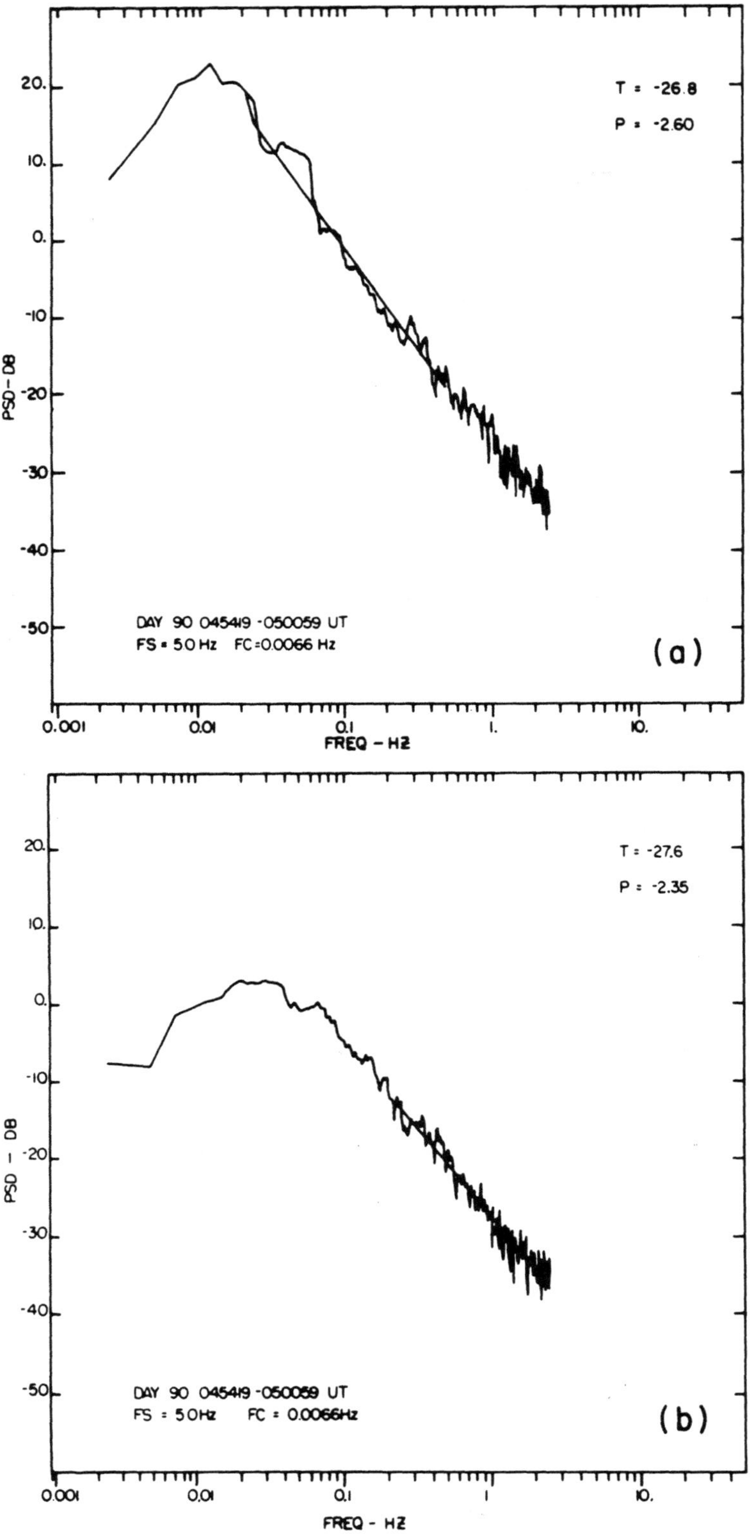

T = -26.8
P = -2.60
DAY 90 045419 -050059 UT
FS = 5.0 Hz FC=0.0066 Hz
(a)
PSD - DB
FREQ - HZ
0.001
0.01
0.1
1.
10.
20.
10.
0.
-10
-20
-30
-40
-50
T = -27.6
P = -2.35
DAY 90 045419 -050059 UT
FS = 5.0 Hz FC = 0.0066Hz
(b)

developed" when the transverse dimension of the ionospheric irregularity equals the size of the first Fresnel zone. Figure 8.5 illustrates the geometry of the irregularity and the observer at ground level. The radius R, of this Fresnel zone is given by

$$R = [(D + \lambda/2)^2 - D^2]^{1/2} \approx \sqrt{D\lambda}. \tag{8.22}$$

The deeper the "phase modulation" of the emergent wavefront, the finer will be the diffraction pattern structure (Ratcliffe 1956). If the phase changes in the wavefront produced by the slab are less than one radian, the scale size of the observed diffraction pattern at the ground is approximately equal to the transverse scale size of the ionospheric irregularities.

Various "scintillation indices" have been derived by workers in the field (Briggs and Parkin 1963; Aarons et al. 1980). In defining the various indices utilized, one usually starts with the autocorrelation function, $R_v(t_1, t_2)$, when the random process has the functional form $v(t) = g(X, t)$; X being an ordinary random variable, then we can define the autocorrelation function as

$$R_v(t_1, t_2) = \int_{-\infty}^{\infty} g(x, t_1)\, g(x, t_2)\, p_x(x)\, dx, \tag{8.23}$$

where:

$p_{x(x)}$ = the probability density function (PDF) of $x = dF_x(x)/dx$
$F(x)$ = the cumulative distribution function $= P(X \leq x)$; $-\infty < X < \infty$
P = the probability of the event.

Briggs and Parkin (1963) defined four amplitude scintillation indices (moments of the distribution) as follows:

$$S_1 = 1/R\,|\overline{R\,\bar{R}}\,| = \text{normalized mean deviation} \tag{8.24}$$

$$S_2 = 1/\bar{R}\,\{R - \bar{R}^2\}^{1/2} = \text{normalized rms deviation.} \tag{8.25}$$

The next two indices are related to power, which is proportional to R^2, as follows

$$S_3 = 1/R^2\,|R^2 - \overline{R^2}\,| \tag{8.26}$$

and

$$S_4 = 1/\overline{R^2}\,\{\overline{R^2 - \bar{R}^2}\}^{1/2}. \tag{8.27}$$

Briggs and Parkin (1963) originally defined S_2 as

$$S_2^2 = \frac{\overline{R^4} - \overline{R^2}^{\,2}}{\overline{(R^2)}^{\,2}}, \tag{8.28}$$

and they denoted R_4 as S. The relationships between these indicies (for a Rayleigh distribution) is:

$S_1 = 0.42\, S_4$
$S_2 = 0.52\, S_4$
$S_3 = 0.73\, S_4$
$S = S_4$.

Table 8.3. Comparison of various scintillation coefficients

S_4	DB_{JSSG}	SI_{JSSG}
0.075	1	11
0.17	3	32
0.3	6	59
0.45	10	81

A less rigorous scintillation index was proposed by Whitney et al. (1969) as

$$\text{S.I.} = \frac{P_{max} - P_{min}}{P_{max} + P_{min}}, \tag{8.29}$$

where:

P_{max} = the power amplitude for the third peak down from the maximum excursion during a 15-min recording

P_{min} = the third peak up from the minimum excursion during a 15 min recording.

The SI defined in Eq. (8.29) is also known as the Joint Satellite Scaling Group (JSSG) index. The most-used ionospheric scintillation index is probably S_4, which is proportional to the RMS electron density variation.

Aarons et al. (1980) used the S_4 and SI_{JSSG} indices to investigate high-latitude scintillation levels and one of their results was the SI equivalence table (Table 8.3). Aarons et al. (1980) also developed quantitative analytical formulas for the auroral latitude RBE receiving stations at Narssarssuaq, Greenland, and Goose Bay, Labrador, and for the Sagamore, Massachusetts, station.

Further work on the statistics of scintillating signals was done by Fremouw et al. (1980) and Secan et al. (1987), and much effort has gone into the development of the theory of scattering and scintillation of trans-ionospheric radio signals (Rino 1979a, b, Parts 1 and 2, 1980; Livingston 1982; also see the survey of scintillation theory by Yeh and Liu 1982), and recent results of a spaced receiver investigation of equatorial scintillations by Vacchione et al. (1987). Another recent and useful addition to the literature on scintillation has been published by Hewish (1989).

8.2 Satellite Topside Sounders

In the late 1950's advances in available satellite payloads, precision of tracking, and electronic miniaturization by use of solid-state electronics made it feasible to design small ionosondes which could be carried by earth-orbiting satellites. Advanced mechanical engineering techniques also made it possible to deploy sufficiently large antennas for ionosondes carried on satellites. Thus, a tool for exploring the ionosphere above the level of maximum ionization on a global scale became available to ionospheric physicists. It is interesting that development of incoherent scatter radars (which also can probe the topside ionosphere) occurred concurrently. Topside sounders have been utilized to investigate the global morphology of the F-region, to stimulate the ambient ionosphere, and to detect natural radiation phenomena (such as Auroral Kilometric Radiation – AKR).

Reviews of topside sounder technology and results of early investigations are reported in the *Special Issue of JATP* of November, 1967, by Chapman and Warren (1968), and in the *Special Issue of the Proceedings of the IEEE* of June 1969 (guest editors, Schmerling and Langille) also by Benson and Bitoun (1979).

Workers in West Germany, the UK, the US and Canada commenced design on topside sounders in 1958, and the first operational topside sounder was the Canadian Alouette I, which was orbited on 29 September 1962 and continued to produce ionograms until 1972.

8.2.1 Sweep Frequency Topside Sounders

In a joint Canadian-US undertaking, the Canadians designed and built (Franklin and Maclean 1969) a sweep-frequency ionosonde covering 0.5–11.5 MHz, using a transmitter pulsed at 63 Hz, with 100-μs pulses and a peak power output of 250 watts into crossed dipoles of 22.9 m and 45.7 m length. This topside sounder (Alouette I satellite) was launched by a US rocket from the Western Test Range into a prograde near-circular orbit with inclination 80.5°, perigee of 993 km and apogee of 1033 km.

The Canadians built and the US launched a second sweep-frequency sounder, Alouette II, on November 28, 1965, followed by ISIS I and ISIS II (Florida 1969; Franklin and Maclean 1969; Jackson 1986, 1988; James 1989). In addition to launching the Canadian topside sounders, the US provided several global tracking and data acquisition facilities. A number of other nations also participated in this international program for ionospheric studies (ISIS) by providing other tracking and data acquisition sites and by sharing in the ionogram filming and analysis. The Alouette/ISIS series of satellites provided 50 satellite-years of measurements of the terrestrial ionosphere and led to the publication of ≈1000 papers (Jackson 1986).

Other topside sounders were built and launched by Japan at the Tanegashima Space Center of the National Space Development Agency (NASDA) in 1976 (ISI-a) and in 1978 (ISI-b), see (Radio Research Laboratories Reports 1978–1982). A "relaxation sounder" was orbited on the European Space Agency's (ESA) GEOS-1 and GEOS-2 satellites (Etcheto and Bloch 1978) and on EXOS-C by Japan (GRL/1988) and a similar sounder using a "mother-daughter" satellite system was flown by the Sun-Earth Explorers ISEE-1 and ISEE-2 (Harvey et al. 1978). Other experiments, such as magnetometers, particle detectors, and probes were also operated on most of the topside sounder satellites.

Table 8.4 shows the salient parameters of the various primary topside sounders which have operated from 1962 until the present time.

Table 8.4. Salient parameters of primary topside sounders satellites

Satellite/type	Frequency range MHz	Sweep rate MHz/s	PRF Hz	Pulse width (μs)	Transmitter power (W)	Remarks
Alouette I sweep	0.5–12	≈1	63	100	100	1962, 1000 km polar orbit
Alouette II sweep	0.12–14.5	0.13–1.0	30	100	300	L 1965, 500–3000 km polar orbit
Explorer XX step	1.50–7.22	0.57–1.1			100	L 1964, 800 km polar orbit
ISIS I step and sweep	Fixed frequency from 0.25–9.3	0.31–1.03	30 60	98	400	L 1969, 565–3500 km polar orbit
ISIS II step	Fixed frequency from 0.1–20	0.376–1.5	45	87	400	L 1971, 1400 km circular polar orbit
ISS-b step and sweep						L 1978, 972–1220 km orbit at 70° inclination

L = "Launched".

Examples of typical results of global ionospheric morphology and physics obtained from topside sounder investigations are given by Jackson et al. (1980); Jackson (1986); Hoegy and Benson (1988); also in four Summary Plot Reports of ISI-b data published between 1978 and 1985 by the Japanese Ministry of Posts and Telecommunications.

Plasma resonance studies have been published by Calvert (1966), Oya (1970, 1971), Benson (1977), Kiwamoto and Benson (1979), Benson (1982), Pulinets and Selegey (1986), Benson and Viñas (1988), and other investigators.

Natural radio emissions from the ionosphere, including Auroral Kilometric Radiation (AKR) have been studied and reported by Calvert (1966), Benson (1981, 1985), Benson and Wong (1987), La Belle (1989); Benson et al. (1988); Nishino and Tanaka (1987); and Tanaka and Nishino (1988).

Researchers outside the United States may request data and documentation on topside sounding from their regional World Data Centers, including:

- World Data Center A
- Rockets and Satellites, Code 630.2
- Goddard Space Flight Center
- Greenbelt, MD 20771 USA

US researchers may obtain the foregoing information from:

- National Space Science Data Center
- Code 633.4
- Goddard Space Flight Center
- Greenbelt, MD 20771

8.3 Capabilities and Limitations of Satellite Radio Techniques

In general, the satellite RBE and Topside Sounder Techniques provide very significant results on physics of the ionosphere and on radio propagation phenomena down to the F-region peak on a global basis. The topside sounders also operate in a much lower environment of radio frequency interference (RFI) than that for the bottomside sounder, thus opening the field of ionospheric resonance investigations. Although satellite-borne equipment is quite expensive, it should be remembered that radio beacons and sounders are usually only *part* of a total payload. Another consideration affecting the economics is that *one* topside sounder can provide *global* information on the topside of the ionosphere – similarly, *one* beacon can be used by *many* observers over the globe.

Some of the specific advantages and disadvantages of radio beacons and topside sounders onboard satellites are given in Table 8.5.

Table 8.5 Advantages and limitations of radio beacons and topside sounders

Technique	Advantages	Limitations
Radio Beacon		
A. TEC studies	Global coverage of the ionosphere from polar orbiting satellites; constant beacon parameters; not-too-complex receiving system; continuous coverage for a large area from geostationary satellites; ability to study TIDS	(1) Relatively expensive; (2) rather complex calibration problems; yields a *vast quantity* of data (sometimes overwhelming!); also, rather painstaking data analysis is required. At the present time, there are few RBE's suitable for ionospheric studies; (3) the polar orbiting satellites are, of course, always moving in reference to the earth station, thus convolving spatial and temporal effects
B. Scintillation	The advantages of global and temporal coverage listed in (A) also apply to the polar-orbiting and geostationary satellites, respectively used in scintillation studies. *Many* earth stations can use the same beacon for TEC and scintillation studies	Interpretation of these data in the context of extant theories is a non-trivial task. Remarks 1, 2, and 3 above also apply
Topside sounders	Since all topside sounders use relatively high inclination orbits, they have good global coverage. They are also free from D-layer absorption effects, and provide much information on the ionosphere above the F2 peak	More complex instrumentation than most beacons

[4] Another increment of expense in the BRE earth station receiving system is required to fully utilize data from the advanced polar-orbitting RBE satellites such as HILAT and Polar Bear.

Chapter 9 Other Radio Techniques

There is no intent to minimize the importance of the radio methods described in this chapter compared to the techniques described in the previous chapters. Several of the methods described in Chapter 9 are simply variants of basic radio systems previously covered, while others are not quite as widely used techniques as ionosondes, ISR's, riometers, etc.

9.1 Ionospheric in Situ Measurements

We will define in situ measurements as radio experiments which measure the ionospheric parameters in the immediate vicinity of the vehicle carrying the experiment. In practice, these experiments utilize electrical probes mounted on rocket or satellite payloads. Reviews on this topic have been published by Florida (1969), Sayers (1970), the *Special Issue of Radio Science* for April 1973, Bauer and Nagy (1975) Meyer–Vernet and Perche (1989), and (pertaining to the Dynamics Explorer satellite) by Hoffman (1981). These reviews describe techniques which measure ionospheric parameters above ≈ 100 km, which is the area where the ionospheric mean-free path for charged and neutral particles exceeds the characteristic dimensions of the probe. The Scandinavian Viking satellite was also a recent successful ionospheric probe (Hultgrist and Lundin 1987).

Probably the most used of these devices are the Langmuir probe (which is basically a direct current device), and the various plasma resonance probes. The Langmuir probe essentially measures the plasma density and temperature by its voltage-current response caused by drawing a DC current from the ionospheric plasma. Good descriptions of the principles of spacecraft charging and operation of Langmuir (and RF) probes are given by Bauer and Nagy (1975), Giraud and Petit (1978), and Hargreaves (1979).

RF techniques, on the other hand, determine ionospheric characteristics from the dispersive properties of the plasma over a larger sampling volume than Langmuir probes. Jackson and Kane (1959) first suggested using the RF Impedance Probe to measure characteristics of the ionospheric plasma from spacecraft, and it has seen much use in this application. The basic operating principle of the RF impedance probe is that the input impedance of an electrically short antenna is capacitive reactance ($1/\Omega C_o$) in free space, and departs from C_o when immersed in plasma. The probe capacitance C, when it is in the ionosphere is given by

$$C = \frac{C_o\{1 + X[1 + G(\phi)Y^2]\}}{(1 - Y^2)}, \tag{9.1}$$

where:

$G(\phi)$ = a probe geometry factor

ϕ = the angle between the electric field of the applied RF and geomagnetic field

x, y, f_B, f_N are magnetoionic parameters defined in Section 2.2.1 of this book.

By varying the frequency applied to the RF probe, one obtains two series resonance values, from which all quantities in Eq. (9.1) including f_N, f_B, and the upper hybrid resonance frequency, $f_T = \sqrt{(f_N^2 + f_B^2)}$, so the electron density, N, can be calculated.

Rather complete descriptions of the methodology of RF probe measurements are given by Bauer and Nagy (1975) and by Benson (1982). There is a discrepancy between ionospheric electron temperatures obtained by Langmuir probes and by ISR's which is probably attributable to anisotropies in ionospheric velocity distributions and to probe calibration considerations (Carlson and Sayers 1970). Figure 9.1 illustrates some of the plasma resonances observed with the Alouette

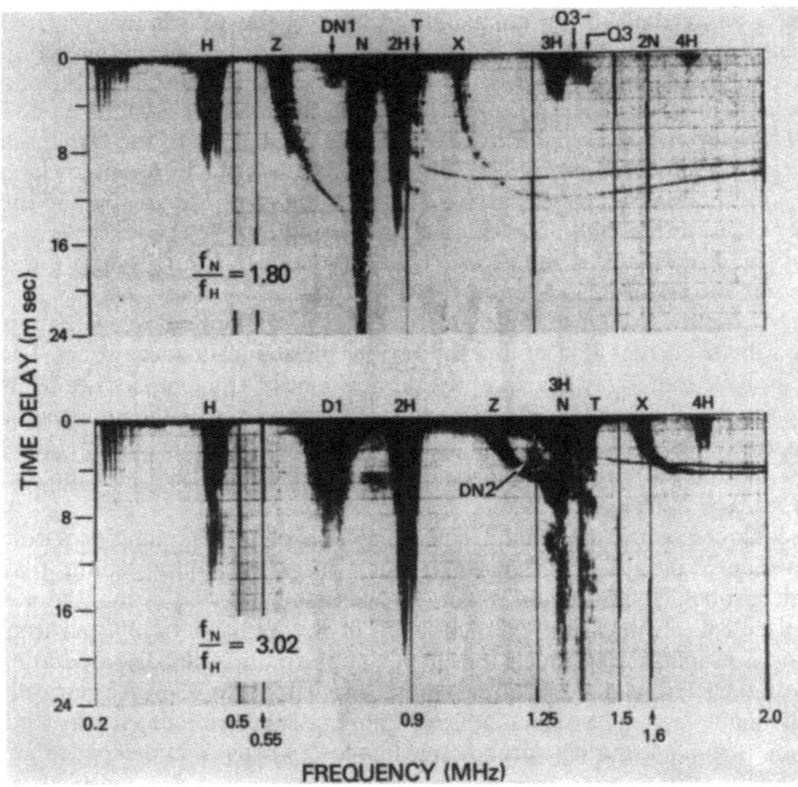

Fig. 9.1. Typical ionospheric resonance signatures observed with the Alouette II topside sounder on April 23, 1966. The resonances are denoted by the *capital letters at the top of the ionograms* and are defined in Benson 1982

II topside sounder (which in this case may be considered to be an in-situ probe, since its antennas are immersed in the ionospheric plasma) (see Calvert and McAfee 1969, and Dougherty and Watson 1971). The topside-sounder-observed resonances represent electrostatic plasma waves whose wave normals are nearly parallel to the magnetic field (f_N and f_B), and wave normals nearly perpendicular to B (f_T), plus the electron cyclotron harmonics (n f_B).

Another in-situ radio technique used aboard satellites is to record *ion-cyclotron whistlers* on a VLF receiver. It is possible to obtain the *crossover* frequency between the electron and the ion whistler and the ion temperature from the difference between the ion whistler at frequency cutoff and the asymptotic ion cyclotron resonance frequency, respectively (Gurnett et al. 1965, and Gurnett and Brice 1966).

9.2 HF Spaced-Receiver and Doppler Systems

There is some confusion concerning the nomenclature of these techniques. Some names utilized to describe the methods of measuring horizontal motions of ionospheric bulk plasma are: HF Doppler sounding, HF spaced receiver and Doppler interferometry.

The groundwork for the Spaced Receiver Technique (SRT) was done by Ratcliffe and Pawsey (1933) and by Pawsey (1935), but the first experimental application of the system was accomplished by Mitra (1949), who measured the drift of ionospheric irregularities.

The SRT usually employs one transmitter and three or more receiving systems (quite often arranged as shown in Fig. 9.2) and if the receiving antennas are suitably spaced, records similar to those shown in Fig. 9.2 will be observed when ionospheric irregularities pass overhead (see Hargreaves 1979). One must assume that the diffraction pattern is spatially isotropic and that fading is due to a steady drift of the irregularities whose structure does not significantly change with time as they drift over the receiving antenna array. Either pulse or CW systems may be used. By a proper choice of operating frequencies, it is possible to study horizontal

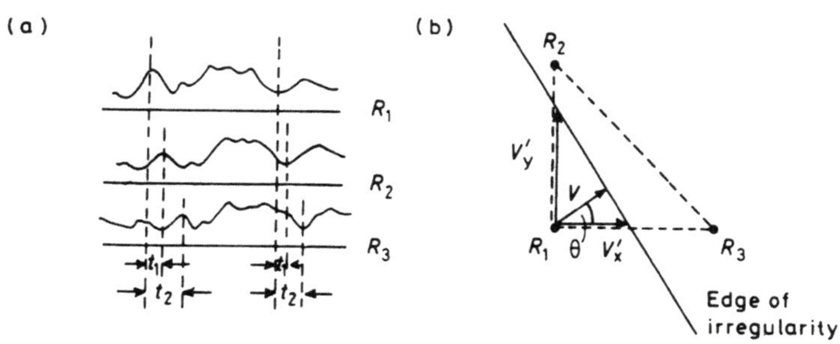

Fig. 9.2 a, b. The spaced receiver technique for observing the movement of a pattern of irregularities over the ground-based observing station is shown in **b**. A typical set of fading patterns observed at these stations is shown in **a**. (After Hargreaves 1979)

plasma motions in both the E- and F-regions and obtain crude estimates of horizontal drift velocities of irregularities.

From Fig. 9.2 one can obtain from the apparent velocities a characteristics velocity, $V_c = d_0/dt_0$, where d_0 is a characteristic distance related to the size of the irregularity, and t_0 is a characteristic fading time as measured by an observer traveling with the drift pattern. If the velocity of the observer is V, the velocity of the observed irregularity movement will be $V_c'^2 = V_c^2 + V^2$, or $V_c'^2 = V'V$ (see Hargreaves 1979, Section 6.3.1).

A refinement of the measurement of the time delay of similar fading patterns uses the autocorrelation and cross correlation functions between the signal received at the antennas to estimate the anisotropy, the true drift, and the apparent velocity of the diffraction pattern.

The relationship between the assumed ionospheric diffraction screen and the fading on the ground as handled by autocorrelation and cross correlation analysis are well covered in a review paper by Ratcliffe (1956), in a book by Al'pert (1963), in a report by Kisabeth (1970), and in papers by Rino and Livingston (1982), Livingston et al. (1982), and Costa and Fougere (1988), and Costa et al. (1988). The basic assumptions and salient conclusions of ionospheric diffraction theory as applied to the SRT will be summarized in this section.

In the simplest terms, pulsed radio waves reflected at vertical incidence form a diffraction pattern on the ground which is then measured by the three or more antennas in the SRT. Both spatial and temporal characteristics of this diffraction pattern can be studied by measuring the amplitude of the reflected radio waves at the three antennas over a period of time. If only the amplitude of the wave front on the ground is measured, the relationships between the amplitude variations on the ground and the physical properties of the ionosphere must be known, a priori, in order to study these properties.

If we treat the ionosphere as a random phase-changing screen at the frequencies used for SRT measurements (Ratcliffe 1956), then the level at which the phase modulation occurs is very important in the determination of horizontal plasma velocity. Jones (1958) confirmed that the phase modulation is, indeed, imposed upon the emerging wavefront close to the level of reflection, by observing a vertical variation of the phase of tidal winds using two closely spaced frequencies and showing that the higher frequency was reflected at a greater height.

The statistical properties of the amplitude variations on the ground were studied by McNichol (1949), who showed that these amplitude variations ranged from completely Rayleigh to displaced Gaussian. The Rayleigh distribution represents diffusely reflected power, and the displaced Gaussian represents both diffusely and specularly reflected power. The ratio of specular to diffuse power varied from 0 to 6, with $\approx 67\%$ of the observations < 0.25. When interpreting SRT data, it is important that two specular echoes are not being received at the same time, since the observed fading at the ground could be the result of interference between the two specular components.

We can use the autocorrelation function (see Section 8.1.5) to describe the statistical properties of the diffraction screen and the resulting diffraction pattern observed on the ground. If the complex amplitude distribution of a one dimensional screen is given by f(x), then the autocorrelation function is given by

$$\rho_f(\xi) = \frac{\int_{-\infty}^{\infty} f(x) f^*(x+\xi)\, dx}{\int_{-\infty}^{\infty} f(x) f^*(x)\, dx}, \tag{9.2}$$

where f* is the complex conjugate.

Ratcliffe (1956) has shown that for Fresnel diffraction, the power spectrum remains the same on all planes parallel to the diffracting screen if the phases of the angular spectrum are completely random – therefore the autocorrelation functions are the same. If g(x) is the complex amplitude on the ground, the equation

$$\rho(f) = \rho_g(\xi) \tag{9.3}$$

holds for plane wave illumination of the diffraction screen. In the SRT case, using a point source for the illuminating waves, we have

$$\rho_f(\xi) = \rho_g(2\xi). \tag{9.4}$$

The structure of the diffraction pattern on the ground, therefore, will be twice the structure size in the ionospheric diffracting screen. For a horizontally moving diffraction screen (or for random motions in the screen), the autocorrelation function is both a spatial and temporal function represented by

$$\rho_f(\xi, \tau) = \rho_g(\xi, \tau) \tag{9.5}$$

for plane wave illumination and

$$\rho_f(\xi, \tau) = \rho_g(z\xi, \tau) \tag{9.6}$$

for point source illumination.

Both Eqs. (9.5) and (9.6) apply for phase deviation < 1 radian in the phase-changing screen and for SNR < 1 (see Ratcliffe 1956).

The one-dimensional analysis of a moving diffraction pattern has been developed by Fedor (1967) and by Briggs (1968a), resulting in the following expression for the spatial-temporal cross correlation function observed at two spaced points separated by a distance d

$$\rho(\xi_o, \tau) = f\left[\frac{(\xi_o - V\tau)^2 + V_c^2\tau^2}{d^2}\right]. \tag{9.7}$$

Equation (9.7) will have a maximum value at some value of d which gives

$$(\xi_o/\tau')V - V^2 + V_c^2, \tag{9.8}$$

where: $V' = (\xi_o/\tau')$ is the apparent velocity; so Eq. (9.8) can be written as

$$V V' = V^2 + V_c'^2, \tag{9.9}$$

where:

V = true horizontal velocity of plasma
V_c = characteristic or random velocity
V' = apparent horizontal velocity

V_c' = fading velocity
VV' = is defined as $(V_c')^2$.

(which is the same result obtained following Fig. 9.2).

The fading velocity Vc′ is found by equating the correlation function $\rho(\xi_o, o)$ and $\rho(o, \tau_E)$ in Eq. (9.7), leading to the equation

$$\xi_o^2/\tau_E^2 = V^2 + V_C^2 = (V_c')^2. \tag{9.10}$$

Since τ_E is obtained from the observed fading records, V_c' can be evaluated. The quantities V and V_c can also be evaluated from the preceding equations. Bowhill (1958) has used the Gaussian autocorrelation function from fading due to ionospheric irregularities to obtain the characteristic time (τ_o), the mean structure size (d) and the lifetime of irregularities. Methods for full autocorrelation analysis of SRT data were developed by Briggs et al. (1950) and Phillips and Spencer (1955), both of whom included a steady drift velocity and random velocities of the ionospheric irregularities.

The interpretation of data from this technique has been critically discussed by various investigators for over three decades (see Hines and Rao 1968, and Briggs 1980). One of the points of interpretation discussed was whether the apparent velocity obtained by this method is actually the true plasma drift velocity at the height of reflection in the ionosphere, or whether it is twice this velocity.

Briggs et al. (1950), Felgate (1970), and Briggs (1980) argue that in the case of mirror reflections of radio waves the apparent velocity is twice the ionospheric drift velocity, while Wright (1968) argues that when one compares simultaneous ionospheric drift data obtained by the SRT and by rocket releases of TMA in the E-region, one obtains the best fit to the two data sets by assuming that the horizontal drift velocity deduced from SRT measurements is the true plasma drift velocity. The latter case represents a diffuse reflection process, and Kent and Wright (1968) also claim that the direction of ionospheric plasma drift at the reflection height is the same direction which is measured by the SRT. Strom et al. (1986) made the assumption that the SRT does, indeed, measure the true plasma drift velocity in their interpretation of SRT observations of F-region plasma drifts from Svalbard.

Another problem in interpreting SRT data in terms of bulk plasma drift has been raised by Kato (1963), Kato and Matsushita (1968), and Bostrom (1973). They argue that the ionospheric drift velocity indicated by radio reflections from irregularities will differ with respect to the background plasma velocity depending on the form of the irregularities.

As of the time of publication of this book, no definitive experiments have been reported which clarify the measurement interpretations mentioned above. Such an experiment could be performed near the Sondrestrom incoherent scatter radar (ISR), for instance; using simultaneous horizontal plasma velocity data acquired by the ISR, an SRT installation, a modern digital ionosonde, the HILAT/Polar Bear beacon satellite interferometer system, and applicable optical techniques (such as a Fabry-Perot Interferometer).

Since from Sondrestrom one can observe the polar, auroral, and sub-auroral ionosphere at various times (depending on the level of magnetic activity, time of day and season), several campaigns taken over an interval of 1–2 years could provide clarification of the relationship between horizontal velocities measured by the different techniques and the true bulk plasma velocity in the high-latitude thermosphere.

Another method of analyzing data from SRT facilities is the dispersion analysis techniques. This is based on the assumption that a given Fourier component in the data arises from the passage over the antennas of a sinusoidal waveform with its own velocity, which is then estimated at each Fourier frequency component (Briggs and Golley 1968, and Wernik et al. 1983). This is accomplished by equating the calculated phase of a cross spectrum to the phase difference between the two corresponding Fourier components. This method used only the phases of the cross spectra to calculate an apparent velocity, hence provides no information on the anisotropy of the ground diffraction pattern.

Costa and Fougere (1988) have proposed a cross spectral analysis which has produced good general agreement with cross correlation analysis data obtained from the HILAT spaced receiver systems at Tromso for a relatively broad frequency band.

HF Doppler Systems

Basically, HF Doppler (HFD) systems utilize a very stable transmitter and one or more stable receivers and local oscillators which heterodyne the received skywave signal; the *beat* signal is usually tape recorded at slow speed. The data tapes are then speeded up by a factor of several thousand and the amplitude and phase of the Doppler variation with time is spectrum analyzed. This technique was pioneered by Davies and Watts at the National Bureau of Standards in Boulder, Colorado in the early 1960's (Watts and Davies 1960; Davies 1962; Davies and Baker 1966). Some ionospheric phenomena studied included sudden-frequency deviation (SFD) (Chan and Villard 1963), sunrise and sunset layer variation, traveling disturbance, magnetic sudden-commencement effects, and equatorial flutter.

HFD systems have been used in two principal modes; the first is the straightforward measurement of the line-of-sight Doppler frequency variation with time caused by ionospheric layer changes, exemplified by the work of Davies and others cited in the preceding paragraph; the second mode utilize three or more spaced receivers to simultaneously obtain the Doppler variation with time, then specific signatures on the individual receiver records are compared, enabling one to obtain the horizontal velocity of traveling ionospheric disturbances (TID's). If one also employs several fixed frequencies, then the vertical characteristics of TID's can also be measured (see Georges 1967, 1968). In the experiment described by Georges, three spaced transmitters and one receiver were utilized, since low-power transmitters were less expensive than receivers. This was a wide-spaced network, with spacings on the order of hundreds of kilometers which, of course, biased the system toward detection of large-scale TID's.

A thorough treatment of the theory of ionospheric phase and frequency variations, including principles of the HFD system, are given in Chapter 9 of

Davies (1969), in Davies and Baker (1966), and in Davies (1990). In developing the theory, one starts with the relationship giving the frequency deviation (f) of an ionospherically propagated wave (Davies 1962)

$$\Delta f = -\frac{f}{c}\frac{dP}{dt}, \tag{9.11}$$

where:

$$P = \int_s \mu \, ds$$

and:

c = free-space velocity of the wave
s = ray path
μ = refractive index
P = the phase path.

Davies (1963) has shown that for mirror reflection from two simple ionospheric models Δf is a function of f as follows:

a) for no change in layer shape, and a small change in height

$$\Delta f \propto f, \tag{9.12}$$

b) for a nondeviating flat slab and a small change in μ below the level of reflection (ignoring collisions and the geomagnetic field),

$$\Delta f \propto f^{-1}, \tag{9.13}$$

If one includes the geomagnetic field and the QL approximation (see Sect. 2.2.1), then

$$\Delta f \propto (f + f_L)^{-1}, \tag{9.14}$$

where f_L is the electron gyrofrequency of the component of the geomagnetic field in the direction of wave propagation, and the + and − signs refer to the ordinary and extraordinary waves, respectively.

Since different frequencies are reflected from different levels in the ionosphere, this can only be accounted for rigorously by ionospheric ray tracing. Simpler corrections to the foregoing equations can be obtained by assuming various layer shapes, i.e., for a layer for which the electron density is zero below some reference height and increases linearly with height above the reference height, it can be shown that

$$\Delta f \propto f^3. \tag{9.15}$$

Agy et al. (1965) have used a somewhat different ionospheric model for vertical reflection which gives

$$f\Delta f \propto \partial N/\partial t(h' - h_0). \tag{9.16}$$

Davies has used the Secant law, Martyn's group path theorem and the result obtained by the Agy et al. (1965) [$\Delta f_o = \Delta f_v$] to obtain the relations

$$f_o \Delta f_o \propto \partial N/\partial t(P' - P_o) \tag{9.17}$$

and

$$P_o(f_o) = 2h_o(f_c) \mathrm{Sec}\, \phi_o \tag{9.18}$$

Equations (9.12) through (9.18) are useful in relating Doppler variations at vertical incidence to Doppler data obtained on oblique paths up to ≈ 2000 km at mid-latitudes, remembering that the following assumptions govern the results:

a) flat ionosphere
b) no geomagnetic field
c) electron collisions are ignored
d) ionospheric layers are horizontally stratified.

HF Doppler data have also been used to estimate time variations of the horizontal vectors of the ionospheric electric field at mid-latitude by Jacobs and Watanabe (1966) and by Tsutsui et al. (1988).

9.3 Whistlers and Controlled Stimulation of VLF Emissions

9.3.1 Whistlers and Other Natural VLF Radiation

Whistlers are bursts of electromagnetic radiation in the VLF range produced by lightning discharges. These bursts travel through the ionosphere and magnetosphere in ducts approximately parallel lines of force in the geomagnetic field and can be detected using low noise audio amplifiers with short antennas. Since about 1951 these signals have been scientifically studied for the information they reveal about the ionospheric and magnetospheric plasma. Other natural VLF emissions (called Dawn Chorus, Risers, Hiss, etc.) which are thought to originate in the ionosphere can also be heard on whistler detection equipment. Some of the fascination with the whistler phenomenon is due to the fact that it is a remarkable sound in the audio range, resembling a human whistle, and can be heard on sensitive audio equipment and on telephone lines under certain circumstances. Helliwell (1965) reported that during the early phases of whistler research at Stanford University, the subject was of great interest to newspaper reporters, and as a result of their stories, to the general public. Some people even attributed the origin of whistlers to people on Mars! At any rate, whistlers and other natural VLF emissions resemble some sounds that people and other animals make (whistles, clicks, hisses, and bridsongs at dawn). Modern scientific descriptions of whistler phenomena are given by Eckersley (1935), Storey (1953), Helliwell (1959, 1965, 1988), Morgan (1959), Gallet 1(964), Davies (1969), Helliwell and Katsufrakis (1978). Probably the most complete description of the history of whistler observations, whistler propagation theory, and global studies of the phenomena has been presented by Helliwell (1965, 1988). This section contains a brief summary of the history and theory of whistlers.

As Helliwell (1965) points out, the study of whistlers and related VLF emissions can be divided into two principal periods; ≈ 1886–1935 and ≈ 1951 until the present time. The lull in whistler studies from ≈ 1935–1951 can mainly be attributed to scientists' interests being diverted into other channels before, during, and shortly after World War II.

The attention of the radio community was directed toward further scientific study of whistlers in a paper by Fuchs (1938) and in a paper presented at the Twelfth General Assembly of URSI in Boulder, Colorado in 1957. Fuchs described whistler observations by Preece (1894) and in Austria in 1896. Preece's paper described the whistler phenomenon as reported by operators at the British Government Post Office during an auroral display on March 30 and 31, 1894.

Another early published study was Barkhausen (1919), who reported on whistler observations by Army signalers using amplifiers to attempt to overhear enemy telephone messages. The WWI inefficient telephone circuits coupled a significant amount of energy into the earth, so high-gain amplifiers connected to metal probes in the ground were employed, which sometimes produced strange whistling sounds in their telephone receivers. Next, Eckersley (1925) described disturbances of a musical nature which had been known to radio workers for some years. In papers by Eckersley (1925, 1928, 1929), he continued his studies on whistlers, describing the frequency variations, possible mechanisms, and duration.

Other pre-World War II papers describing whistler characteristics were published by Barkhausen (1930), Eckersley (1931), Burton and Boardman (1933a, b), etc.. Eckersley (1935) apparently was the first to enunciate the dispersion theory explanation of the relation between frequency and time for whistlers.

After the end of World War II there was renewed effort on whistler research, starting with a paper by Potter (1951), who applied a new technique to analysis of whistler data – the sound spectrograph. Improvements in this equipment and in tape recorders contributed greatly to whistler research since that time. Storey (1953) made many new and significant contributions to the field, including the use of existing direction-finding equipment for cause-effect studies. Results of Storey's many contributions to whistler research were presented by J.A. Ratcliffe at the 10th URSI General Assembly in Sydney, Australia in 1952, and served to stimulate more worldwide research on whistlers.

Typical descriptions of this new burst of activity in this field were published by Crary et al. (1956), Helliwell (1956), Alcock and Martin (1956), Hines (1957), Morgan (1957), and others. These studies merged into the IGY/IGC/IQSY and the present era, and references to published works are contained in the reviews listed in the first part of Section 9.3.1. Reviews of more recent whistler and other VLF research in Antarctica have been published by Park and Carpenter (1978) and by Carpenter (1988).

Whistler Theory

The starting points for whistler theory are Appleton's equations for dispersion and polarization and the QL approximation (see Section 2.2.1). A complete development of the theory was given by Helliwell (1965) and a more concise treatment by Davies (1969), and we will present some of the results of the theory from the latter work.

Figure 9.3 is a simplified presentation of the basic whistler mode of propagation, with typical whistler signatures obtained near the source and near the source conjugate area.

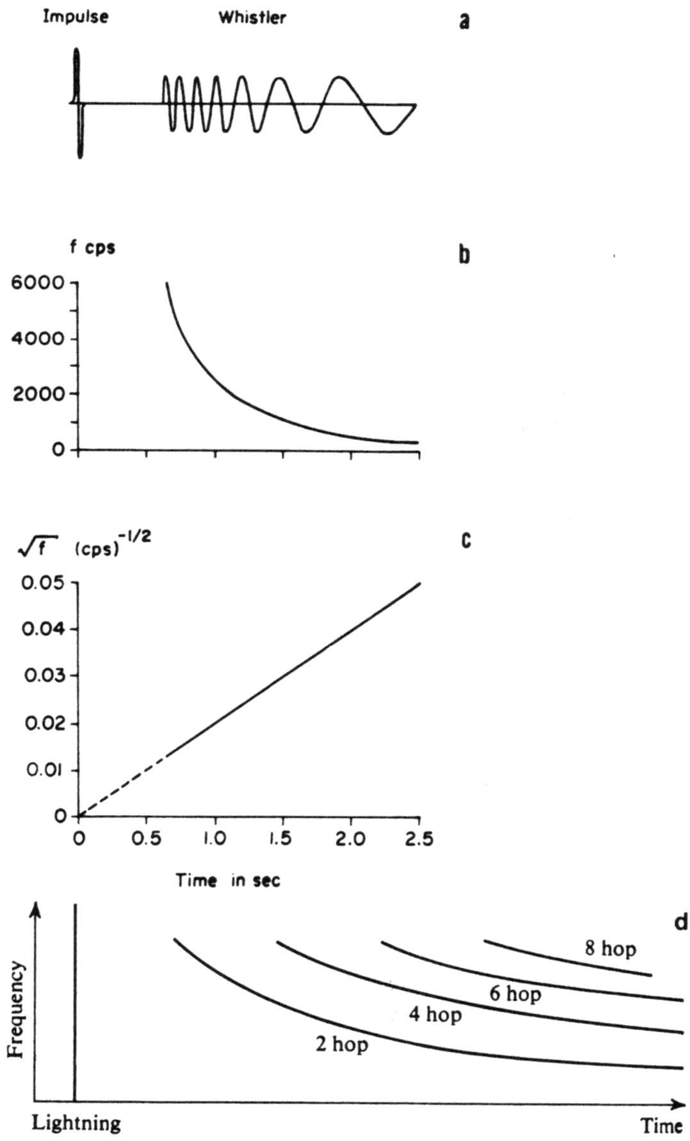

Fig. 9.3 a–d. Sketch of basic manifestations of a "whistler" and its initiating disturbance. **a** The frequency spectrum. **b** Frequency-time curve of a typical whistler. **c** Curve 1/f with time. **d** Initiating disturbance and multiple hops when the source and receiver are at the same end of a line force. (After Davies 1969 and Helliwell 1956)

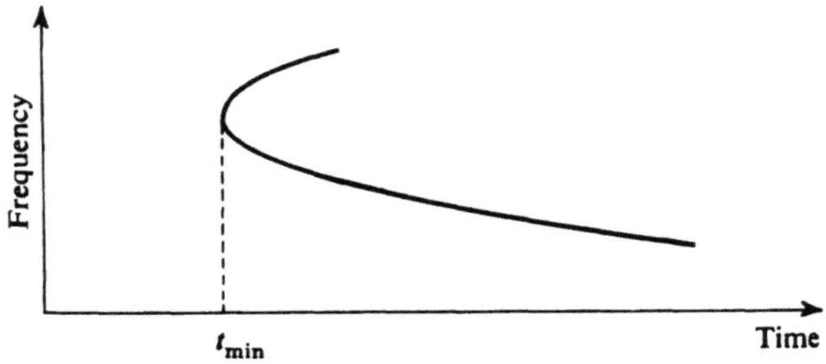

Fig. 9.4. Idealized sketch of the frequency versus time characteristics of a "nose Whistler". (Davies 1969)

Another basic whistler signature is the nose whistler (Helliwell 1956) illustrated in Fig. 9.4.

Frequency Dispersion and Other Whistler Characteristics

The whistler branch of the magnetoionic dispersion curves is the branch of the + curve for values $X > 1$ (see Davies 1969, Fig. 5.7). When collision are included, that is, after the wave enters the ionosphere, energy is coupled into the whistler wave from particles in the ionosphere. Thus, as Helliwell (1987) has pointed out, the whistler phenomenon can be considered to be an ionospheric cooling process – as contrasted to the HF heating experiments described in Chapter 6!

From an examination of the magnetoionic polarization equation [Eq. (2.31) of Section 2.2.1] as applied to whistler propagation in the ionosphere, we find that the wave angle θ (angle between the wave normal and the imposed magnetic field) $\approx 88°$ (QL propagation), and is valid for a large range of propagation angle (θ). Under the conditions stated above, the refractive index is

$$\mu^2 = 1 + \frac{f_N^2}{f|f_L| - f^2} \tag{9.19}$$

For $f_L \gg f$, the refractive index can be simplified to

$$\mu \approx f_N/\sqrt{ff_L}. \tag{9.20}$$

For values of f_N, f, and f_L of whistler waves in the ionosphere, $\mu >$ unity, phase propagation is nearly vertical, and μ approaches infinity as $Y \to 0$ and as $Y \to 1$. The curve of μ vs. $1/Y$ then is symmetrical with a minimum at $Y_m = 2\cos\theta$.

We can obtain another approximation to Eq. (2.19) by dropping the 1 on the right side, yielding

$$\mu^2 \approx f_N^2/(f|f_L| - f^2). \tag{9.21}$$

Since μ^2 cannot be negative, Eq. (9.21) gives the range of frequencies for which it applies, namely for $0 < f_H < f\cos\theta$, implying an upper limit (f_H) for whistler propagation. Also for frequencies less then f_H, the propagation angles are limited to $0° < \theta < \theta_{max}$, where

$$f_H \cos\theta_{max} = f \tag{9.22}$$

the wave normals must lie within a cone of semi-angle θ_{max}. Whistler energy flow will be guided along ducts in the geophysical magnetoplasma by the following relation,

$$\tan(\theta - \alpha) \approx 0.5 \tan\theta/(1 + \tfrac{1}{2}\tan^2\theta), \tag{9.23}$$

where α = the angle between the whistler raypath and the wave normal, and upon differentiating Eq. (9.23) we find that the maximum value of $(\theta - \alpha)_{max}$ occurs for a value of a θ given by $\tan\theta' \approx \sqrt{2}$, which when substituted back into Eq. (9.23) gives $(\theta - \alpha)_{max} \approx 19° 28'$ so that over a wide range of θ, whistler energy is constrained to flow within a cone with a semi-angle of 19° 28′. Further discussion of the behavior of the quantity $(\theta - \alpha)$ with θ can be found in Helliwell (1965) and in Davies (1969).

Another important characteristic of a whistler wave packet is the group velocity, v_g, (Section 2.1.1) which Davies (1969) derives, and for suitable frequency approximation is

$$v_g = 2c\frac{f^{1/2}(|f_L| - f)^{3/2}}{|f_L| f_N} \tag{9.24}$$

when $f_L \gg f$, Eq. (9.24) simplifies

$$v_g = 2c\frac{f^{1/2} f_L^{1/2}}{f_N}. \tag{9.25}$$

Furthermore, if the direction of phase propagation is close to that of the magnetic field ($\theta \approx 0°$ and $\alpha \approx 0°$), the group ray velocity is

$$v_{gr} \approx 2c\frac{f^{1/2}(f_H - f)^{3/2}}{f_H f_N} \tag{9.26}$$

and the minimum frequency of the nose whistler is found to be

$$f_{min} = \tfrac{1}{4} f_H \tag{9.27}$$

The time of flight of a pulse of EM energy traveling along a geomagnetic line of force from a point in one hemisphere to its conjugate point is

$$T = \int_{path} ds/v_g \tag{9.28}$$

substituting v_g from Eq. (9.25) gives Eckersley's dispersion law,

$$T = \frac{1}{2c}\int_s f_N\, ds/(f^{1/2} f_L^{1/2}) = \frac{D}{f^{1/2}} \tag{9.29}$$

where:

$$D = \frac{1}{2c}\int_s f_N/f_L^{1/2}\, ds, \tag{9.30}$$

where D is called the dispersion.

As mentioned earlier in this section, there are several other naturally occurring VLF emissions, in addition to whistlers. These emissions are thought to originate in the ionosphere and/or the magnetosphere and are observed in the frequency range from ≈ 200 to 30 000 Hz. They are known by descriptive names like chorus, tweeks, and auroral radio noise, and have been reported in the literature since the IGY. An Atlas of VLF emissions was published by Helliwell (1965), and some recent results and reviews are presented by Benson et al. (1988), Carpenter (1988), and LaBelle (1989).

9.3.2 Controlled Stimulation of VLF Emissions

Helliwell (1965) pointed out in his monograph that Morse code transmissions from VLF stations had also been observed to excite whistlers. This discovery led to initiation of studies of controlled stimulation of VLF emissions using a

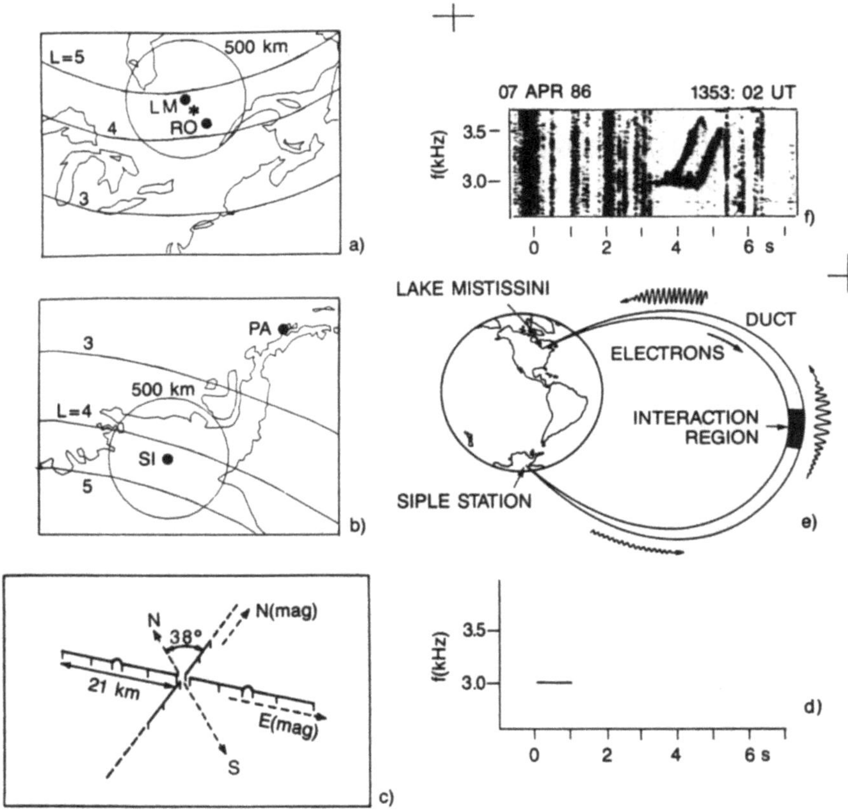

Fig. 9.5 a–d. Siple/Lake Mistissini experiment plan. **a** Map showing the new station, Lake Mistissini (*LM*), previous station, Roberval (*RO*), and estimated Siple Station conjugate point (*) for 1986. **b** Map of Antarctic peninsula, showing Siple Station (*SI*) and Palmer Station (*PA*). **c** Crossed dipole antenna at Siple, with planned extensions shown as *dashed lines*. **d** f–t Diagram of 1-s pulse at 3 kHz. **e** Diagram of duct propagation path from Siple to conjugate region, showing wave growth near equatorial plane. **f** f–t Diagram of two one-hop amplified signals with associated emissions that propagated to LM in two ducts that were excited by the 1 s pulse of **d**. (Courtesy of R.A. Helliwell 1988)

transmitter located at Siple Station, Antarctica (Helliwell and Katsufrakis 1978), Helliwell (1983, 1988). Figure 9.5 is a simplified sketch of the basic propagation mode between the transmitter at Siple and the receiver at Roberval, Quebec, Canada.

There are two suggested interaction mechanisms between the ducted waves and the counter-streaming energetic electrons on the magnetic field line over the equator; Pitch angle scattering (Kennel and Petschek 1966); and phase bunching of the interacting electrons (Helliwell and Inan 1982).

Helliwell (1988) has pointed out that the study of naturally occurring whistlers (in progress some 50 years now) is still alive and provides very complementary information to the newer field of artificially stimulated VLF radiation. Helliwell's review (1988) concentrates on the results of *controlled stimulation* of VLF emissions.

9.4 VLF and HF High Resolution Sounders

Utilization of the VLF sounder technique has been described by Lewis et al. (1973), Rasmussen et al. (1975), and Kossey et al. (1983). They employed a steep-incidence sounder sweeping over ≈ 5–30 kHz using pulse lengths $< 100\,\mu s$, to obtain data on the lower ionosphere at low, mid and high latitudes. The essential elements of the system are shown in Fig. 9.6, with some typical data shown in Fig. 9.7 for a quiet day and in Fig. 9.8 for a disturbed day.

Another type of sounder (HF Channel Prober), has as its main function the determination of the characteristics of a particular HF oblique path, but since it is a high resolution system, it can also provide information on the structure and dynamics of the ionospheric reflection point. One operational system of this type is the NRL Channel Prober (Wagner et al. 1983, 1988) which spectrum analyzes short coded pulses transmitted over short (≈ 50–150 km) oblique paths in the range of operating frequencies from ≈ 2–14 MHz to obtain the temporal fine structure of the E- and F-layers.

The NRL Channel Prober uses a maximal length pseudo-noise binary sequence to bi-phase modulate a high frequency carrier. The in-phase and quadrature components of the received signal are processed by a coherent detector and an on-line high-speed digital correlator operating in real time, it then cross correlates the received signals with a replica of the transmitted signal. The end result is an enhanced sensitivity, high resolution pulse sounder system. The basic system block diagram is shown in Fig. 9.9, and Fig. 9.10 shows some typical data obtained by this technique.

Another HF Channel Prober has been developed by Stanford Research Institute-International (SRI-I) and deployed in polar and equatorial regions (Basler et al. 1988a, b). It also employs pseudo-random noise (PRN) bi-phase modulation in order to determine delay and Doppler spread of the received signal, but uses a frequency-modulated continuous (FMCW) signal for the ionospheric sounding mode.

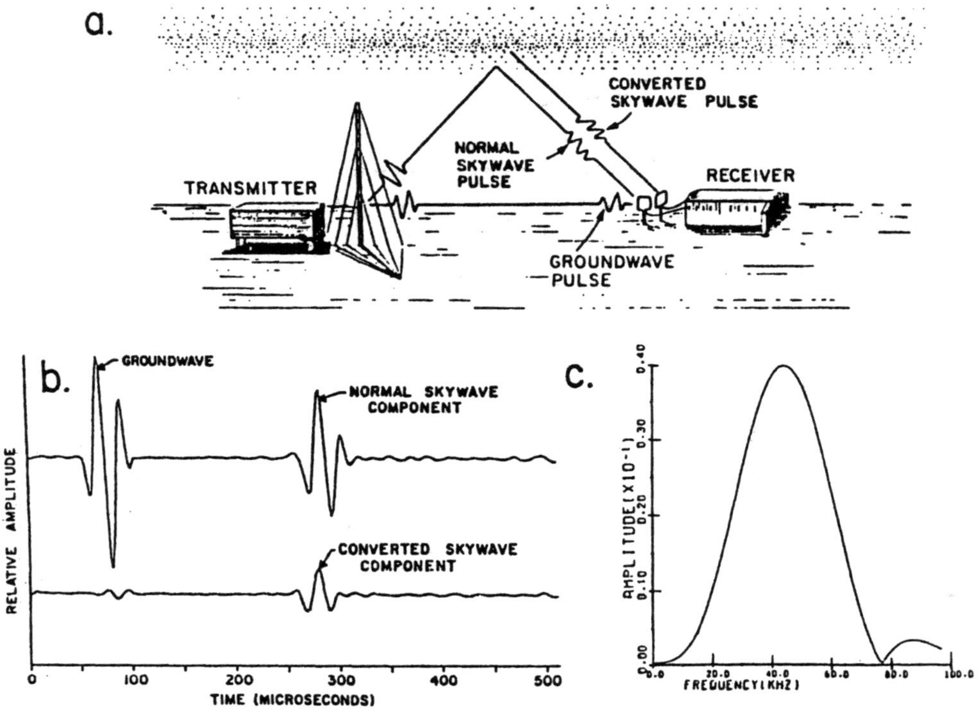

Fig. 9.6. **a** VLF pulsed ionosonde technique. **b** Example of typical observed waveforms. **c** Spectrum of typical transmitted pulse. (After Kossey et al. 1983)

Figure 9.11 shows typical high resolution sounder data obtained by the NRL channel probe. Figures 9.12 through 9.15 show block diagrams and typical data obtained by the SRI-I/HF channel Prober (Basler et al. 1988a, b).

As mentioned previously, the HF Channel probers are used primarily to assess the transfer characteristics of a particular HF ionospheric communications channel, especially during disturbed conditions; but they also can provide some information on ionospheric physics at the reflection point of the ionospheric path.

9.5 Ionospheric Imaging by Radio

Ionospheric physicists have attempted for several decades now to use radio waves to provide two-dimensional images of portions of the ionosphere projected onto ground-based sensors. Some success in these endeavors has been achieved in the 1980's by adapting techniques of holography, computerized tomography, and aperture synthesis imaging from other disciplines to radio ionospheric studies.

Probably the first suggestion of using the wave front reconstruction method applied to radio waves ionospheric studies was made by Rogers (1956). Schmidt

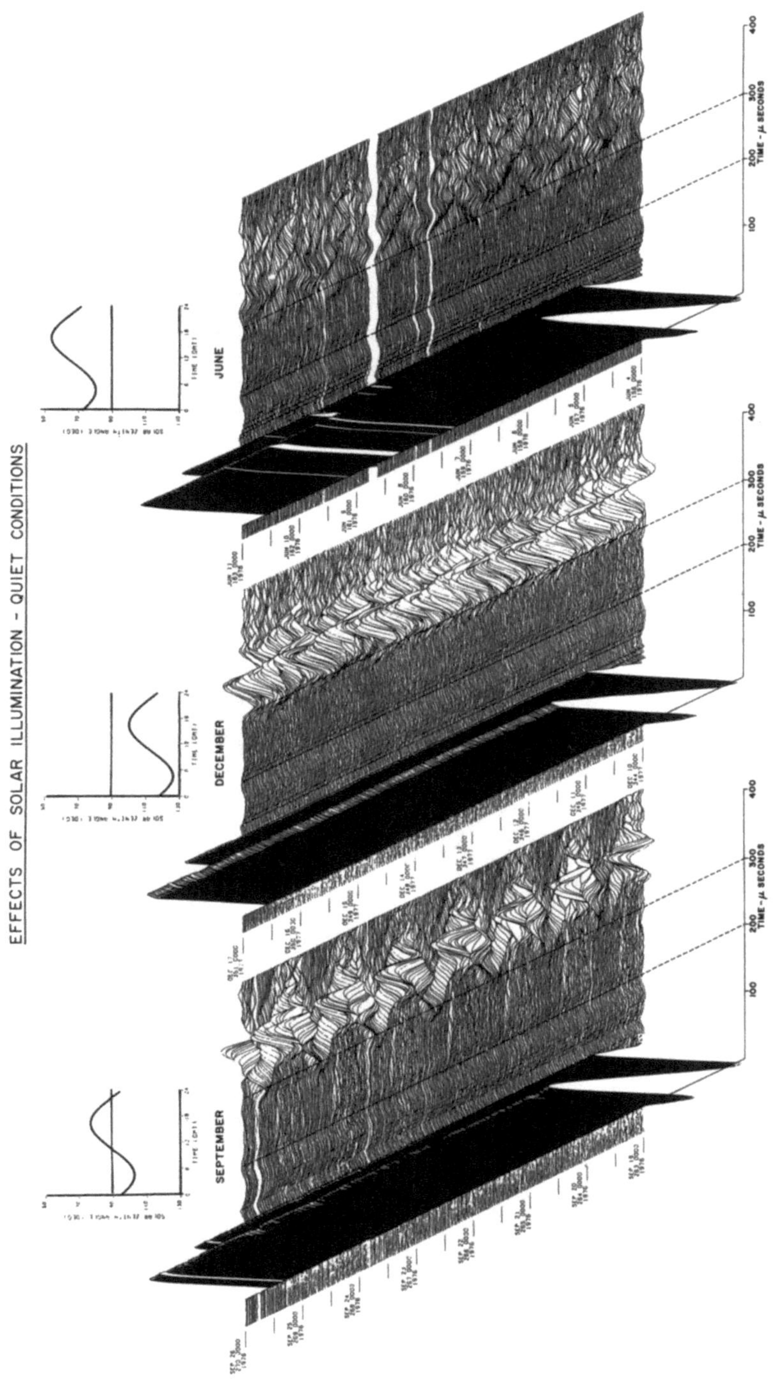

Fig. 9.7. VLF pulse reflection data received under quiet polar conditions. (Kossey et al. 1983)

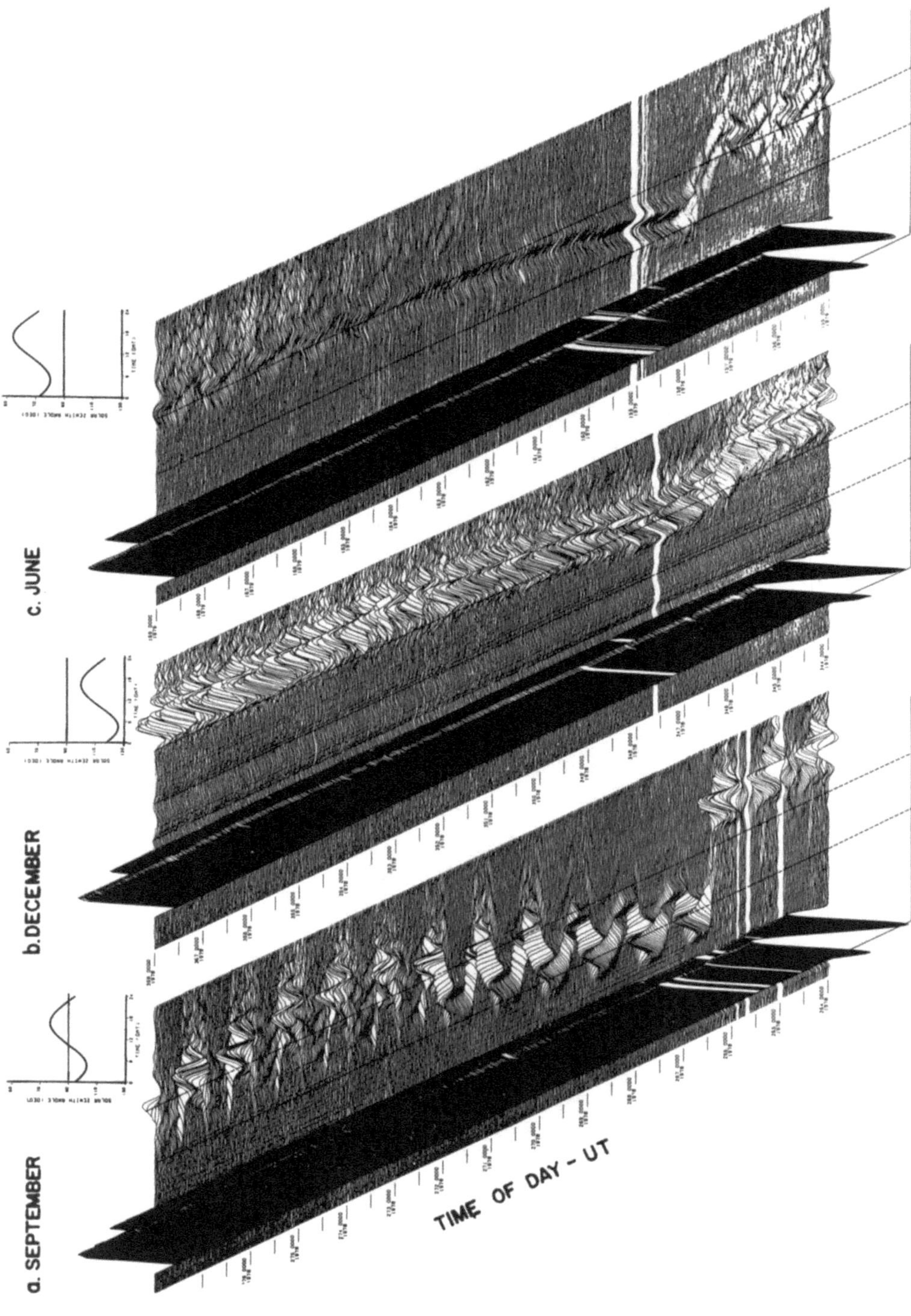

Fig. 9.8 a–c. VLF pulse reflection data for disturbed polar period. (Kossey et al. 1983)

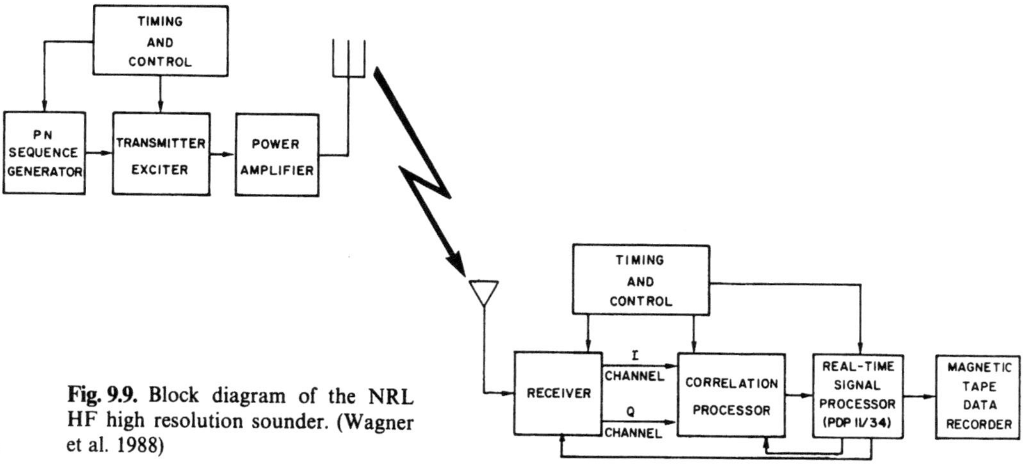

Fig. 9.9. Block diagram of the NRL HF high resolution sounder. (Wagner et al. 1988)

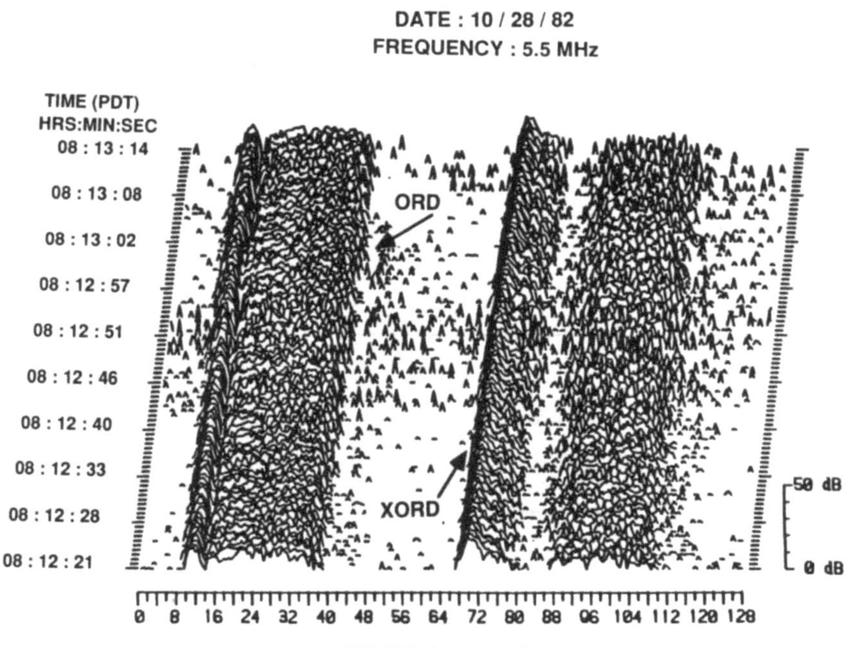

Fig. 9.10. NRL HF high resolution sounder data for a 1-hop, F2 layer mid-latitude path – quiet conditions, transmitter pulse width = 1 s, signal bandwidth was 1 MHz. (Wagner et al. 1988)

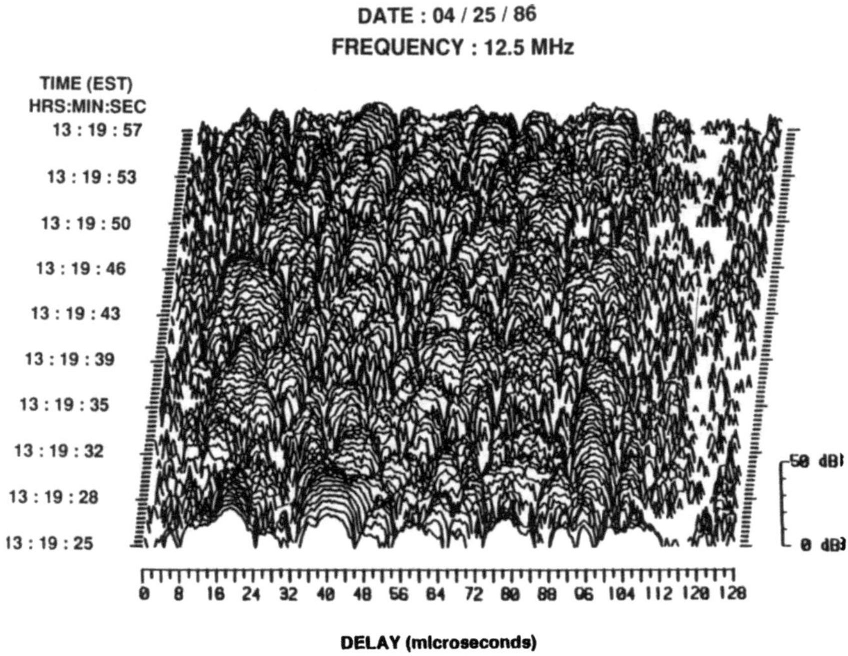

Fig. 9.11. NRL HF high resolution sounder data for a long (2300 km) auroral path, 1-hop F1 layer mode, moderately disturbed. (Wagner et al. 1988)

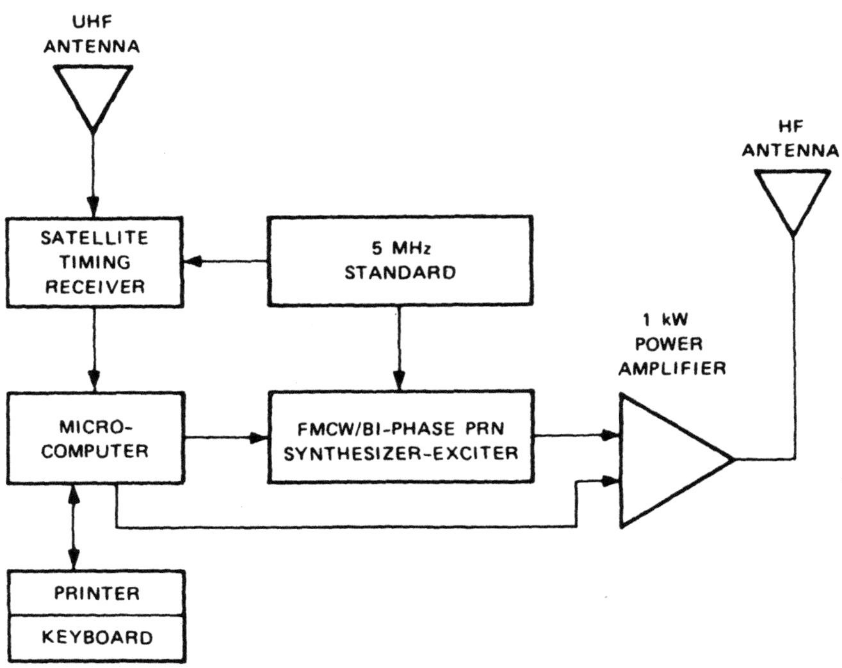

Fig. 9.12. Block diagram of SRI-I HF channel prober transmitter. (Courtesy of R.P. Basler)

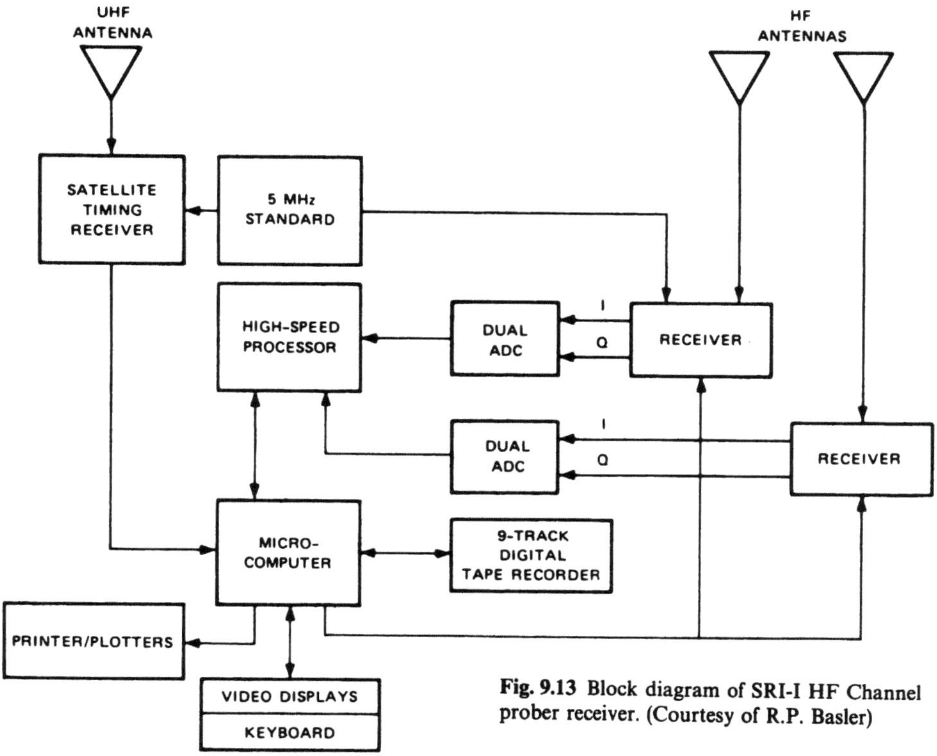

Fig. 9.13 Block diagram of SRI-I HF Channel prober receiver. (Courtesy of R.P. Basler)

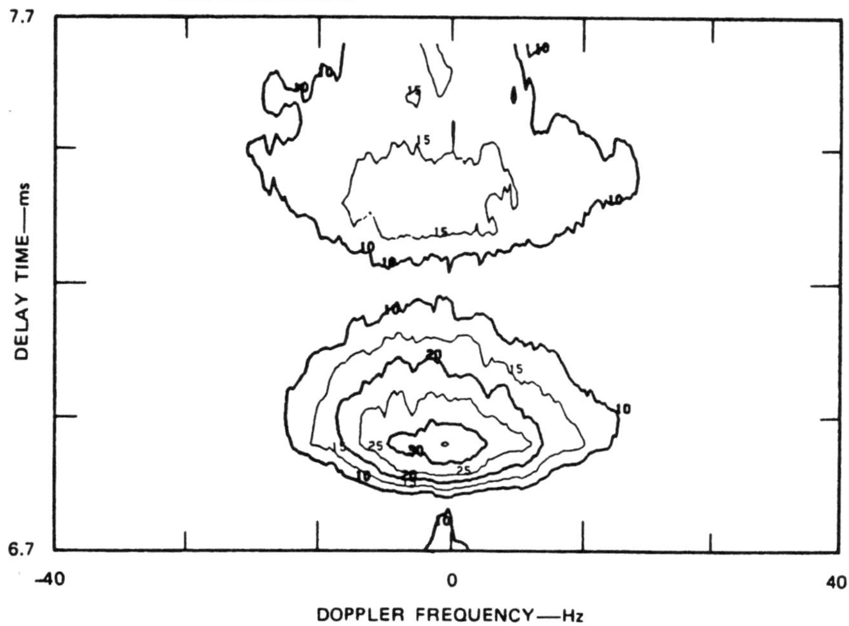

Fig. 9.14. Observed channel scattering function from a relatively undisturbed channel-1723 UT 5 October 1984 at 10.935 MHz SRI-I HF channel prober. (Courtesy of R.P. Basler)

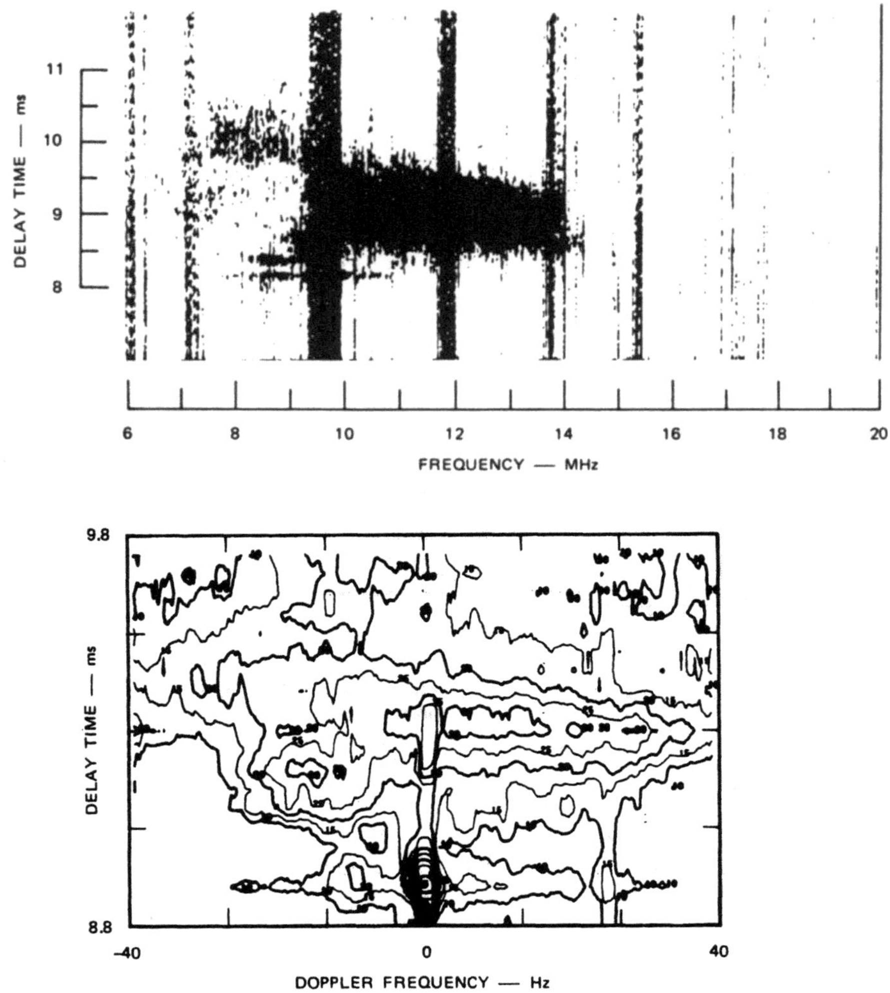

Fig. 9.15. Observed channel scattering function for an auroral path and disturbed conditions (Kp = 5-) on 7 March 1986 at 1916 UT, f = 10.705 MHz. (Courtesy of R.P. Basler)

(1972) proposed carrying out a holographic experiment using VHF signals transmitted by a satellite to localize intense ionospheric irregularities, and a description of the technique used to reconstruct the wavefront in two dimensions was presented by Schmidt and Tauriainen in 1975. A more sophisticated type of holographic radio camera was developed by Stone (1976, 1978) using a 32-element linear antenna array perpendicular to the satellite path which had the capability of three-dimensional reconstruction.

In the 1980's advances in the theory and in a simulation for ionospheric holography were reported by Rogers and Ireland (1980a, b) and by the late Anti Tauriainen (1982), but as of the time of publication of this book, no fully operational systems have been reported.

Recently two scientists familiar with radio astronomical interferometry have reviewed the concept and proposed experiments in ionospheric radio holography (Knowles and Andrews 1987). They concluded that: (1) at least in radio astronomy, the Aperture Synthesis Imaging (ASI) technique is well proven. The ASI method is defined as the phase differencing of several subapertures of an antenna array separated spatially. It is a type of radio holography with a locally generated reference beam and corresponds to the diffuse case of holograms. (2) Only a thoroughly rough ionosphere will produce an image of the reflecting layer. (3) Some 20–30 receivers are probably adequate for holography of the ionosphere illuminated by an HF transmitter. (4) The radio holographic technique similar to present use in radio astronomy is capable, when properly interpreted, of providing significant information about the structure of the ionosphere under quite and disturbed conditions.

Another technique – this one borrowed from medical technology – being applied to radio studies of the ionosphere is Computerized Tomography (CT). The use of CT to produce a two-dimensional image of the electron density in the ionosphere has been presented by Austen et al. (1988). The authors show that the CT problem requires that the measured data be the line integral through the medium of the unknown parameter. They further show that total electron content (TEC) measured using Faraday rotation and differential Doppler techniques on transionospheric paths meet this requirement. The theory of CT ionospheric imaging and simulation results from a polar-orbitting satellite in a 100-km orbit using three to five ground stations are presented, and it is concluded that CT radio imaging of at least large-scale ionospheric structure is feasible. Research in this area is continuing (Yeh and Raymund 1991; Fremouw et al. 1991).

Yet another application using an adaptation of the ASI technique described in the section on radio holography, is the Imaging Riometer Ionospheric Studies (IRIS) system developed by Rosenberg in 1990 and described briefly in Section 7.2 of this monograph.

9.6 Backscatter from Meteoric Ionization

The Ionization trails of small meteors which "burn up" in the height range of ~75–105 km can be used as "tracers" of neutral winds in that region of the thermosphere (McKinley; Rishbeth and Garriott 1969; Giraud and Petit 1978). Forward scatter from these meteoric ionization trails is also used as a propagation mechanisms in "meteor burst" communication systems (Vogan and Campbell 1957; Sugar 1964; Cannon et al. 1985; Ostergard et al. 1985; Roberts and Kokjer 1986; Weitzen 1986, 1989).

The backscatter from meteoric ionization trails can either be partial reflection from underdense trials or total reflection from overdense trials, and is strongly aspect angle-dependent. Many investigators have used this technique to measure lower-thermospheric winds and turbulence (Booker and Cohen 1957; Greenhow and Neufeld 1961; Hook 1972; Revah 1969). Gilbert (1977) traced the rise and eventual decline of the use of meteoric backscatter to deduce ionospheric winds and pointed out that, although the technique is still utilized, it has blended into some of the other more powerful techniques. More recently, the meteoric-ion-backscatter technique has been utilized to study the "middle atmosphere" (Wang et al. 1988).

This concludes the chapter on *Other Radio Techniques*, and it is hoped that no significant techniques have been overlooked. There will be no section on *Capabilities and Limitations* for this chapter, since some of that information is included in the description of each of the methods. It is also rather obvious that some of the techniques described have been subsets of previously discussed techniques.

Chapter 10 Epilog

This chapter provides a synthesis of the material presented in the first nine chapters. The goals of the chapter are:

1. To summarize information on which particular observational techniques are applicable to the investigation of specific ionospheric phenomena.
2. To emphasize the complementarity of various techniques.
3. To point out the synergistic effects of coordinated multi-technique campaigns.

10.1 "Where We Have Been and Where We Are Now"

The history of the development of radio techniques for probing the earth's ionosphere can be summarized by the phrase... "in the beginning was the ionosonde..." – which basically was the device which was used to demonstrate convincingly the existence of the ionosphere and to describe its gross structure. It seems quite remarkable that this basic technique (HF swept-frequency pulse reflection at vertical incidence) is still the most ubiquitous of the radio techniques used for ionospheric research! In fact, the current worldwide deployment of digital ionosondes (or Advanced Ionospheric Sounders – AIS), in addition to the standard ionosondes still usable, will probably bring the global network to ≈ 150 sounders within this decade.

Most of the basic radio techniques for ionospheric research were invented and deployed during the period of about 1925 to 1965. Variants of these techniques, plus adaptation of technologies used in other fields, i.e., computer tomography, holography, etc., were increasingly employed from about 1965 until the present time. One would be rather foolish to claim that no new radio techniques will be developed in the future, but since there are only a finite number of electromagnetic wave properties (frequency, amplitude, phase, polarization, and direction of propagation, experimenter's minds will most likely be strained to invent a truly new technique). Nevertheless, the author would certainly like to encourage continued creativity on the part of the researcher!

Two points should be emphasized when examining Table 10.1: first, the dates are estimates of first use of these particular techniques as cited in the open literature; second, much theoretical and considerable experimental work preceded the first reported use of each particular technique.

Table 10.1. Approximate year of first use of some selected ionospheric radio techniques

Year	Techniques	Reference
1925	Frequency-change method ($\Delta p/\Delta f$) of radio reflection	Appleton and Barnett (1926)
1925	Ionosonde[a]	Breit and Tuve (1926)
1926	HF pulsed groundscatter sounder	Mögel (1930); Taylor and Young (1928)
1938	VHF direct E-region backscatter	Harang and Stoffregen (1938, 1961)
1947	Ionospheric cross modulation	Ratcliffe and Shaw (1948)
1950's	A1 and A3 absorption techniques	Piggott (1953)
1956	HF direct backscatter from F-region	Bates (1959)
1957	HF radio satellite beacons	Al'pert (1958); Aitchison et al. (1958)
1958	Incoherent scatter radar	Gordon (1958); Bowles (1958)
1953	Cosmic noise ionospheric absorption	Mitra and Shain (1953)
1958	Riometer (practical application of the Mitra and Shain principles)	Little and Leinbach (1959)
1962	Topside sounding	Florida (1969); Franklin and Maclean (1969)
1969	HF resistive heating of D-region	Utlaut (1970)
1975	VHF/UHF geostationary satellite beacon	Davies and Hartman (1976)

[a] This refers to the ionosonde principle of measuring the reflecting properties of the ionospheric layers by measuring the characteristics of radio echoes from the layers.

10.2 Specific Techniques and Their Applications

This section summarizes in Table 10.2 various ionospheric properties and phenomena, together with specific techniques which are applicable to measuring them. This will not necessarily be a complete list, but should at least point in the right direction toward specific radio techniques for measuring the major ionospheric features.

At the risk of committing a tautology, it should be pointed out that (even in the macroscopic world) the measurement of a physical quantity using *any* experimental technique involves a convolution of the response curve of that particular instrument with the characteristic curve of the quantity. Figure 10.1 is an illustration of one example of the difficulty in determining the true nature of a specific phenomenon, i.e., atmospheric gravity waves (AGW) (see bibliography in Hunsucker 1982.)

It is also possible that when one employs certain techniques, the exploring signal may affect the phenomenon under investigation in a measurable way. One such case might occur when using an OTH radar with very high radiated power levels – the backscattered signals may contain components produced by the modification of the ionosphere by the transmitted signal, as well as the backscatter from ionospheric and other types of targets.

On a somewhat different subject, it is tempting to try to contrive an objective method for determining the relative cost-effectiveness of the various radio

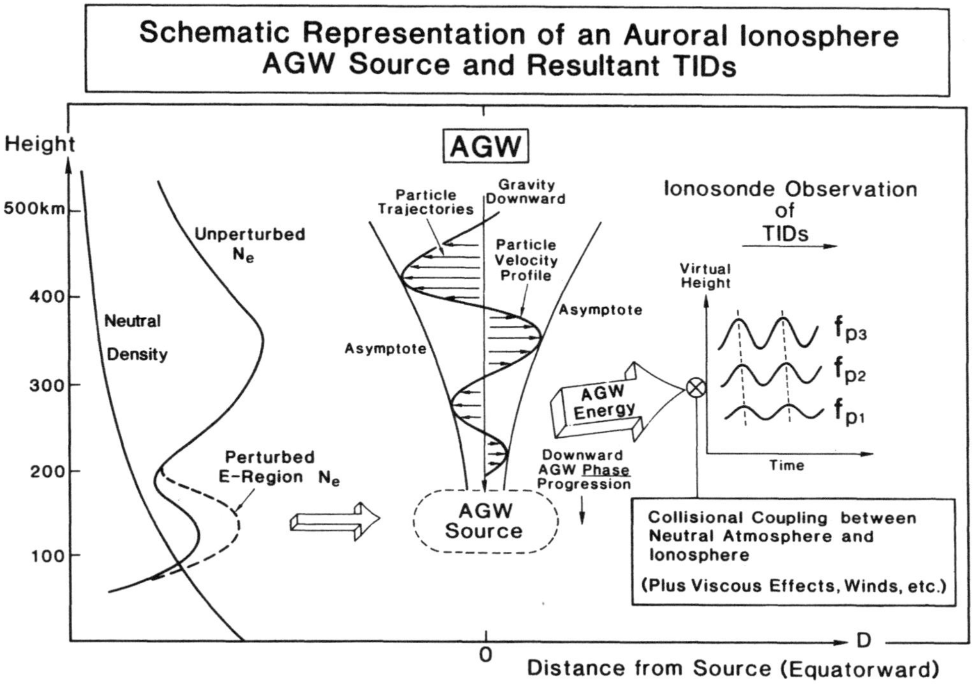

Fig. 10.1. An example of the difficulty of determining the nature of an ionospheric anomaly from remote probing. In this idealized sketch, a specific radio technique (an ionosonde) detects traveling ionospheric disturbances (TID's) produced by an atmospheric gravity wave (AGW) which is, in turn, generated by a joule heating event in the auroral ionosphere. The ionosonde measures only one specific feature of the TID, leaving most of the physics of the TID, AGW, and source unknown. Obviously, several independent simultaneous techniques located in the AGW source region and "downstream" from the source will reveal much more about the phenomena in question

techniques. Such an equation would supposedly provide quantitative information on the effectiveness per unit cost, so that research program managers could better allocate resources. Deriving such an equation, however, is fraught with almost insurmountable difficulties!

Supposedly, at the risk of oversimplification, one would started with Information Theory and with total capital expenditure for a particular technique – something like Information (I) in bits/channel divided by cost (for instance US 1991 dollars), or I/$. Heuristically, the numerator would probably be some function of the data rate, the different quantities measured as a function of spatial coordinates and the quality of the data (error bars on determined quantities, etc.). Probably a Usefulness factor (U) should also be included in the numerator, which, however, would be rather subjective. However subjective, the factor U would be very important – having to do with the relative importance of the scientific results and the age of the technique (a few new results are usually more useful than many results from an old technique).

Table 10.2. Summary of ionospheric phenomena and pertinent radio techniques

Phenomena to be investigated	Applicable radio techniques	Described in this book	Remarks
D-Region			
Electron density profile $[N_e(h)]$	ISR	Chap. 5	Mainly applicable to region where $N_e > 10^4$ el/cm^3
$N_e(h)$ and $\nu(h)$	Radiowave Interaction	Sects. 6.1 and 6.3.2	Only applicable over a limited height range
Absorption	URSI A1 and A3 methods	Sects. 7.1 and 7.3	Mainly applicable to low and mid-latitude regions
Absorption	URSI A2 methods	Sect. 7.2	Auroral and polar ionosphere
E-region			
$N_e(h)$ profile	ISR	Chap. 5	Limited geographical coverage
Profile of $N_e(h)$, f_0E, fE_s, f_{min}, large-scale irreg. structure, etc.	Digital Ionosondes (AIS)	Sect. 3.3	Possibility for good geographical and temporal coverage
Small- and medium-scale irregularity structure and motion	Auroral radars	Sect. 4.3.2	Requires approximate normality between radar line-of-sight and magnetic field at h. ≈ 100 km
F-region			
A. Entire height region			
Total electron content (TEC)	Satellite RBE	Sect. 8.1	Global coverage, but rather sparse data (includes protonosphere)
Integrated electron content	ISR	Sect. 2.3.2	Ionosphere only (Faraday rotation of signal)
B. Topside region			
Irregularity structure (medium scale)	Topside sounders	Sect. 8.2	Possibility for good global coverage. Poor time resolution at a given station (it takes ≈ 3 months to get a global map)
Small-scale irregularities	RBE on satellites	Sect. 8.1.5	"

Table 10.2. (*Continued*)

Phenomena to be investigated	Applicable radio techniques	Described in this book	Remarks
C. Low and Middle F-region (up to F2 peak)			
Profiles of N_e, T_e, and T_i, E-field, composition	ISR	Chap. 5	ISR's usually operate ≈ 100–200 h/month at intermittent times, and there are only ≈ six of them at the present time. They are not limited by a severely disturbed ionosphere
$N_e(h)$, f_0F1, f_0F2, spread-F, TID's, tilts, etc.	AIS	Sect. 3.3	Limited during aurorally disturbed conditions
TID's	HF ground-scatter sounders	Sect. 4.1	Tveten (1961)
TID's	HF spaced-receiver and Doppler sounders	Sect. 9.2	Davies (1969)

The denominator is also difficult to evaulate, since there is a tendency for research program directors to sometimes not include *all* funding received for a particular project (probably due to multiple sponsors, in-house support, donations of land, etc.); also because true believers in certain techniques sometimes advance rather tendentious arguments for the technique use and/or funding!

Once the form of the equation was determined, it might not be difficult to evaluate relatively straightforward techniques such as riometers and standard ionosondes, but if one were to try to evaluate techniques as sophisticated as advanced ionosondes, Doppler interferometers or incoherent scatter radars, the numerator would probably become very difficult to evaluate. The derivation of the cost-effectiveness equation and its application to state-of-the-art radio techniques is left to the advanced, objective, motivated (and funded) investigator!

10.3 Complementarity

The simultaneous use of *several* different radio techniques to investigate specific ionospheric phenomena is a very powerful approach, one which has been employed since the IGY. A good example of complementarity has been given recently by Wright et al. (1988), who illustrate how the simultaneous use of the ISR and digital ionosonde techniques together can result in a very good description of the complete electron density profile – with each instrument providing information to complement the other one. Figure 10.2 is an illustration

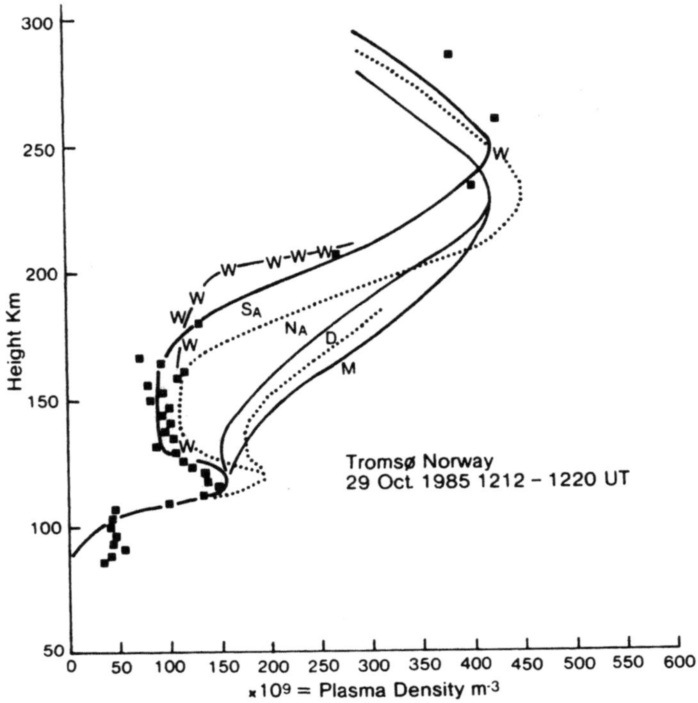

Fig. 10.2. An example of using two complementary techniques (ISR and Digital ionosonde) to determine an ionospheric parameter (electron density profile). The *solid black squares* are the EISCAT electron densities, the *solid lines* are from the southward (near the EISCAT beam), *dotted lines* are farther from the EISCAT beam, the *upper portion of the lines* are obtained from the POLAN ionosonde interpretation program. (Wright et al. 1988)

of a composite profile obtained using simultaneous data from these two instruments.

Other examples of the use of complementary techniques include simultaneous topside/bottomside backscatter sounding (Lund et al. 1967), V/I-Topside Sounding (Jackson 1986, 1988), and recent large area campaigns such as SUNDIAL (Szuszewicz et al. 1987), CEDAR (Romick et al. 1987), VIKING (Hultqvist 1987), WAGS (Argo and Hunsucker 1988), GITCAD (Wickwar and Smith 1988), GIZMOS (Fejer et al. 1990), and WITS (1988).

Not to belabor a point, but consider the investigations of small-scale ionospheric irregularities in the F-region by isolated techniques (*not* simultaneous multiple techniques): an ionosonde alone could indicate a spread-F signature and produce good statistics on occurrence behaviour, whereas amplitude and phase measurements from a satellite beacon would provide good information on the scale size of the irregularities, and an ISR could provide the behaviour of the electric field as a function of time and height – but taken together, these three techniques could be very synergistic for advancing knowledge of the physics of the phenomenon.

10.4 Concluding Remarks

10.4.1 The Present Status of Radio Instrumentation

When one examines most of the radio techniques used to study our ionosphere, it is obvious that they have reached a fairly high level of maturity and sophistication. In most cases, it seems that new technologies and theoretical advances have driven the scientific breakthrough, i.e., solid-state devices, sensitive receivers, satellites, information theory, magnetoionic theory, applied mathematics, antenna array theory, etc. have led to topside sounders, incoherent scatter radars, radio interferometric sounders, digital ionosondes, etc. Probably one reason for the present level of sophistication of ionospheric radio techniques is the close coupling between the disciplines of applied mathematics, plasma physics, and electrical engineering at various universities and at some laboratories.

Ionospheric scientists have also adapted selected techniques from other fields – two examples which appear nearly ready to yield dividends are holography and computed tomography (see Section 9.5).

10.4.2 Possible Future Trends

Some promising technologies for continued application to ionospheric studies are: digital-signal-processing (DSP), large-scale integration (LSI) of solid-state devices, and parallel computer processing for rapid analysis of large amounts of data. Acoustooptics is also becoming an attractive to DSP because of its great bandwidth, dynamic range, resolution, and speed (see the IEEE Spectrum for August 1988).

10.4.3 Excelsior!

The present and the incoming generation of experimental ionospheric physicists and engineers are cordially challenged to carefully plan their investigations to best utilize the arsenal of available radio technologies to study specific phenomena; and perhaps even invent new and better methods for exploring the terrestrial plasma envelope! It is the author's hope that this book may be of some use in the continuing adventure of understanding our plasma universe.

Appendix A.1 List of Principal Symbols, Constants, and Conversion Factors

All of the symbols used in this monograph have been defined as they were introduced, so this list is limited to those most often used. In some instances, one symbol may have more than one meaning, but the context of its use should clarify its meaning. The SI units for the quantities are also given. Short listings of constants and conversion factors, hopefully of use to radio scientists, are also included.

A.1.1 Principal Symbols

Roman Letters

A	Area, amplitude of a quantity, a constant
a	Radius of the earth ($\simeq$ 6370 km), various constants
B	Magnetic flux density, a constant
C	Capacitance (farad)
c	Velocity of an EM wave in free space
D	Electric flux density (coulomb/m^2) also region of the ionosphere, Debye length
d	Distance
E	Electric field strength (volts/m), also region of the ionosphere
e	Charge on electron, base of natural logs
F	Force in newtons (vector), region of ionosphere, shape factor, noise factor
f	Frequency, noise figure
G	Gain, constant
g	Acceleration due to gravity
H	Magnetic intensity in ampere-turns/m), scale height in the ionosphere
h	Height, Planck's constant
I	Magnetic dip angle or inclination, electric current
i	Electric current, angle of incidence
J	Electric current density
j	"Imaginary" operator, electric current density
K	Ionospheric absorption coefficient, wave number, magnetic disturbance index, Boltzmann's constant
k	Propagation vector, various constants, unit vector
L	Inductance, loss
l	length
M	Molecular weight
m	Mass (kg)

N	Number density of particles or ions, electron content
n	Complex refractive index, number density of neutral particles, a real integer
P	Electric polarization, Poynting vector, electric power in watts, probability of an event
p	Pressure
Q	Production rate, electric charge
q	Electric charge (coulombs), neutral production rate
R	Relative sunspot number, reflection coefficient
r	Radius vector
S	Surface, area, scintillation index
s	Distance along a path, polarization parameters
T	Temperature, period, cross modulation
t	Time
V	Plasma drift velocity, volume, electric potential
v	Velocity
W	Work in joules
X	Statistical variable, magnetoionic parameter
x	Cartesian coordinate (usually independent variable)
Y	Magnetoionic parameter
y	Cartesian coordinate (usually dependent variable)
Z	Magnetoionic parameter
z	Cartesian coordinate

Greek Letters

α	Ionospheric attachment coefficient, various angles, attenuation
β	Loss coefficient for F-region, phase-shift constant
γ	Complex propagation constant
Δ	Small change in quantity
δ	Small change in a quantity
ε	Electrical permittivity (in farads/m) also various constants
θ	Various angular measurements
κ	Constant
λ	Wavelength
μ	Real part of refractive index, magnetic permeability in henry/m
ν	Particle collision frequency
ζ	A shifting function
π	Ratio of circumference to diameter of a circle (3.1419)
ρ	Cross correlation function
σ	Electrical conductivity, scatter cross-section
τ	Ionospheric slab thickness
Φ	Various angular measurements
χ	Solar zenith angle, a variable in the autocorrelation function
Ω	Rotation of plane of polarization, ohms
ω	Angular frequency

A.1.2 Selected Physical Constants

Boltzmann's constant; $k = 1.3807 \times 10^{-23}$ joule/kelvin
Classical electron radius; $r_e = 2.8179 \times 10^{-15}$ m
Elementary charge; $e = 1.6022 \times 10^{-19}$ coulomb
Electron mass; $m_e = 9.1095 \times 10^{-31}$ kg
Gravitational acceleration at sea level; $g = 9.8067$ m/s^2
Impedance of free space (resistive); $R = 376.73\,\Omega$
Permittivity of free space; $\varepsilon_0 = 8.8542 \times 10$ farad/m
Proton mass; $m_p = 1.6726 \times 10^{-27}$ kg
Planck constant; $h = 6.6262 \times 10^{-34}$ joule/s
Skin depth in a lossy medium; $\delta = (\pi f \mu \sigma)^{-1/2}$
Speed of light in free space; $c = 2.9979 \times 10^8$ m/s
Thomson cross-section $(8\pi/3)r_e = 6.6524 \times 10^{-29}$ m^2
Wave-impedance in a lossy medium; $Z = [\mu/(\varepsilon + J\sigma/\Omega)]^{1/2}$
Wavelength associated with 1 eV = hc/e

A.1.3 Some Selected Conversion Factors

To convert (1)	into (2)	multiply (1) by
1	2	
Acres	hectares	0.4047
Acres	square feet	43560
Ampere-hours	coulombs	3600
Angstrom unit	meter	1×10^{-10}
1 Atmosphere	cm of Hg	76.0
1 Atmosphere	in of Hg	29.92
Btu	joules	1054.8
Btu/h	watts	0.2931
Centigrade	Fahrenheit	(°C × 9/5) + 32°
Centimeters	inches	0.3937
Days	seconds	86400
Degrees	radians	0.01745 (angle)
Degrees/s	revolutions	0.1667 per min (rpm)
Dyne/cm^2	atmospheres	9.869×10^{-7}
Ergs	joules	10^{-7}
Fathom	meter	1.828804
Feet	meters	0.3048
Furlongs	feet	660
Gallons	cm^3	3785
Gallons	litres	3.785
Gallons (Brit Imperial)	gallons (US)	1.20095

Some selected conversion factors (*Continued*)

To convert (1)	into (2)	multiply (1) by
1	2	
Gallons of water	pounds of water	8.3453
Gausses	webers/m^2	10^{-4}
Grams	ounces (avdp)	0.03527
Hectares	acres	2.471
Horsepower U.S. and Brit)	Btu/min	42.44
(metric) Horsepower	(U.S. and Brit) horsepower	0.9863
Horsepower	watts	745.7
Joules	Btu	9.480×10^{-4}
Kilograms	dynes	980665
Kilograms	joules/m (newton)	9.807
Kilolines	maxwells	1000
Kilometers	feet	3281
Kilometers	miles	0.6214
Kilometers/h	feet/s	0.9113
Kilowatt-h	joules	3.6×10^4
Knots	mi/h nautical	1.0
Knots	km/h	1.8532
Knots	ft/s	1.689
Maxwells	webers	10^{-8}
Meters	feet	3.281
Meters	inches	39.37
Meters	yards	1.094
Meter-kg	lb-ft	7.233
Micron	meters	1×10^{-6}
(naut) Miles	feet	6080.3
(naut) Miles	km	1.853
Miles	km	1.609
Mils	in	0.001
(angle) Minutes	degrees	0.0167
Nepers	decibles	8.686
Ounces	grams	28.349
Ounces (fluid)	liters	0.029
Parsec	km	3.084×10^{13}
Pounds	kg	0.454
Pounds of water	gallons	0.1198
Pounds/m^3	kg/m^3	16.02
Pounds/ft^2	kg/m^2	4.882

Some selected conversion factors (*Continued*)

To convert (1)	into (2)	multiply (1) by
1	2	
Quarts (liq)	liters	0.9463
Radians	degrees	57.30
Sphere	steradians	12.57
Tons (long)	pounds	2240
Tons (long)	kg	1016
Tons (metric)	lb	2205
Tons (short)	lb	2000
Tons (short)	kg	907.185
Watts	ergs/s	107

Appendix A.2 Acronyms

This appendix contains, hopefully, all the acronyms used in this book which apply to *Radio Techniques*. Most of these abbreviations have also been defined when they were introduced in the text. The projects (CEDAR, WITS, SUNDIAL, etc. were in progress as of 1989).

AFGL – Air Force Geophysics Laboratory – U.S. research center at Hanscom Air Force Base, Massachusetts (the name has recently been changed to Geophysics Laboratory-GL).

AG – Annales Geophysicae – Official Journal of the European Geophysical Society; Part A – Upper Atmosphere and Space Sciences (Series A); Editorial Office; Bruxelles, Belgium.

AGU – American Geophysical Union – Headquarters in Washington, DC

AGW – Atmospheric Gravity Waves – sometimes also known as Acoustic Gravity Waves.

AIS – Advanced Ionospheric Sounder – the present generation of digital ionosondes.

AKR – Auroral Kilometric Radiation.

APL/JHU – Applied Physics Laboratories/Johns Hopkins University Laurel, Maryland USA.

ASC – All Sky Camera – a very wide field of view camera designed primarily to observe the horizon to horizon structure and motion of the visible aurora.

BARS – Bistatic Auroral Radar System – a Canadian VHF auroral radar system.

BMEWS – Ballistic Missile Early Warning Radar.

CASS – Center for Atmospheric and Space Science – research center located at Utah State University in Logan, Utah.

CCIR – International Radio Consultative Committee – the part of the ITU which concentrates on the radio spectrum.

CEDAR – Coupling, Energetics, and Dynamics of Atmospheric Regions.

CNET – Centre National d'Etudes des Télécommunications.

CNRS – Centre National de la Recherche Scientifique.

CRPE – Centre de Recherches Physique de Environment Terrestre et Planetaire.

CRPL – Central Radio Propagation Laboratory – a US NBS Department of Commerce Lab at Boulder, Colorado, in the 1950's and 1960's.

CT – Computerized Tomography – specifically applied to ionospheric imaging by recording RBE signals at several suitably placed earth stations.

CUPRI – Cornell University Portable Radar Interferometer.
CW – Continuous Wave – an unmodulated "carrier wave".
DAR – Dual Auroral Radar – a USSR twin VHF auroral radar system.
EISCAT – European Incoherent Scatter Facility.
ELF – Extra Low Frequency – that part of the radio frequency spectrum designated by the ITU from 300 to 3000 Hz.
EM – Electromagnetic
ESSA – Environmental Sciences Services Administration – a US Federal Laboratory in the 1960's.
fE_s – The highest frequency of the sporadic-E (E) trace on an ionogram.
f_0F2 – The "critical frequency" of the F2 layer as measured by an ionosonde.
FM – Frequency Modulated wave.
FCC – Federal Communications Commission – US regulating agency for wire and radio communications.
GaAsFET – Gallium Arsenide Field Effect Transistor.
GITCAD – Global Ionosphere-Thermosphere Coupling and Dynamics.
GISMOS – Global Ionospheric Simultaneous Measurement of Substorms.
GI/UAF – Geophysical Institute/University of Alaska, Fairbanks.
GRL – Geophysical Research Letters – main editorial office in Washington, DC.
GSFS – Goddard Space Flight Center – NASA/US research center at Greenbelt, Maryland.
HF – High Frequency – the 3–30 MHz segment of the radio spectrum.
HFD – High Frequency Doppler technique for observing ionospheric irregularities.
HFPL – High Frequency Plasma Line – as observed on an ISR.
HILAT – High Latitude multi-frequency beacon and multi-sensor polar-orbiting satellite.
HIPAS – High Power Auroral Stimulation – an HF ionospheric "heating" facility, operated by UCLA near Fairbanks.
IEEE – Institute of Electrical and Electronic Engineers – a trans-national technical society with headquarters in New York.
IGY – International Geophysical Year – an 18-month period during 1957–1958 when most of the nations on earth cooperated in a vast study of earth sciences.
IGC – International Geophysical Cooperation – a continuation of the studies of the IGY for 12 months (1959).
INAG – International Network Advisory Group of URSI – provide advice, coordination, and advocacy for the worldwide network of ionosondes.
IMS – International Magnetospheric Study – another international cooperative geophysical study (1976–79). One feature was the establishment of chains of small observatories along several magnetic meridians.
IQSY – International Year of the Quiet Sun – an intensive 12-month international cooperative geophysical study during 1964 – 65.

IRIS – Imaging Riometer Ionospheric Studies – a very high resolution riometer system.
ITS – Institute for Telecommunication Sciences – a US Department of Commerce Laboratory in Boulder, Colorado.
ITU – International Telecommunications Union – an international organization headquartered in Geneva, Switzerland, which is primarily concerned with worldwide efficient use of the electromagnetic spectrum for communications.
IZMIRAN – Institute of the Earth's Magnetism, Ionosphere, and Radio Propagation, Headquarters in Moscow, U.S.S.R.
JATP – Journal of Atmospheric and Terrestrial Physics – Main Editorial Office in Aberystwyth, UK, and published by Pergamon Press.
JGR – Journal of Geophysical Research – main editorial office in Washington, DC.
LANL – Los Alamos National Laboratory – a U.S. research facility addressing nuclear physics and EM radiation; located in Los Alamos, New Mexico.
LF – Low frequency – the 30 to 300 KHz segment of the radio spectrum.
LHC – Left Hand Circular (sense of polarization of an EM wave).
LHI – Linear, Homogeneous, Isotropic medium.
LOF – Lowest Observed Frequency on an HF oblique sounding system.
LORAN – Long Range Radio Aid to Navigation – a global electronic navigational aid operating in the LF range.
LTF – Least Time Focused – ionospheric focusing mechanisms dominant in the ground scatter mode.
MF – Medium Frequency – the 300 kHz to 3 MHz segment of the radio spectrum.
MIT – Massachusetts Institute of Technology – Boston, Mass, USA.
MST – Mesosphere-Stratosphere-Troposphere radar system.
MPAe – Max-Planck-Institut fur Aeronomie, Research Institute located at Katlenburg–Lindau, Germany.
MTI – Moving Target Indicator – type of radar display.
MW – MegaWatt.
NAIC – National Astronomy and Ionosphere Center – a US research facility at Arecibo, Puerto Rico.
NASA – National Aeronautics and Space Agency – national agency; located at various places in the US.
NCAR – National Center for Atmospheric Research – US research center at Boulder, Colorado.
NGC – Non Great Circle HF Propagation mode.
OBS – Oblique Backscatter Sounder.
NBS – National Bureau of Standards – a US research center with main labs located in Washington, DC and Boulder, Colorado. Recently renamed National Institute for Standards and Technology (NIST).
NOSC – Naval Ocean Systems Center – US Navy research complex in San Diego, California.

NRL – Naval Research Laboratory – US Navy research Lab in Washington, DC.

NSF – National Science Foundation – US scientific research agency with headquarters in Washington, DC.

OTH – Over The Horizon Radar – an HF OBS radar utilizing ionospheric reflection to extend its range.

PACE – Polar Anglo-American Conjugate Experiment – ionospheric research program using two HF radars, located conjugately at Halley Station, Antarctica and Goose Bay, Labrador.

PARAMP – Parametric Amplifier

PL – Radar Pulse Length (usually given in microseconds).

PRE – Partial Reflection Experiment – for measuring D-region absorption and dynamics.

PRF – Pulse Repetition Frequency.

PSS – Planetary and Space Science – monthly journal published by Pergamon Press with editorial office in Belfast, Northern Ireland.

QDC – Quiet Day Curve of riometer data.

QL – Quasi-linear approximation in magnetoionic theory.

RBE – Radio Beacon Experiment – ground-based observation of radio beacons on satellites, primarily to observe ionospheric electron content and scintillation.

RFI – Radio Frequency Interference.

Riometer – Relative Ionospheric Opacity Meter Extra Terrestrial Ionization – a self-calibrating total power receiver used to measure ionospheric nonderivative absorption.

RHC – Right Hand Circular – (sense of propagation of an EM wave).

RS – Radio Science – research journal of US National Committee of URSI; published by AGU in Washington, DC.

SAFARI – Sweden and France Auroral Radar Investigation – a VHF E-region direct backscatter system.

SABRE – Scandinavian And British Radar Experiment – a VHF E-region direct backscatter system.

SFD – Sudden Frequency Deviations – observed on HF Doppler systems.

SHERPA – System HF d'Etude Radar Polaires Aurorale – a French and Canadian phased array radar.

SMY – Solar Maximum Year – cooperative geophysical studies during 1979–1981.

SRT – Spaced Receiver Technique – used to measure motion of ionospheric irregularities.

SRI-I – Stanford Research Institute-International – main office in Menlo Park, California USA.

STARE – Scandinavian Twin Radar Experiment – a VHF E-region direct backscatter system.

STAR – Space, Telecommunications, Aeronomy, and Radar Research labs located on the campus of Stanford University; Stanford, California.

STEP – Solar Terrestrial Energy Program – a US research program for the early 1990's.

SUNDIAL – an international solar-terrestrial research effort which combines theoretical and empirical modeling with coordinated ground-based and satellite-borne measurement programs.

TID – Traveling Ionospheric Disturbance – a perturbation in electron density usually moving through the F-region, with a horizontal component of velocity with respect to a ground-based radio observatory.

TLC – "Tender Loving Care".

TEC – Total Electron Content of the ionosphere.

URSI – International Union of Radio Science – one of 18 unions which adhere to the International Union of Scientific Unions (ICSU), founded in 1919 to encourage "scientific studies of radio telegraphy". interests span all the scientific aspects of telecommunication and radio studies of planetary atmospheres.

VLF – Very Low Frequency – the frequency range from 3–30 kHz

VHF – Very High Frequency – the 30 to 300 MHz segment of the radio spectrum.

UHF – Ultra High Frequency – the 300 to 3000 MHz (3 GHz) portion of the radio spectrum.

ULCAR – University of Lowell Center for Atmospheric Research – research Lab located in Lowell, Massachusetts.

UT – Universal Time – a reference time often used in global measurement of geophysical parameters. Formerly called Greenwich Mean Time (GMT), it refers to the local solar time at 0° longitude.

VLSI – Very Large Scale Integration – microminiaturized electronics circuit.

WAGS – Worldwide Atmospheric Gravity Wave Study.

WDC – World Data Center.

WITS – Worldwide Ionosphere Thermosphere Study.

Appendix A.3 A Radio Frequency Spectrum Chart Showing the ITU Nomenclature and Principal Uses

Figure A.3.1 is a somewhat expanded version of the spectrum shown in Fig. 1.1 of Chapter 1 of this book, showing some of the services that share the spectrum. Much more detailed information on radio spectrum allocations may be obtained from CCIR publications, Radio Engineering Handbooks, and from the Telecommunication Authorities of each country.

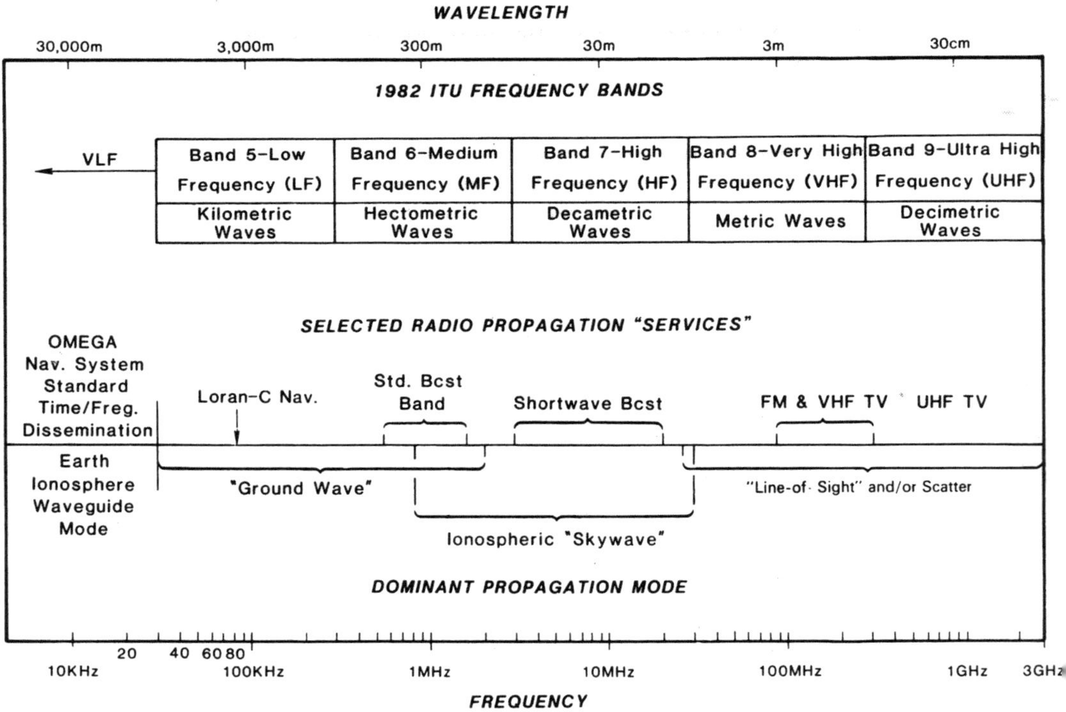

Fig. A.3.1. A truncated frequency spectrum

Appendix A.4 Annotated Bibliography

The books, reports, and articles included in this annotated bibligraphy have been chosen according to one or more of the following criteria:

1. The importance of the article to an understanding of the subjects presented in this book.
2. They offer a lucid presentation or a detailed derivation of pertinent mathematical relations.
3. They contain an account of the historical development of the idea or concept.
4. They are good sources for seeking further information on a topic.

The descriptive paragraph for each item in this bibliography is a summary of the contents of the article or book and an indication of why it may be of interest to readers of this book.

These annotations express the opinions of the author of this book and will, of course, be colored by my own biases. I have attempted to select references which should be obtainable in a university library system or in the book shelves of active workers in this field.

Akasofu S-I, Kamide Y (eds) (1987) The solar wind and the earth. Reidel, Dordrecht Boston

An excellent update and/or introduction to solar-terrestrial relations, intended for graduate students in astronomy and geophysics and for other students.

Appleton EV, Barnett MAF (1926) On some direct evidence for the downward atmospheric reflection of electric rays. Proc R Soc, 109A (752): 621–641

An account of one of the first experiments which proved the existence of the ionosphere.

Banks PM, Kockarts G (1973) Aeronomy, Parts A and B. Academic Press, New York, Vol A, 430 pp, Figures, Vol B, 355 pp, Figures

This two-volume set gives an encyclopedic coverage of the basic processes involved in terrestrial aeronomy. The fundamental character of the earth's upper atmosphere is stressed without going into a detailed description of the global morphology. The best quantitative values (as of 1973) of reaction rates, cross-sections, and interaction coefficients are given, along with very adequate mathematical treatment of the physical processes. These volumes are probably best suited to the needs of advanced graduate students and active researchers in aeronomy.

Volume A is principally concerned with describing the atmosphere as a whole and with a presentation of the outstanding problem of chemical aeronomy, while Volume B is concerned with phenomena which occur mainly above 100 km.

Bauer SJ (1973) Physics of planetary ionospheres. Vol 6. Physics and chemistry in space. Springer, Berlin Heidelberg New York, 230 pp, 89 Figs. Bauer's monograph, addressed to graduate students and researchers who are "getting acquainted with the field", appraises "where we stand, what we know, and what we still need to know" (as of 1973) about the fundamental physical processes in an idealized planetary ionosphere. Actual ionospheres are described as special cases. The book provides the intended audience with an excellent background on the subject in a very logical sequence of topics.

Beckmann P (1967) Probability in communication engineering. Harcourt, Brace World, Inc, New York, 511 pp

An excellent introduction to probability, information, noise and reliability theory with radio propagation applications.

Bernstein J (1984) Three degrees above zero – Bell Labs in the Information Age. Scribner's Sons, New York, 241 pp

This popular book is quite valuable for its insight into the creative processes of individual scientists and managers and the importance of a well-defined management research philosophy in a unique and very successful internationally famous laboratory. Much of the theory and some of the basic techniques used by ionospheric physicists were products of this laboratory.

Born M, Wolf E (1975) Principles of optics, 5th ed Pergamon Press, Oxford, A classic work on electromagnetic wave behavior.

Budden KG (1985) The propagation of radio waves – the theory of radio waves of low power in the ionosphere and magnetosphere. Cambridge Univ Press, Cambridge CB2 2RU, England, 669 pp

A comprehensive treatment of the topic – the magnum opus of Professor Budden's extensive studies. Covers most of the mathematical topics pertinent to radio waves in the ionosphere and contains a bibliography of over 600 entries. Some of the topics included are: magnetoionic theory, ray tracing in the ionosphere, full wave methods for anisotropic stratified media, etc.

CCIR Report 322 (1985) International Telecommunications Union. Geneva, Switzerland

The most comprehensive reference on galactic, atmospheric, and man-made radio noise in the terrestrial environment.

Chamberlain JW (1978) Theory of planetary atmospheres – an introduction to their physics and chemistry. Academic Press, London, New York, Int Geophys Ser 22, 330 pp

This is a monograph written for graduate students and research scientists engaged in planetary or atmospheric studies. The approach is mainly theoretical and descriptive, and a rather complete summary of the historical development of each topic is included as Bibliographical Notes at the end of each chapter.

Collin RE (1985) Antennas and radiowave propagation. McGraw-Hill, New York, 508 pp

A modern textbook for junior or senior level electrical engineering students which provides an introduction to the fundamental principles of antennas and propagation with a communications emphasis. It treats wire and aperture antennas as well as including a valuable coverage of antenna array principles in the first 336 pages of the book. Radiowave propagation is covered in the last chapter (117 pages).

Davies K (April 1, 1965) Ionospheric radio propagation. Nat Bur Stand Monograph 80, 470 pp

Dr. Davies writes that this book is for use by "research workers and communications engineers who already have some background knowledge of radio propagation and the ionosphere." The book draws on material from the published literature through 1964 (post-IGY and IQSY) and on a series of lectures given at CRPL/Boulder in 1961 and 1962. Topics covered include: electron-layer production, magnetoionic theory, results of synoptic studies of the ionosphere, oblique propagation, ionospheric disturbances, predictions, and LF and VLF wave behavior. This book is out of print, but most libraries have copies.

Davies K (1969) Ionospheric radio waves. Blaisdell, Waltham, Massachusetts, 460 pp

This textbook is based on lecture courses given by Dr. Davies at the University of Colorado and at the Institute for Telecommunications Science and Aeronomy at Boulder, Colorado. The book is intended to bridge the gap between ionospheric research workers and radio communication engineers and is suited for introductory courses (graduate or undergraduate) in aeronomy and ionospheric radio engineering. Basic magnetoionic theory is treated without mathematical rigor in the first half of the book, and ionospheric radio phenomena (absorption, polarization, fading, scintillations, etc.) are covered in the second half.

Davies K (1990) Ionospheric radio. Peregrinus Press, London

Intended for scientists and engineers working in ionospheric research and propagation, as well as for students in the field.

de Coulon F (1986) Signal theory and processing. Artech House, Dedham, Massachusetts

A textbook on mathematical fundamentals of signal processing theory, with problem assignments.

di Franco JV, Rubin WL (1980) Radar detection. Artech House, Dedham, Massachusetts, 645 pp

A comprehensive tutorial exposition on signal processing as applied to radar systems, using techniques of mathematical statistics. Intended as a text for graduate students and for practicing radar engineers.

Drabowitch S, Ancona C (1988) Antennas. Vol 2: Applications. Hemisphere, Washington (First Publ in French as Antennas: applications. Masson, Paris)

Intended to follow the book by Roubine and Bolomey, this book treats the use of small and large antennas, including arrays, in considerable detail.

Eaves JL, Reedy EK (eds) (1987) Principles of modern radar. Van Nostrand Reinhold, New York

A collection of articles by experts in the basic principles and state-of-the-art radar technology. The book is based on a course offered at Georgia Tech since 1969 and is well suited to be both a formal textbook and a reference work for practitioners.

Flock WL (1979) Electromagnetics and the environment: Remote sensing and telecommunications. Prentice-Hall, Englewood Cliffs, New Jersey, 325 pp
Based on an undergraduate elective course on *Environmental electromagnetics* in the electrical engineering department of the University of Colorado. Emphasis is on interdisciplinary treatment of the terrestrial environment using electromagnetic remote sensing. Includes descriptive treatment of the electromagnetic spectrum, the atmosphere, atmospheric effects on EM waves, blackbody and solar radiation, and remote sensing in general.

Giraud A, Petit M (1978) Ionospheric techniques and phenomena. Reidel, Dordrecht, Holland, 264 pp
This monograph is intended for "graduate students and research scientists in atmospheric, ionospheric and magnetospheric physics," and is based on lectures given at the University of Paris and at the International School of Atmospheric Physics (Erice, Sicily). The work is an updated translation of the original French monograph. Part One (p 55) covers "The ionospheric environment," Part Two covers "The techniques of ionospheric measurements" (p 74) [some nine or so techniques], and Part Three discusses "The interpretation of ionospheric phenomena" (p 113).

Hargreaves JK (1979) The upper atmosphere and solar-terrestrial relations (An introduction to the aerospace environment). Van Nostrand Reinhold, New York, 298 pp
This book is based on lectures presented in a course *Physics of the Environment* at the University of Lancaster, United Kingdom. The book is intended to serve as an introductory test for undergraduates who have basic knowledge of classical physics. Included in the book are 26 pages on *Observational Techniques*, 54 pages describing the vertical, geographical, and temporal structure of the terrestrial ionosphere, and 29 pages devoted to winds, currents, irregularities, and waves. The remainder of the book (152 pages) discusses the structure of the magnetosphere, dynamics of the substrom, waves in the magnetosphere, solar flares, ionospheric storms, etc.

Johnk CTA (1975) Engineering electromagnetics fields and waves. Wiley New York, 655 pp
Written for junior or senior electrical engineering students. This book starts with Maxwell's equations for free space, then develops them to include material regions. The wave equation is developed on p. 100, then applied to various and sundry situations including quite complete treatments of antennas and wave guides. The book is characterized by the use of clear illustrations and excellent didactics.

Johnson RC, Jasik H (1984) Antenna engineering handbook, 2nd edn McGraw-Hill, New York
A very complete engineering handbook on antennas. Suitable for graduate students and practicing radio scientists.

Johnson RC, Jasik H (1987) Antenna applications reference guide. McGraw-Hill, New York
A rather succinct summary of the basic characteristics and essential design criteria for the most often used antennas spanning the spectrum from LF through microwaves.

Jordan EC, Balmain KG (1968) Electromagnetic waves and radiating systems, 2nd edn, Prentice-Hall, Englewood Cliffs, New Jersey, 753 pp
A basic electrical engineering text covering the principles of electromagnetic theory, wave guides, antenna fundamentals, antenna array theory, broadband antennas, ground wave and ionospheric propagation, plus other related topics. Suitable as a text for advanced or beginning graduate EE students, and as a reference for experimental ionospheric physicists.

Kelley MC (1989) The earth's ionosphere. Plasma physics and electrodynamics. Academic Press, London New York San Diego, Int Geophys Ser 43
A recent, thorough treatment of ionospheric plasma physics intended for active ionospheric research and graduate students.

Kelso JM (1964) Radio ray propagation in the ionosphere. McGraw-Hill, New York, 408 pp
This work is primarily intended as a textbook for researchers who have had an introduction to EM theory and who are starting to specialize in ionospheric propagation. It starts with the propagation of waves in a homogeneous medium, followed by a short treatment of ionospheric physics, then the dispersion equations in the ionosphere, and finally ray theory, both with and without the magnetic field.

Kolosov AA et al. (1987) Over the horizon radar. Artech House, Boston London
This monograph describes quite adequately the topic of OTH radar; including the radar equation, path loss, target cross-section HF interference, signal detection, and signal estimation, etc. Approximately 90% of the references are to Soviet literature.

Krall NA, Trivelpiece AW (1973) Principles of plasma physics. McGraw-Hill, New York, 674 pp
Intended to provide graduate students in physics and astronomy a comprehensive background in plasma physics. Contains a fairly rigorous treatment of incoherent scatter from plasma density fluctuations in the ionosphere.

Kraus JD (1966) Radio astronomy. McGraw-Hill, New York, 481 pp
An introduction to radio astronomy for undergraduate students in electrical engineering Chapter 7 on radio-telescope receivers by Marti Tiuri contains a thorough discussion of noise temperature and noise figure of receivers.

Kraus JD (1984) Antennas, 2nd edn McGraw-Hill, New York, 553 pp
A classic treatment of the theory and practice of radio antennas starting with point sources, covering the dipole, loop, linear, helical antennas, arrays of linear antennas, other miscellaneous antennas, and basic antenna measurements.

Lewis BL, Kretschner FF, Sheltion WW (1986) Aspects of radar signal processing. Artech House, Norwood, Massachusetts, 554 pp
Written for engineering students and radar engineers. Covers basic radar theory, signal enhancement techniques, clutter suppression, adaptive processing, etc. – practically oriented.

Lo YT, Lee SW (eds) (1988) Antenna handbook: theory, applications and design. Van Nostrand Reinhold, New York, 28 Chap.

A truly encyclopedic work, covering antenna fundamentals, mathematical tools, basic design, and related topics such as propagation, transmission lines, near-field/far-field measurements, and pulse response. A very useful book for the student or practicing radio scientist.

Mayaud PN (1980) Derivation, meaning, and use of geomagnetic indices. Geophys Monogr 22, Am Geophys Union, Washington DC
The definitive source of information on K, R, Q, AU, Dst, and all other magnetic indices of importance to the geophysicist.

Ramo S, Whinnery JR, Van Duzer T (1965) Fields and waves in communication electronics.
A modern textbook intended for use as an intermediate level textbook on electromagnetic fields and waves in the electrical engineering curriculum. It includes a very complete treatment of guided EM wave theory, characteristics of transmission lines and wave guides, resonant cavities and antennas.

Ratcliffe JA (1959) The magnetoionic theory and its applications to the ionosphere – a monograph. Cambridge Univ Press, Lond, 206 pp
Ratcliffe uses both the macroscopic and microscopic approaches to give a very detailed picture of physical reasons for the behavior of radio waves in the earth's ionosphere in Part 1 of this book. In Part 2 the results are summarized in many graphs describing in exquisite detail the special cases and approximations of the Appleton–Hartree equations. Part 3 is devoted to applications of the magnetoionic theory to the terrestrial ionosphere. Part 4 completes the treatment by covering miscellaneous items such as the Lorentz term, waves in a mixture of ions and electrons, and inhomogeneous media.

Rees MH (1989) Physics and chemistry of the upper atmosphere. Cambridge Univ Press, Cambridge, UK
A new book, including recent results. Intended for graduate students and scientists in this field.

Roubine E, Bolomey JC (1987) Antennas. Vol 1: General principles. Hemisphere Washington (publ first in French as Antennes: introduction générale. Masson, Paris)
Written for "practicing engineers... as well as students". A summary of key electromagnetic principles applicable to antennas, presented in a modern fashion.

Rudge AW, Milne K, Oliver AD, Knight P (eds) (1982, 1983) The handbook of antenna design. Vols 1 and 2, respectively, Peregrinus, Lond, UK, 1: 708, 2: 945
An encyclopedic reference work covering a very wide range of antennas operating from VLF through microwaves. Starts with a brief outline of basic antenna theory and includes details of design and measurement of most antennas used for communication, radar, and research.

Rulf B, Robertshaw GA (1987) Understanding antennas for radar, communications, and avionics. Van Nostrand Reinhold, New York, 335 pp
Written for the non-specialist, this book gives a nonmathematical treatment of wire, aperture, reflector and lens, traveling-wave, broadband, surface wave, and array antennas. Geometrical optics principles are used extensively in

explanations of antenna principles. There is a short annotated bibliography and a rather inadequate index.

Schey HM (1973) Div, Grad, Curl and all that – an informal text on vector calculus. Norton, New York, 163 pp
A relaxed treatment of the basic mathematical principles of vector calculus and the gradient, divergence, and curl operators for undergraduate students taking a course in electricity and magnetism.

Skolnik MI (1980) Introduction to radar system, 2nd edn McGraw-Hill, New York, 581 pp
Intended as a reference for those seeking a comprehensive introduction to basic radar principles and to modern radar systems. Covers coherent radar detection of hard targets.

Spaulding AD, Washburn JS (1985) Atmosphere radio noise: worldwide levels and other characteristics. NTIA Report 85–173. Natl Telecom Information Administration, 325 Broadway, Boulder, Colorado
An updated, worldwide compilation of atmospheric radio noise in graphical and numerical form including an outline of the pertinent noise measurement theory. This report indicates that corrections up to 20 dB are required for some locations where 1 MHz noise was measured and reported in CCIR Report 322.

Stix TH (1962) The theory of plasma waves. McGraw-Hill, New York, 283 pp
The author's twofold aim is to give a unified presentation of the branch of plasma physics theory which deals with the solutions of the linearized, collision-free Boltzmann equations and to provide a text for graduate students in plasma physics. This book should provide the basic background necessary to understand current work in plasma waves and instabilities both in the laboratory and in the ionosphere.

Volland H (ed) (1982) CRC Handbook of atmospherics. CRC Press, Boca Raton, Florida, 1, 2: 704
This is an exhaustive treatment on the general subject of electromagnetic pulses of cosmic, terrestrial atmospheric, or man-made origin. It is obvious that "one man's noise is another man's signal" from perusing the Table of Contents of these volumes, whose chapters include such topics as: Schumann resonances, atmospheric radio noise, man-made radio noise, sferics, whistlers, theory of low-frequency wave propagation, and acoustic radiations from lightning.

Wait JR (1985) Electromagnetic wave theory. Harper & Row New York, Hagerstown Lond, 308 pp
Designed for an advanced course in electromagnetic waves at the senior or beginning graduate level in electrical engineering. Follows the point of view of Schelkunoff and Booker in using the impedance concept to describe wave phenomena. Emphasizes currently relevant topics in radio wave propagation and antennas.

Appendix A.5 Selected Global Maps and Other Information on Electrical Noise

Electrical noise from natural and man-made sources and radio frequency interference (RFI) effects have been discussed in Section 1.3 of this monograph. This Appendix contains a few selected maps, charts, and tables which may be of use when selecting a receiver site.

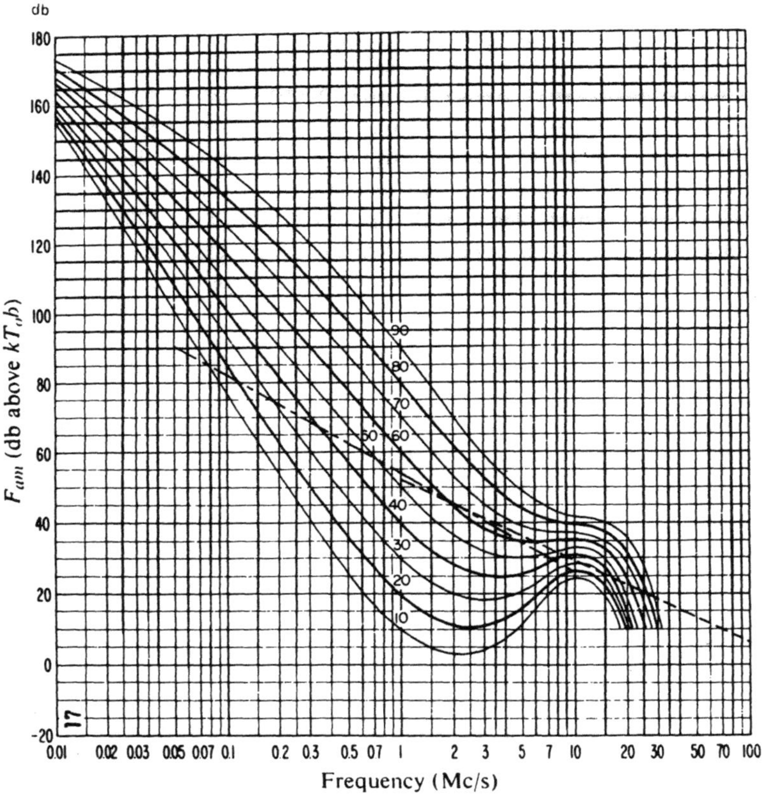

Variation of radio noise with frequency
(Summer; 1200–1600 h.)

——— Expected values of atmospheric noise
—·—·- Expected values of man-made noise at a quiet receiving location
- - - - Expected values of galactic noise

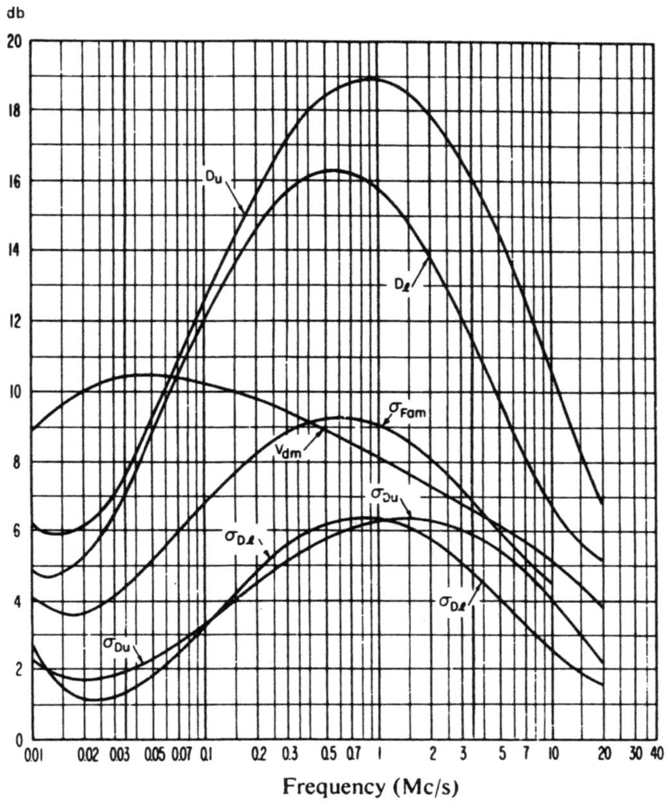

Data on noise variability and character
(Summer; 1200–1600 h.)

σ_{Fam} = Standard deviation of values of F_{am}
D_u = Ratio of upper decile to median value, F_{am}
σ_{Du} = Standard deviation of values of D_u
D_l = Ratio of median value, F_{am}, to lower decile
σ_{Dl} = Standard deviation of value of D_l
V_{dm} = Expected value of median deviation of average voltage.
The values shown are for a bandwidth of 200 c/s.

Fig. A.5.2. Statistics of radio noise behavior

◄

Fig. A.5.1. Typical variation of radio noise

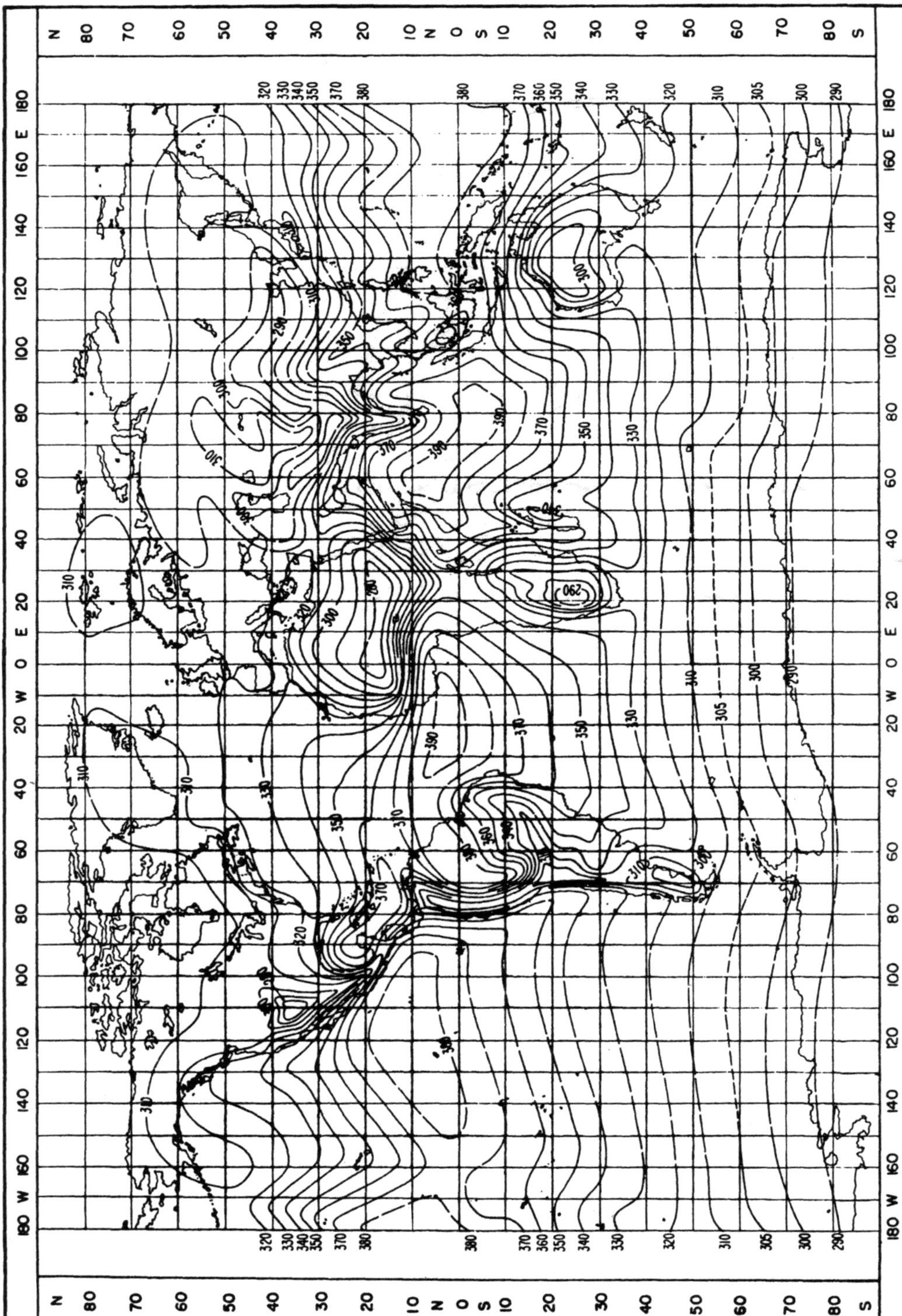

Fig. A.5.3. Typical global radio noise map from CCIR Report 322

Appendix A.6 Some Representative Values of Ground Electrical Characteristics

Figure A.6.1 illustrates some electrical conductivity and permitivity values for different types of terrain. Much more detailed information may be found in the various electrical engineering handbooks listed in the Annotated Bibliography and references in this book.

Type of Surface	Conductivity, σ in mhos (or Siemens) per meter	Permitivity (ε) (Relative dielectric) constant
Coastal dry sand	0.002	10.0
Flat, wet coastal	0.01 to 0.02	4.0 to 30.0
Rocky land (steep hills)	0.002	10.0 to 15.0
Highly moist soil	0.005 to 0.02	30.0
Marshy	0.1	30.0
Hills (to $\sim$ 1000 m)	0.001	5.0
Freshwater	0.001	80.0 to 81.0
Sea water	3.0 to 5.0	80 to 81.0
Sea ice	0.001	4.0
Polar ice (free)	0.000025	3.0
Polar ice (cap)	0.0001	1.0
Arctic Land	0.0005 to 0.001	23–34 for silts $\sim$12 for dry sand
*Tundra underlain with permafrost surface	$\sim 10^{-3}$ to 10^{-2}	$\sim 5-70$

*Aquired in 1988/1989 in Central Alaska from 2–30 MHz by G. Hagn of SRI International.

Fig. A.6.1. Electrical conductivity and permitivity for various types of terrain

Appendix A.7 Fresnel Zone Antenna Siting Fundamentals

A.7.1 Introduction

There is a brief discussion of an application of the Fresnel zone as applied to the study of ionospheric irregularities associated with the scintillation of beacon satellite signals in Section 8.1.5 of this monograph.

The geometrical optics description of the diffraction of electromagnetic waves by a straight edge or a slit is given in detail by Born and Wolf (1970), Stratton (1941), and Jordan and Balmain (1968). In 1678 Huygens analyzed the diffraction of light by a point and stated that "every point exposed to electromagnetic radiation may be regarded as the source of a secondary wave". Fresnel in 1810 then improved Huygen's principle by eliminating secondary waves by destructive interference using the superposition principle. The individual secondary waves are very weak, and a radiation field may only be observed when there is constructive interference of the field due to superposition.

Consider Fig. A.7.1 below, depicting the diffraction of plane, parallel electromagnetic waves incident upon a slit of width D, showing the Fresnel and Fraunhofer diffraction regions. As may be seen, the Fresnel region is considered to extend to a distance $2D^2/\lambda$ from the secondary source, and it defines the

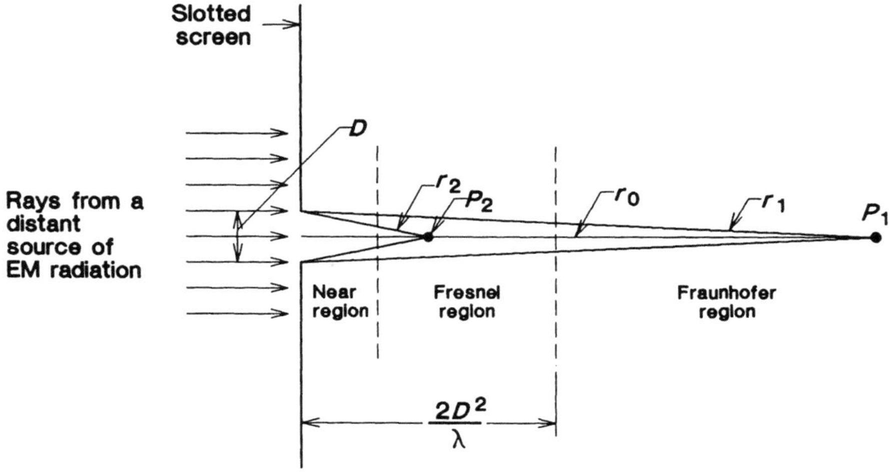

Fig. A.7.1. Geometry of a simple diffraction problem. The diffraction of a plane, parallel wave front by a slit whose width $D < \lambda$

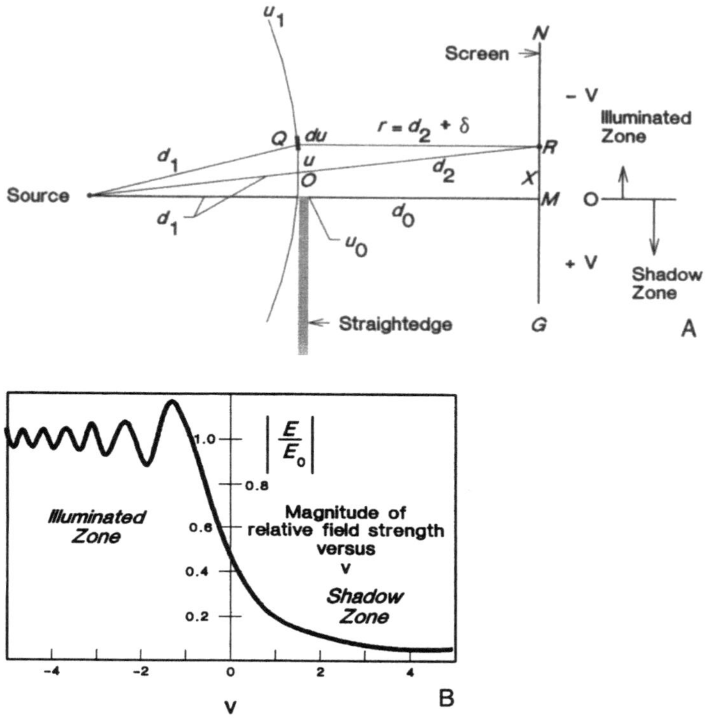

Fig. A.7.2. A Geometry of Fresnel diffraction of a spherical wavefront by a straightedge. **B** Plot of relative field strength from the scattering problem shown in **A**

boundary between Fraunhofer and Fresnel diffraction. Figure A.7.1 is, of course, idealized and rays in the Fraunhofer region are really considered to be parallel.

To illustrate the Fresnel diffraction principle, consider the diffraction of a spherical wave front by a straight edge as shown in Fig. A.7.2.

If we assume that elemental strip du of the wave front produces an effect at R of

$$dE = \frac{k_1 \, du \, e^{-j\beta r}}{f(r)}, \tag{A.7.1}$$

where $\beta = \dfrac{2\pi}{\lambda}$ and k_1 is a constant.

Following the development given by Jordan and Balmain (1968), one can obtain the square of the field strength at R as

$$|E|^2 = k_1/f^2(d_2)\left[\left(\int_{u_0}^{u_1} \cos \beta\delta \, du\right)^2 + \left(\int_{u_0}^{u_1} \sin \beta\delta \, du\right)^2\right]. \tag{A.7.2}$$

The standard Fresnel integral is used

$$C(v) - jS(v) = \int_0^v e^{-j(\pi/2)v^2} dv \tag{A.7.3}$$

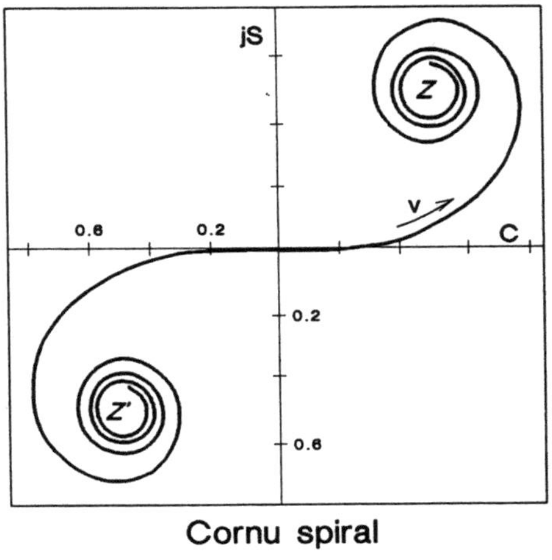

Fig. A.7.3. The Cornu spiral, depicting the diffraction field of scattering of a spherical wave front by a straightedge

to evaluate (A.7.2), which can then can be plotted as the Cornu spiral displayed in Fig. A.7.3.

Cornu's spiral has some very important properties which describe diffraction phenomena. If we define the following terms

$$C = \int_0^v \cos(\pi v^2/2)dv; \quad s = \int_0^v \sin(\pi v^2/2)dv \quad \text{and}$$

$$\delta s = \sqrt{(\delta C)^2 + (\delta C)^2} = \delta v; \quad v = s, \text{ then}$$

$$\tan\phi = \delta S/\delta C = \tan \pi v^2/2 \quad \text{and} \quad \phi = sv^2/2 = \pi s^2/2, \quad \text{and}$$

$$C(\pm\infty) = \pm\tfrac{1}{2}; \quad S(\pm\infty) = \pm\tfrac{1}{2}.$$

The following properties of the Cornu spiral are applicable to analysis of diffraction fields:

1. The length of arc s along the Cornu spiral measured from the origin = v.
2. As v approaches + or − infinity, the spiral winds an infinity of times about the points $(\frac{1}{2}, \frac{1}{2})$ or $(-\frac{1}{2}, -\frac{1}{2})$.
3. The magnitude and phase values of the Fresnel integral are defined by a vector drawn from the origin to any part of the curve.
4. Maximum values of the magnitude $\sqrt{C^2 + S^2}$ of the integral occur when $\Phi = 3\pi/4$, or at $v = \sqrt{3/2}$. Minima occur at $v = \sqrt{7/2} + 4m$, $(m = 0, 1, 2, 3, \ldots)$.

Other properties of Fresnel integral may be deduced for limiting conditions when u_0 = large negative values.

A.7.2 HF Antenna Siting Example

The importance of accurate determination of the vertical antenna patterns in long distance HF propagation and the application of Fresnel zones to this problem was pointed out by Utlaut (1961). An accurate determination of the vertical radiation pattern of antennas used to probe the ionosphere is, of course,

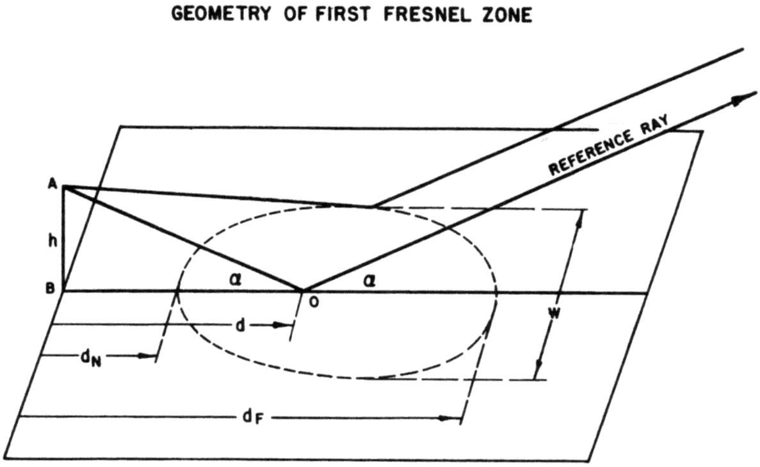

Fig. A.7.4. Two-dimensional geometry of the first Fresnel zone for the ground reflection region of an antenna. (After Utlaut 1962)

extremely important, and Fresnel diffraction theory is very applicable. The following example of the application of Fresnel zones to HF antennas, hopefully will illustrate this principle.

The area required in front of an antenna to effectively form the lowest ground reflection lobe is described by the first Fresnel zone. Figure A.7.4 shows the two-dimensional geometry of the first Fresnel zone. This treatment assumes *horizontal wave polarization*; smooth, perfectly reflecting surface, a point-source antenna and flat-earth.

Using the ground reflection point, o, for the reference ray, as the origin of a Cartesian coordinate system, the following simple equations apply:

The *height* of the antenna above the plane of the Fresnel zone for maximum radiation in the direction of the ionospheric reflection is

$$h = \lambda/4 \sin \alpha. \tag{A.7.4}$$

The *distance* of the ground reflection point, o, from the base of the antenna, B, is

$$d = h/\tan \alpha. \tag{A.7.5}$$

Similarly, the distance from the antenna base, B, to the *near* edge of the Fresnel zone is

$$d_N = h/\tan \alpha (3 - 2\sqrt{2}/\cos \alpha), \tag{A.7.6}$$

and the distance to the *far* edge of the Fresnel zone is

$$d_F = (h/\tan \alpha)(3 + 2\sqrt{2}/\cos \alpha). \tag{A.7.7}$$

The maximum *width* of the zone is given by

$$w = 4\sqrt{2}h = 5.66\,h. \tag{A.7.8}$$

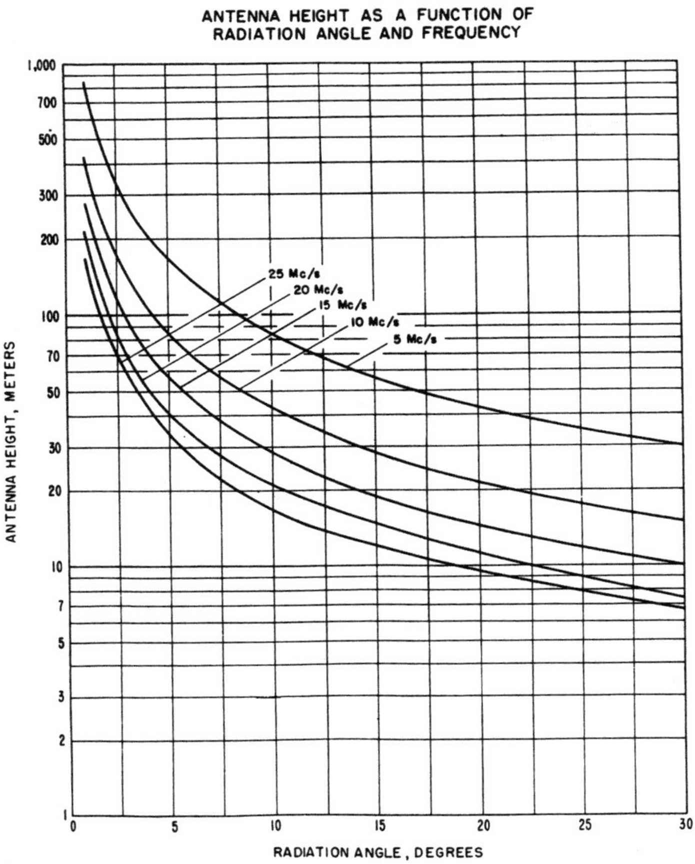

Fig. A.7.5

Figs. A.7.5.–A.7.9. Fresnel zone parameters for a horizontally polarized antenna over a flat, perfectly conducting surface for selected HF frequencies. (After Utlaut 1962)

Figures A.7.5 through A.7.9 illustrate the relation between the elevation takeoff angle and the parameters defined by Eq. (A.7.4) through (A.7.8) for various frequencies in the HF bands. The foregoing treatment of Fresnel zone reflection is subject to some caveats.

For low takeoff angles (below $\approx 3°$) the curvature of the earth has the effect of reducing the height, h, required for formation of a first lobe at a specified angle.

In the real world, one does not find plane-earth, perfectly smooth and infinitely conducting reflecting surfaces; so some siting criteria must be used. If the reflecting surface is irregular in contour and has poor conductivity, as much as a 2-dB degradation of performance may result. In order to keep the loss in performance from exceeding ≈ 2 dB, the Fresnel zone requirements may be compromised as listed:

a) d_N may be increased in length from 1.0 to 1.6 times full N value.
b) d_F may be decreased in length from 1.0 to 0.6 times the F full value.
c) w may be reduced from 1.0 to 0.7 times full value.

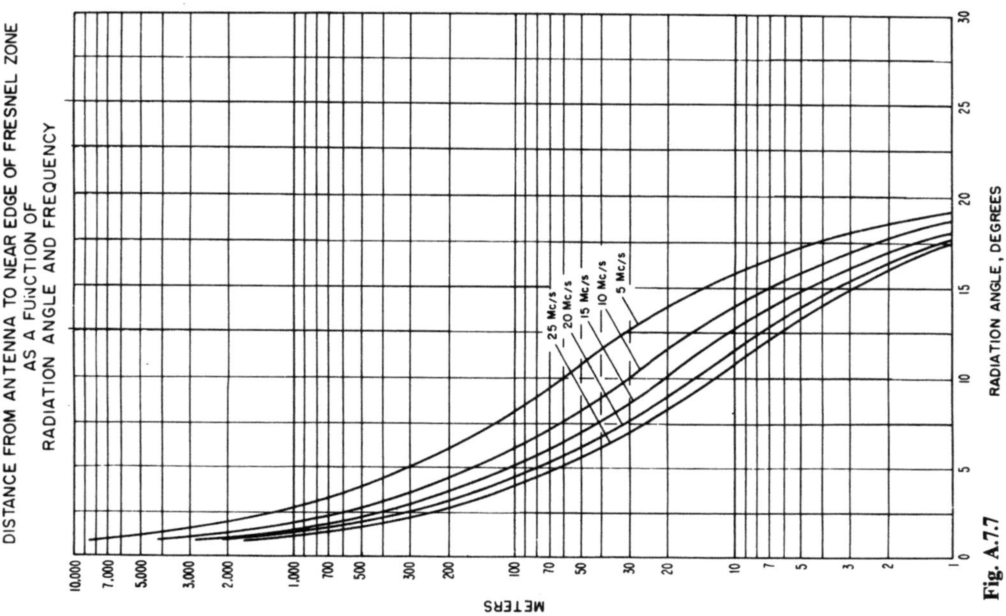

Fig. A.7.7

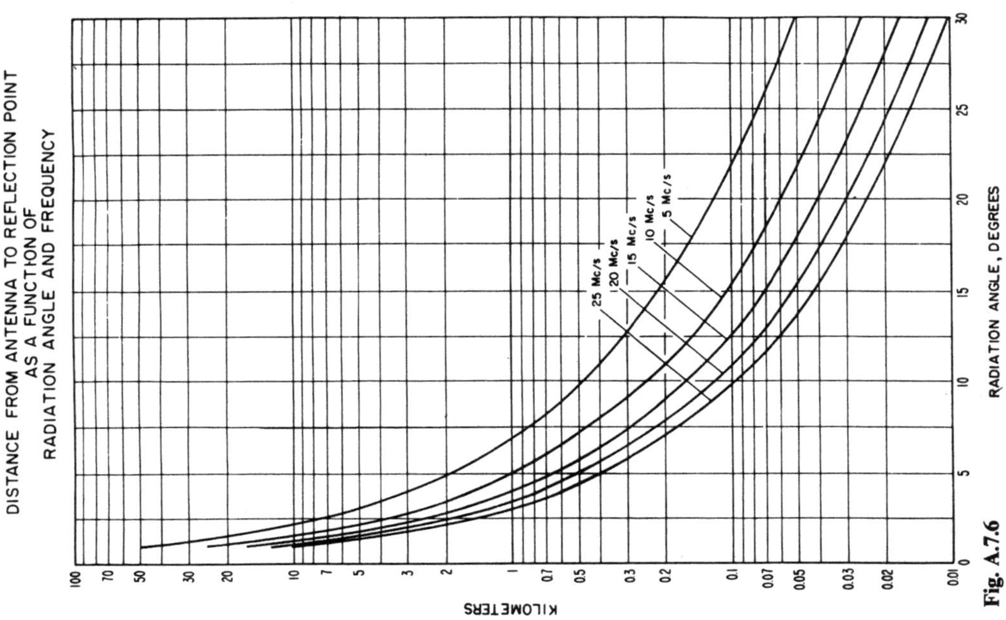

Fig. A.7.6

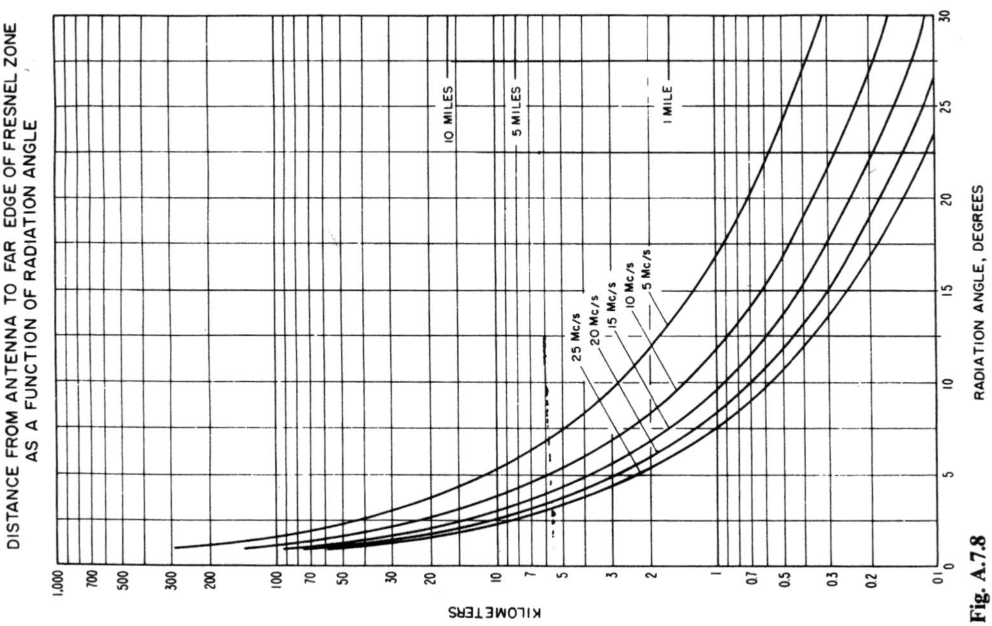

Fig. A.7.8

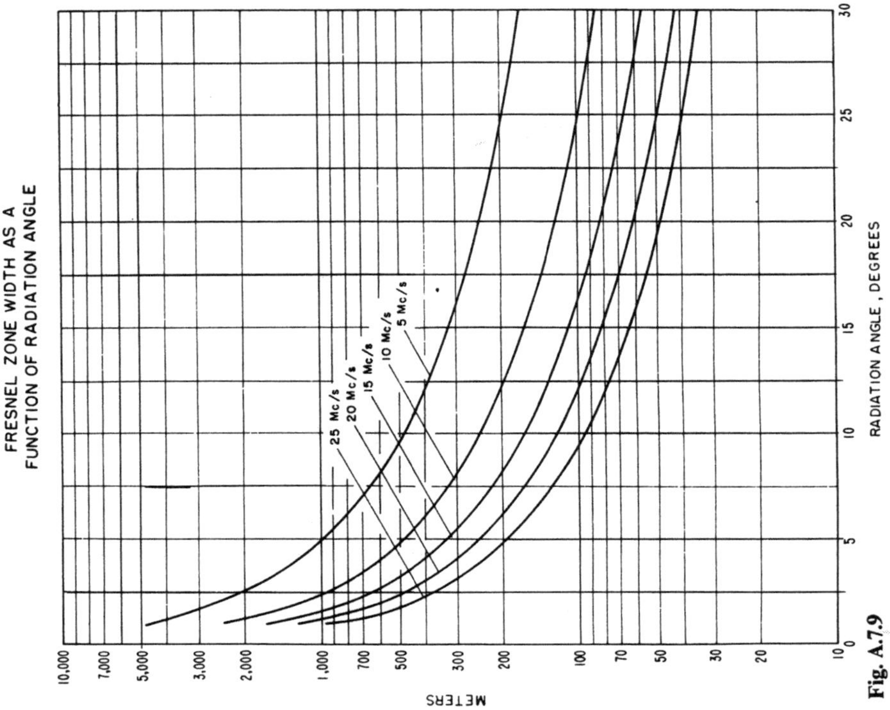

Fig. A.7.9

If the reflecting surface is of low conductivity, a metallic ground screen should be used with spacing of elements $\leq \lambda/10$.

Rayleigh's criteria states that the transition between specular reflection and scattering takes place at a height of an irregularity above the reflecting plane (for the geometry of Fig. A.7.4) of

$$H = \lambda/16 \sin \alpha. \tag{A.7.9}$$

We can obtain a rule of thumb for height (H) of an irregularity above the average terrain profile, of one-quarter the antenna height

$$H = h/4. \tag{A.7.10}$$

Since Eqs. (A.7.9) and (A.7.10) define the *transition* region between specular reflection and scattering, a more conservative engineering compromise is to keep the irregularities of terrain below $\approx \lambda/12$. The antenna engineering books in the Annotated Bibliography of this monograph contain a much more complete treatment of antenna siting than presented here.

Remember that all the foregoing treatment of antenna beam forming by reflection was for *horizontal* polarization. For *vertical* polarization there are dramatic changes in the magnitude and phase of the reflected wave at low takeoff angles i.e., Brewster angle effects.

A.7.3 The Fresnel Zone in Three Dimensions

The discussion in Section 8.1.5 and Appendices A.7.1 and A.7.2 describe two-dimensional Fresnel zones, so Fig. A.7.10 is included in this discussion to illustrate the essential features of the three-dimensional Fresnel zone.

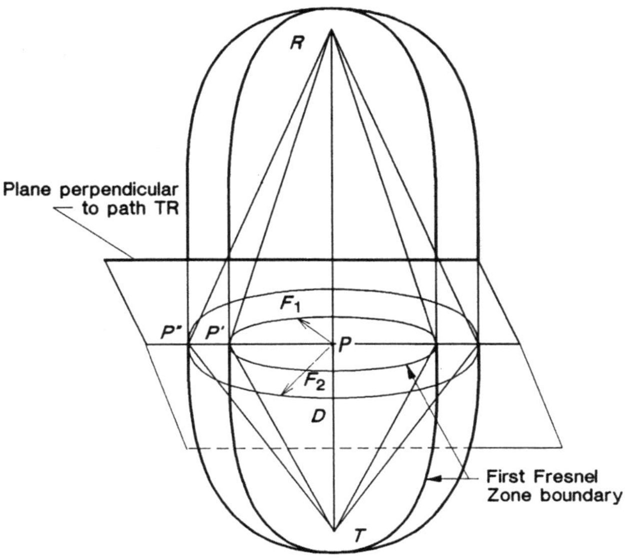

Fig. A.7.10. Geometry of a three-dimensional Fresnel zone

The distance from P along the major axis of the ellipse in Fig. A.7.10 defines $\frac{1}{2}$ the width of the first Fresnel zone, and is given by

$$F = 17.3\sqrt{d_1 d_2/fD}, \tag{A.7.11}$$

where: all distances are in km

d_1 = distance from transmitter to Fresnel object 1
d_2 = distance from fresnel object to receiver 2
$D^2 = d_1 + d_2$
f = frequency in MHz.

High-order Fresnel zone (nth) dimensions can be obtained from

$$F_n = \sqrt{n} F_1, \quad \text{where} \quad n = 2, 3, 4, \ldots..$$

References

Aarons J (ed) (1970) A survey of scintillation data and its relationship to satellite communication. Rep AFCRL-70-0053, Spec Rep No 106, US Air Force

Aarons J (1982) Global morphology of ionospheric scintillations. Proc IEEE 70: 360–378

Aarons J, Allen RS (1971) Scintillation boundaries during quiet and disturbed magnetic conditions. J Geophys Res 76: 170–177

Aarons J, MacKenzie E, Bhavnani K (1980) High-latitude analytical formulas for scintillation levels. Rad Sci 15: 115–127

Abel WG, Edwards LC (1951) The source of long distance backscatter. Proc. IRE 39: 1538–1541

Agy V, Baker DM, Jones RM (1965) Studies of solar flare effects and other ionospheric disturbances with a high frequency Doppler technique. NBS Tech Note 306

Aitchison GJ, Thomson TH, Weekes K (1958). J Atmos Terr Phys 14: 244

Aitken Hugh GJ (1985) Syntony and Spark – the origins of radio. Princeton Univ Press, Princeton NJ

Aitken Hugh GJ (1985) The continuous wave technology and American radio, 1900–1932, Princeton Univ Press, Princeton NJ

Alcock G McK, Martin LH (1956) Simultaneous occurrence of "Dawn Chorus" at places 600 km apart. Nature (Lond) 178: 937–938

Allen EM, Thome GD, Rao PB (1974) HF phased array observations of heater-induced spread-F. Rad Sci 9: 905–916

Al'pert YL (1976) On ionospheric investigations by coherent radiowaves emitted from artificial earth satellites. Space Sci Rev 18: 551–602

Al'pert YL (1958) Method of investigating the ionosphere with the aid of artificial earth satellites. Usp Fiz Nauk (UFN, in Russian) 64: Nos 1,3

Al'pert YL (1963) Radio wave propagation and the ionosphere. Consultants Bureau, New York

Amayenc P (1968) Caracteristiques des résultant de mesure de l'absorption ionosphérique a l'aide d'un satellite balise. Plant Space Sci 16: 163–219

Anderson SJ, Lees ML (1988) High-resolution synoptic scale measurements of ionospheric motions with the Jindalee sky wave radar. Rad Sci 23: 265–272

Andre D, McNamara AG, Wallis DD (1988) Diurnal radio aurora variations at 50 MHz measured by the bistatic auroral radar system radars. J Geophys Res 93: 8651–8661

Anonymous (1939) QST 23: 78

Ansari ZA (1963) The spatial and temporal variations in high latitude cosmic noise absorption and their relation to luminous aurora. PhD Dissertation, University of Alaska, Fairbanks Alaska

Ansari ZA (1964) The aurorally associated absorption of cosmic noise at College, Alaska. J Geophys Res 69: 4493

Ansari ZA (1965) A peculiar type of daytime absorption in the auroral zone. J Geophys Res 70: 3117

Appleton EV (1925) Geophysical influence on the transmission of wireless waves. Proc Phys Soc (Lond) 37(2): 16D–22D

Appleton EV, Barnett MAF (1926) On some direct evidence for the downward atmospheric reflection of radio waves. Proc R Soc 109A: 621–641

Appleton EV, Piggott WR (1954) Ionospheric absorption measurements during a sunspot cycle. J Atmos Terr Phys 5: 141

Argo P, Hindman M (1987) PC radar – a data acquisition tool. Pers Eng Instrum News, August 1987

Argo PE, Hunsucker RD (1988) Foreword: The worldwide atmospheric gravity wave study (WAGS). Radio Sci 23: 865

Armstrong RJ, Berkey FT, Melbye T (1977) The day to night absorption ratio in auroral zone riometry measurements. Planet Space Sci 25: 1193

Austen GL (1971) A direct measuring differential phase experiment. J Atmos Terr Phys 33: 1667
Austen JR, Franke SJ, Liu CH (1988) Ionospheric imaging using computerized tomography. Rad Sci 23: 299–307
Bailey VA (1937a) On some effects caused in the ionosphere by electric waves. Part I. Phil Mag 23: 929–960
Bailey VA (1937b) Interaction by resonance of radio waves. Nature (Lond) 139: 68–69
Bailey VA (1937c) Resonance in the interaction of radio waves. Nature (Lond) 139: 838–839
Bailey VA, Martyn DF (1934) The influence of electric waves on the ionosphere. Phil Mag 18: 369–386
Bailey VA, Smith RA, Landecker K, Higgs AJ, Hibberd FH (1952) Resonance in gyro-interaction of radio waves. Nature (Lond) 169: 911–913
Baker KB, Greenwald RA, Walker ADM (1986) A case study of plasma processes in the dayside cleft. J Geophys Res 91: 3130–3141
Baker KB, Greenwald RA, Ruohoniemi JM (1989) PACE: polar Anglo-American conjugate experiment. EOS, 22: 785, 799
Banks PM, Kockarts G (1973) Aeronomy, Parts A and B. Academic Press, Lond New York, A: 430, B: 355
Barke W, Willing W, Geisweid KH (1974) The southern boundary region of the winter anomaly in ionospheric absorption in winter 1971/72 observed on board a ship between 10° and 55° N. Mitt Max-Planck-Inst Aeronomie 51: 1
Barkhausen H (1919) Zwei mit Hilfe der neuen Verstärker entdeckte Erscheinungen. Phys Z 20: 401–403
Barkhausen H (1930) Whistling tones from the earth. Proc IRE 18(7): 1155–1159
Barnes C (1965) Xelodop antenna pattern measuring equipment, 2 to 50 MHz. SRI Rep Stanford Res Inst, Menlo Park, CA
Barnum JR (1986) Ship detection with high-resolution HF skywave radar. IEEE, J Oceanic Eng OE-11: 196
Baron MJ (1984) The EISCAT facility. J Atmos Terr Phys 46: 469–472
Baron MJ (1986) EISCAT progress 1983–1985. J Atmos Terr Phys Spec Issue 48: 9–10
Baron MJ, Watt T, Bates HF (1970) Project 617 radar readiness achievement program. Part B: Ionospheric measurements. SRI Rep DASA 2519-2
Baron MJ, de la Beaujardiere O, Craig B (1970) Project 617 Radar readiness achievement program. Part A: Data processing and analysis. Final Rep SRI Proj 6291, SRI Internat, Menlo Park, CA
Barr R, Rietveld MT, Stubbe P, Kopka H (1988) Ionospheric heater beam scanning: A mobile source of ELF/VLF radiation. Rad Sci 22: 1073–1083
Barrington RE, Thrane E (1962) The determination of D-region electron densities from observations of cross modulation. J Atmos Terr Phys 24: 31–42
Bartholomew RR (1966) Results of a high-latitude HF backscatter study. Sci Rep No 2, SRI Proj 5538, Stanford Res Inst, Menlo Park, CA
Basler RP (1963) Radio wave absorption in the auroral ionosphere. J Geophys Res 68: 4665
Basler RP, Bentley PB, Price GH, Tsunoda RT, Wong TL (1988a) Ionospheric distortion of HF signals. Radio Sci 23: 569–579
Basler RP, Price GH, Tsunoda RT, Wong TL (1988b) Ionospheric distortion of HF radio signals. Rad Sci 23: 569–579
Basu Su, Basu Sa, Stubbe P, Kopka H, Waaramaa J (1987) Daytime scintillations induced by high power HF waves at Tromsø, Norway, JGR 92: 11, 149–11, 157
Basu Su, Basu Sa, Weber EJ, Coley WR (1988) Case study of polar cap scintillation modeling using DE2 irregularity measurements at 800 km. Rad Sci 23: 545–553
Bates HF (1959) The height of F-layer irregularities in the arctic ionosphere. J Geophys Res 64: 1257–1265
Bates HF (1961a) An HF sweep-frequency study of the arctic ionosphere. Rep No 1, UAF-R115 Geophys Inst, Univ Alaska, College, AK
Bates HF (1961b) An HF sweep frequency study of the arctic ionosphere. Diss Univ Alaska Fairbanks, Sci Rep No 1, April 1961, UAF-R115, Geophys Inst, Univ Alaska, College, AK
Bates HF (1969) HF propagation characteristics through auroral ionosphere. Final Rep on SRI Project 7539. SRI, Menlo Park, CA

Bates HF, Albee PR (1970) Aspect sensitivity of F-layer backscatter echoes. J Geophys Res 75: 165
Bates HF Baron M (1969) F-region temperatures over Palo Alto in 1967 and 1968. J Geophys Res 74: 6497
Bates HF, Hunsucker RD (1974) Quiet and disturbed electron density profiles in the auroral zone ionosphere. Radio Sci 9: 455–467
Bates HF, Hunsucker RD, Owren L (1961) Experimental proof of focusing at the skip distance by backscatter records. Proc IRE 49: 369
Bates HF, Belon AE, Romick GJ, Stringer WJ (1965) On the continuous observation of the auroral belt by radio means. Nature (Lond) 207: 1081–1082
Bates HF, Albee PR, Hunsucker RD (1966) On the relationship of the aurora to non-great-circle high frequency propagation. J Geophys Res 71: 1413–1420
Bauer P, Waldteufel P, Vialle C (1974) The French quadristatic incoherent scatter facility. Rad Sci 9: 77–83
Bauer SJ (1973) Physics of planetary ionospheres, Vol 6. Phys Chem Space, Springer, Berlin Heidelberg New York, p 230
Bauer SS, Daniels BF (1959) Measurements of ionospheric electron content by the lunar radio technique. JGR 64: 1371–1376
Bauer SJ, Nagy AF (1975) Ionospheric direct measurement techniques. Proc IEEE 63: 230–249
Beaujardiere O de la, Wickwar VG, Caudal F (1985) Universal time dependence of night time F-region densities at high latitudes. J Geophys Res 90: 4319–4332
Beckmann P (1968) The depolarization of electromagnetic waves. Golem Press, Boulder Co
Beckmann P, Spizzichino A (1963) The scattering of electromagnetic waves from rough surfaces. Pergamon/Macmillan, London
Bell Telephone Laboratories (1982) Transmission systems for communications, 5th edn.
Belrose JS (1970) Radio wave probing of the ionosphere by the partial reflection of radio waves (from heights below 100 km). J Atmos Terr Phys 32: 567–596
Belrose JS, Burke MJ (1964) Study of the lower ionosphere using partial reflection. J Geophys Res 69: 2799–2818
Belrose JS, Hewitt LW, Bunker R (1968) The partial reflection experiment as a tool for synoptic D-region research; and results of recent experiments. Conf Digest, 3rd Conf Meteorol Chem Fact D-region Aeronomy-Univ Illinois
Benson RF (1962) Cross-modulation of cosmic noise. J Geophys Res 67: 2569–2572
Benson RF (1963) The cross modulation of cosmic noise: a technique for investigating the disturbed D-region. Rep UAF-R137, Geophys Inst Univ Alaska, Fairbanks, Alaska
Benson RF (1977) Stimulated plasma waves in the ionosphere. Rad Sci 12: 861–878
Benson RF (1981) Auroral kilometric radiation source region observations from ISIS 1. In: Kan JR, Akasofu SI (eds) Phys of auroral arc formation. Geophys Monogr Ser 25: 369–379
Benson RF (1982) Stimulated plasma instability and nonlinear phenomena in the ionosphere. Rad Sci 17: 1637–1659
Benson RF (1985) Auroral kilometric radiation: wave modes, harmonics and source region electron density structures. J Geophys Res 90: 2753–2784
Benson RF, Bitoun J (1979) Interpretation of satellite gyroharmonic resonance observations. Radio Sci 14: 113–123
Benson RF, Calvert W (1979) ISIS 1 observations at the source of Auroral Kilometric Radiation. Geophys Res Lett 6: 479–482
Benson RF Viñas F (1988) Plasma instabilities stimulated by HF transmitters in space. Radio Sci 23: 585–590
Benson RF, Wong HK (1987) Low-Altitude ISIS-1 observations of auroral radio emissions and their significance to the cyclotron maser instability. J Geophys Res 92: 1218–1230
Benson RF, Desch MD, Hunsucker RD, Romick GJ (1988) Ground-level detection of low- and medium-frequency auroral radio emission. J Geophys Res 93: 277–283, 2767
Berkey FT, Parthasarathy R (1964) An investigation of selected types of radiowave absorption events in the auroral zone. Univ Alaska Rep UAG-R151
Berkner LV, Wells HW (1934) F-region ionospheric investigations at low latitude. Terr Magn Atmos Elec 39: 215
Berkner LV, Wells HW (1937) Abnormal ionization of the E-region of the ionospheric. Terr Magn Atmos Elec 42: 73

Bernhardt PA, Duncan LM, Tepley CA (1988) Artificial airglow excited by high-power radio waves. Science 242: 1022–1027
Bertel L, Bertin F, Testud J (1976) De la mesure du contenu électronique intègre appliquée a l'observation des ondes de gravité de moyenne écelle. JATP 38: 262–270
Bertin F, Papet-Levine J (1970) Latitudinal variation of total electron content in the winter at middle latitudes. Rad Sci 5: 899–906
Best JE, Ratcliffe JA, Wilkes MV (1936) Experimental investigations of very long waves reflected from the ionosphere. Proc R Soc A 156: 614
Beynon WJG, Brown GM (1956) The ionosphere, annals of the international Geophysical Year, Vol III. IGY Instruction Manual, Pergamon Press, New York
Bhonsle RV, da Rosa AV, Garriott OK (1965) Measurements of the total electron content and the equivalent slab thickness of the mid-latitude ionosphere. Radio Sci 690: 939–937
Bibl K, Patenaude JA, Reinisch BW (1970) Digital integrating goniometric ionospheric sounder-Digisonde 128. Fin Rep AFCRL-71-0002, Air Force Geophys Lab, Hanscom Air Force Base, MA
Bibl K, Reinisch BW (1978) The universal digital ionosonde. Rad Sci 13:
Bibl K, Reinisch BW, Kitrosser DF (1981) Digisonde 256 – general description of the compact digital ionospheric sounder. Rep Univ Lowell/Cent Atmos Res 450 Aiken St Lowell, MA 01854
Blair JC, Melanson LL, Tveten LH (1969) HF ionospheric radar ground-scatter map showing land-sea boundaries by a spectral-separation technique. Electronics Lett 5: 75–76
Blum EJ, Denisse JF, Steinberg JL (1954) Influence de l'ionosphere sur la réception du rayonnement galactique de frequency 29.5 Mc/s. C R Acad Sci 238: 1095
Boithias L (1987) Radio wave propagation. McGraw-Hill, New York. Original French language edition, Propagation des ondes radio electriques: Dans l'environment terrestre. BORDAS and CNET, Paris 1984
Booker HG (1956a) A theory of scattering by nonisotropic irregularities with application to radar reflections from the aurora. J Atmos Terr Phys 8: 204–221
Booker HG (1956b) Turbulence in the ionosphere with applications to meteor trails, radio-star scintillation, auroral radar echoes and other phenomena. J Geophys Res 61: 673–705
Booker, HG (1979) The role of acoustic gravity waves in the generation of spread F and ionospheric scintillation. J Atmos Terr Phys 41: 501–515
Booker HG (1960) Radar studies of the aurora. Physics of the upper atmosphere. (ed JA Ratcliffe) Academic Press, New York
Booker HG (1984) Cold plasma waves. Martinus Nijhoff, the Hague
Booker HG, Cohen R (1957) A theory of long-duration meteor echoes based on atmospheric turbulence with experimental confirmation. J Atmos Terr Phys (Spec Suppl) 20: 171–194
Booker HG, Gordon WE (1950) A theory of radio scattering in the troposphere. Proc Inst Radio Engr 38: 401–402
Born M, Wolf E (1970) Principles of optics 4th edn. Pergamon Press, New York
Bostrom R (1973) Electrodynamics of the ionosphere. In: Egeland A, Holter O, Omholt A (eds) Cosmical geophysics. Universitetsforlaget Oslo, Norway, Ch 12: 181
Bourdillon A, Gauthier F, Parent J (1987) Use of maximum entropy spectral analysis to improve ship detection by over-the-horizon radar. Rad Sci 22: 313–320
Boutacoff DA (1985) Backscatter radar extends early warning times, Defense. Electronics 17: 71–83
Bouvet J, Lafeuille M (1982) Bibliographie du sondeur français à diffusion incohérente. CRPR Note Tech CRPE/124, PER/EMI 38–40 rue du General Leclerc, 92131 Issy Les Moulineaux, France
Bowhill SA (1955) The fading of radio waves of frequencies between 16 and 2400 kc/s. J Atmos Terr Phys 8: 129–145
Bowhill SA (1958) The Faraday-rotation rate of a satellite radio signal. JATP 13: 175–176
Bowles KL (1958) Observations of vertical incidence scatter from the ionosphere at 41 Mc/sec. Phys Rev Lett 1: 454
Bowles KL (1961) Incoherent scatter by free electrons as a technique for studying the ionosphere and exosphere: Some observations and theoretical considerations. J Res NBS, 65D: 1–14
Bowles KL (1963) Measuring plasma densities in the magnetosphere. Science 139: 389–391
Bowman GG, Hainsworth DW, Dunne GS (1986) A research-oriented ionosonde with directional capabilities. Rad Sci 21: 21–297
Bradley PA (1976) A new computer-based method of HF sky-wave signal prediction using vertical-incidence ionosonde measurements. Radio Systems and the ionosphere, AGARD Conf Proc No 173

Breit G, Tuve MA (1926) A test of the existence of the conducting layer. Phys Rev 28: 554
Brekke A (ed) (1977) Radar probing of the auroral plasma. Proc EISCAT summer school, Tromsø, Norway, Universitesforlaget, Tromso, Oslo, Bergen
Briggs BH (1968a) On the analysis of moving patterns in geophysics I. Correlation analysis, J Atmos Terr Phys 30: 1777–1788
Briggs BH (1968b) On the analysis of moving patterns in geophysics. II Dispersion analysis. J Atoms Terr Phys 30: 1789–1794
Briggs BH (1980) Radar observations of atmospheric winds and turbulences: a comparison of techniques. J Atmos Terr Phys 42: 823–833
Briggs BH and Golley MG (1968) A test for dispersion in F-region drifts observed by the radio star scintillation method. J Atmos Terr Phys 30: 963–973
Briggs BH, Parkin IA (1963) On the variation of radio star and satellite scintillations with zenith angle. J Atmos Terr Phys 25: 339–365
Briggs BH, Phillips CJ, Shinn DH (1950) The analysis of observation on spaced receivers of the fading of radio signals. Proc Phys Soc B63: 106–121
Brillouin L (1960) Wave propagation and group velocity. Academic Press, Lond New York
Brookner (1987) Array radars: an update. Microwave J Feb 117–137
Browne IC, Evans JV, Hargreaves JK, Murray WAS (1956) Radio echoes from the moon. Proc Phys Soc B69: 901–920
Buchau J, Reinisch BW, Anderson DN, Weber EJ, Dozois C (1988) Polar cap plasma convection measurements and their relevance to the modeling of the high-latitude ionospheric. Radio Sci 23: 521
Budden KG (1961) Radio waves in the ionosphere. Cambridge Univ Press, Cambridge, UK
Budden KG (1985) The propagation of radio waves – the theory of radio waves of low power in the ionosphere and magnetosphere. Cambridge Univ Press, Cambridge, UK
Buneman O (1962) Scattering of radiation by the fluctuations in a non-equilibrium plasma. J Geophys Res 67: 2050–2053
Buneman O (1953) Excitation of field aligned sound waves by electron streams. Phys Rev Lett 10: 285–287
Burke MJ, Hara EH (1963) Tables of the semiconductor integrals Cp(x) and their approximation for use with the generalized Appleton-Hartree magnetoionic formulas. Rep No 113, DRTE Radio Phys Lab Ottawa, Canada
Burton ET, Boardman EM (1933a) Effects of solar eclipse on audio frequency atmospherics. Nature (Lond) 131: 81–82
Burton ET, Boardman EM (1983b) Audio frequency atmospherics. Proc IRE 21 (10): 1476–1494
Butt AG (1933) World Radio, April 28
Calvert W (1966) Oblique-z-mode echoes in the topside ionosphere. J Geophys Res 71: 5579–5583
Calvert W, McAfee JR (1969) Topside sounder resonances. Proc IEEE 57: 1089–1096
Carlson HC, Duncan LM (1977) HF excited instabilities in space plasmas. Radio Sci 12: 1001
Carlson HC, Sayers J (1970) Discrepancy in electron temperatures deduced from Langmuir probes and from incoherent scatter radars. J Geophys Res 75: 4883–4886
Carlson HC, Gordon WE, Showen RL (1972) High frequency induced enhancements of the incoherent scatter spectrum at Arecibo. J Geophys Res 77: 1242–1250
Carpenter DL (1988) Remote sensing of the magnetospheric plasma by means of whistler mode signals. Rev Geophys 26: 535–549
CCIR Report 413 (1966) Operating noise-threshold of a radio receiving system. Internat Telecomm Union, Geneva, Switzerland
CCIR Report 258 (1978) Man-made radio noise. Study Group 6, Internat Radio Consultative Comm Geneva, Switzerland
CCIR Report 322-3 (1985) World distribution and characteristics of atmospheric radio noise. In Telecomm Union, Geneva, Switzerland
Chamberlain JW (1961) Physics of the aurora and airglow. Academic Press, Lond New York
Chamberlain JW (1978) Theory of planetary atmospheres – an introduction to their physics and chemistry. Academic Press, Lond New York, p 330
Chan KL, Villard OG (1963) Sudden frequency deviations induced by solar flares. J Geophys Re 68: 3197–3224
Chapman JH, Warren ES (1968) Topside sounding of the earth's ionosphere. Space Sci Rev 8: 846–865

Clark C, Peterson AM (1956) Motion of sporadic-E patches determined from high-frequency backscatter records. Nature (Lond) 178: 486–487
Clarke RH (1971) Preliminary observations with a directional ionosonde. Planet Space Sci Rev 19: 1387–1390
Cole AR, Ferraro AJ, Lee HS (1969) Comparison of two ground-based D-region experiments. Nature (Lond) 222: 761–763
Cole KD, Liu C-H (1987) The worldwide ionosphere thermosphere study. EOS 68: 1594–1596
Collin RE (1985) Antennas and radiowave propagation. McGraw-Hill, New York
Collins C, Forsyth PA (1959) A bistatic radio investigation of auroral ionization. JATP 13: 315–345
Cones HN, Cottony HV, Watts JM (1950) A 600-ohm multiple wire delta antenna for ionospheric studies. NBS Res Pap RP2094, 44: 475–488
Conkright RO, Brophy HI (1982) Catalog of ionosphere vertical soundings data. Rep UAG-85, WDC-A, NOAA/EDS, US Dept Comm, Boulder, CO
Conkright R, Ertle M, Hapgood M, Feldstein A, Matuura N (1984) Combined catalog of ionospheric vertical sounding data. World Data Center A for Solar – terrestrial physics, NOAA/EDS, 325 Broadway Boulder. Colorado
Costa E, Fougere PF (1988) Cross-spectral analysis of spaced-receiver measurements. Rad Sci 23: 129–139
Costa E, Fougere PF, Basu S (1988) Cross-correlation analysis and interpretation of spaced-receiver measurements. Rad Sci 23: 141–162
Crary JH, Helliwell RA, Chase RF (1956) Stanford-Seattle whistler observations. J Geophys Res 61: 35–44
Croft TA (1967) Computations of HF ground backscatter amplitude. Rad Sci 2: 739–746
Croft TA (1968) The influence of ionospheric irregularities on sweep-frequency backscatter. J Atmos Terr Phys 30: 1051–1063
Croft TA (1972) Skywave backscatter: a means for observing our environment at great distance. Rev Geophys Space Phys 10: 73–155
Cutolo M (1951) Self-interaction of radio waves in the ionosphere. Nature (Lond) 167: 314–315
Davis JR, Headrick WC, Ahearn JL (1964) A HF backscatter study of solar eclipse effects upon the ionosphere. J Geophys Res 69: 190–193
Davies K (1962) The measurement of ionospheric drifts by means of a Doppler shift technique. J Geophys Res 67: 4909–4913
Davies K (1963) Doppler studies of the ionospheric effects of solar flares. Proc Internatl Conf Ionosphere 76, Inst Phys Soc Lond
Davies K (1965) Ionospheric radio propagation. Natl Bur Stand Monogr 80, Washington DC
Davies K (1969) Ionosphere radio waves. Blaisdell, Watham, MA
Davies K (1980) Recent studies in satellite radio beacon studies with particular emphasis on the ATS-6 radio beacon experiment. Space Sci Rev 25: 357–430
Davies K (1990) Ionospheric radio. Peregrinus, London
Davies K, Baker DM (1966) On frequency variations of ionospherically propagated HF radio signals. Rad Sci 1: 545–556
Davies K, Hartmann GK (1976) Short period fluctuations in total columnar electron content. J Geophys Res 81: 3431–3434
Davies K, Fritz RB, Grubb RN, Jones JE (1975) Some early results from the ATS-6 Radio Beacon Experiment. Rad Sci 10: 785–799
De La Beaujardiere O, Wickwar V, Leger D, McCready M, Baron M (1984) The software system for the Chatanika incoherent-scatter system. SRI-I Tech Rep Proj 8358: 21 SRI-I 333 Ravenswood Dr Menlo Park California 94025
Detrick DL (1984) A multiple-beam phased array antenna for riometer studies of auroral absorption. Univ Maryland Inst Phys Sci Tech Tech Note Sn-1032
Detrick DL, Rosenberg TJ (1988) IRIS: An imaging riometer for ionospheric studies. Antarctic J US XXIII:5
Devlin T, Hargreaves JK, Collis PN (1986) EISCAT observations of the ionospheric D-region during auroral radio absorption events. J Atmos Terr Phys 48: 795–805
Dieminger W (1951) The scattering of radio waves. Proc Phys Soc B64: 142–158
Dieminger W (1974) Early ionospheric research in Germany. J Atmos Terr Phys 36: 2085–2093
Dieminger W (1975) 20 years of cooperation in ionospheric research with Finland. Veroff Geophys Obs Finn Acad Wiss 56/3: 61–72

Dieminger W, Möller HG (1956) Echo sounding experiments with variable frequency at oblique incidence. Nuovo Cimento (Suppl) 4X: 1532–1545

Dieminger W, Geisweid K-H, Möller HG (1955) Echolotungen der Ionosphäre mit veränderlicher Frequenz bei schrägem Einfall. NTZ 11: 578–581

Djuth FT (1984) HF enhanced plasma lines in the lower ionosphere. Rad Sci 19: 383–394

Djuth FT, Gonzales CA, Ierkic HM (1986) Temporal evolution of the HF enhanced plasma line in the Arecibo F region. J Geophys Res 91: 12089–12107

Donnelly RF (ed) (1979) Solar-terrestrial predictions proceedings, Vol II. Superintendent Doc, US Gov Printing Office, Washington DC

Dougherty JP, Farley DT (1960) A theory of incoherent scatter of radio waves by a plasma. Proc R Soc Lond A 259: 79

Dougherty JP, Watson SR (1971) The interpretation of plasma resonances observed by ionospheric topside sounders. In: Simon H, Thompson WB (eds) Advance plasma physics, Vol 4. Wiley, New York

Drevin GR, Stoker PH (1990) Riometer quiet day curves determined by the maximum density method. Rad. Sci 25: 1159–1166

Dudeney JR, Jarvis MJ (1986) A simple graphical method for dealiasing digital ionosonde echo location data. Radio Sci 21: 101–105

Dueño B (1962) Sporadic-E as observed from Mayaguez, Puerto Rico, by backscatter sounder. In: Smith ED, Matsushita S (eds) Ionospheric sporadic-E. Pergamon Press, New York London, pp 110–122

Earl GF, Ward BD (1987) The frequency management system of the Jindalee over-the-horizon backscatter HF radar. Rad Sci 22: 275–291

Eccles WH (1912) Proc R Soc A87:79

Eckersley TL (1925) Note on musical atmospheric disturbances. Phil Mag 49: (5) 1250–59

Eckersley TL (1928) Letter to the editor. Nature (Lond) 122: (3081) 768–768

Eckersley TL (1929) An investigation of short waves. J Inst Elec Engrs 67: 992–1032

Eckersley TL (1931) 1929–1930 developments in the study of radio wave propagation. Marconi Rev 5 (31): 1–8

Eckersley TL (1932) Studies in radio transmission. J Inst Elec Engrs 71: 434–443

Eckersley TL (1935) Musical atmospherics. Nature (Lond) 135: 104–105

Eckersley TL (1937) Irregular ionic clouds in the E-layer of the ionosphere. Nature (Lond) 140: 846

Eckersley TL (1939) Scattering of wireless waves in the ionosphere. Nature (Lond) 143: 33–34

Eckersley TL (1940) Analysis of the effect of scattering in radio transmission. J Elec Engr 86: 548–563

Edwards CF, Jansky KG (1941) Measurements of the delay and direction of arrival of echoes from nearby short wave transmitters. Proc IRE 29: 322–329

Egan RD, Peterson AM (1962) Backscatter observations of sporadic-E. In: Smith EK, Matsushita J (eds) Ionospheric Sporadic-E. Pergamon Press, New York London, pp 89–109

Elford WG (1959) A study of winds between 80 and 100 km in medium latitudes. Planet Space Sci 1: 94–101

Erickson WC, Mahoney MJ, Jacobson AR, Knowles SH (1988) Ionospheric wave and irregularity measurements using passive radio astronomy. Radio Sci 23: 273–282

Essex EA (1977) Equinoctial variations in the total electron content of the ionosphere at northern and southern hemisphere stations. J Atmos Terr Phys 39: 645–656

Etcheto J, Bloch, JJ (1978) Plasma density measurements from the GEOS-1 relaxation sounder. Space Sci Rev 22: 597–610

Evans JV (1956) The measurement of electron content of the ionosphere by the lunar radar method. Proc Phys Soc Lond B69: 953–955

Evans JV (1957) The electron content of the ionosphere. J Atmos Terr Phys 11: 259–271

Evans JV (1969) Theory and practice of ionospheric study by Thomson scatter radar. Proc IEEE 57: 496

Evans JV (1972) Ionospheric movements measured by incoherent scatter: a review. J Atmos Terr Phys 34: 175

Evans JV (1977) Satellite beacon contributions to studies of the structure of the ionosphere. Rev Geophys Space Phys 15: 325–350

Evans JV, Lowenthal M (1964) Ionospheric backscatter observations. Planet Space Sci 12: 915–944

Evans JV, Wand RH (1983) Traveling ionospheric disturbances detected by UHF angle-of-arrival measurements. J Atmos Terr Phys 45: 255–265

Evans JV, Oliver WL, Salah TE (1979) Thermospheric properties as deduced from incoherent scatter measurements. J Atmos Terr Phys 41: 259–278
Faraday M (1845) Selected correspondence of Michael Faraday. 1812–1848, 3–538
Farley DT (1963) A plasma instability resulting in field-aligned irregularities in the ionosphere. J Geophys Res 63: 6083–6097
Farley DT (1971) Radio wave scattering from the ionosphere. In: Loveberg RH, Grien HR (eds) Methods of experimental physics. Academic Press, Lond New York, 9B: 139–186
Farley DT, Ierkic HM, Fejer BG (1981) Radar interferometry: A new technique for studying plasma turbulence in the ionosphere. J Geophys Res 86: 1467–1472
Fedor LS (1967) A statistical approach to the determination of three-dimensional ionospheric drifts. J Geophys Res 72: 5401–5415
Fejer BG, Kelley MC (1980) Ionospheric irregularities. Rev Geophys Space Phys 18: 401–454
Fejer BG, Kelley MC, Senior C, et al. (1990) Low- and mid-latitude ionosphere electric fields during the January 1984 GISMOS Campaign. J Geophys Res 95: 2367–2377
Fejer JA (1955) The interaction of pulsed radio waves in the ionosphere. J Atmos Terr Phys 7: 322–332
Fejer JA (1960) Scattering of radiowaves by an ionized gas in thermal equilibrium. J Geophys Res 65:2635
Fejer JA (1979) Ionospheric modification and parametric instabilities. Rev Geophys Space Sci 17: 135
Fejer JA, Gonzales CA, Ierkic HM et al. (1985) Ionospheric modification experiments with the Arecibo heating facility. J Atmos Terr phys 47: 1165
Felgate DG (1970) On the point source effect in the measurement of ionospheric drifts. J Atmos Terr Phys 32: 241–245
Ferraro AJ, Lee HS (1966) Electron density and collision frequency measurements of the D-region with radio wave phase and amplitude interaction. In: John Frihagen (ed) Electron density profiles in ionosphere and exosphere. Proc NATO Advan Study Inst Finse, Norway, pp 69–82
Ferraro AJ, Lee HS (1968) Capability of high-power wave interaction facility. J Geophys Res 73: 4427–4429
Ferraro AJ, Lee HS, Weisbrod S (1963) Phase interaction: A new tool for D-region studies. J Geophys Res 68: 1169–1171
Ferraro AJ, Lee HS, Allshouse K, Carroll K, Lunnen R (1984) Characteristics of ionospheric ELF radiation generated by HF heating. J Atmos Terr Phys 46: 855–865
Fialer PA (1974) Field-aligned scattering from a heated region of the ionosphere – observations at HF and VHF. Rad Sci 9: 923–940
Flock WA (1979) Electromagnetics and the environment. Remote sensing and telecommunications. Prentice-Hall, Englewood, NJ
Flock WF (1987) Propagation effects on satellite systems at frequencies below 10 GHz – a handbook for satellite systems design. NASA Ref Pub 1108
Flock WF, Slohin SD, Smith EK (1982) Propagation effects on radio range and noise in earth-space telecommunications. Radio Sci 17: 1141–1144
Flock WL (1962) Environmental studies for radar operations in the auroral zone, Part 1. Univ Alaska Geophys Rep UAG-R130
Flock WL, Hunsucker RD (1968) The auroral zone as a source of traveling ionospheric disturbances. In: Georges TM (ed) Symposium proceedings. Acoustic gravity waves in the atmosphere. Boulder, Colorado, 15–17 July 1968, US Govt Printing Office Washington
Flood WA (1968) Revised theory for partial reflection D-region measurements. J Geophys Res 73: 5585
Flood WA (1969) Reply (from discussion of previous paper Holt [1969]). J Geophys Res 74: 5183–5186
Florida CD (1969) The development of a series of ionospheric satellites. Proc IEEE 57: 867–875
Folkestad K, Hagfors T, Westerlund S (1983) EISCAT: an updated description of technical characteristics and operational capabilities. Rad Sci 18: 867–880
Foppiano AJ, Bradley PA (1985) Morphology of background auroral absorption. J Atmos Terr Phys 47: 663–674
Franklin CA, Maclean MA (1969) The design of swept-frequency topside sounders. Proc IEEE 57: 897–929

Fremouw EJ (1966) Aberrations of VHF/UHF signals traversing the auroral ionosphere. Final Rep Contract NASA-3940, Geophys Inst Univ Alaska College, AK

Fremouw EJ (1985) Recent HILAT results. AGARD Conf Proc CP-382: 2.1–1 through 2.1–14

Fremouw EJ, Bates HF (1971) Worldwide behavior of average VHF-UHF scintillation. Rad Sci 6: 863–869

Fremouw EJ, Rino CL (1971) Development of a worldwide model for F-layer-produced scintillation. Final Rep Contract NASA-21551 (SRI Proj 1079)

Fremouw EJ, Secan JA (1984) Modeling and scientific application of scintillation results. Rad Sci 19: 687–694

Fremouw EJ, Leadabrand RL, Livingston RC et al. (1978) Early results from the DNA wideband satellite experiment – complex signal scintillation. Rad Sci 13: 167–187

Fremouw EJ, Livingston RC, Miller DA (1980) On the statistics of scintillating signals. J Atmos Terr Phys 42: 717–731

Fremouw EJ, Secan JA, Lansinger JM (1991) On the feasibility of ionospheric tomography. Proc of the First Annual Conference Prediction and Forecasting of radio propagation at high latitudes for C^3 systems. Monterey CA, 12–14 Feb 1991

Fritz RB (1976) ATS6 radio beacon electron content measurements at Boulder. July 1974–May 1975, Rep UAG-58, WDC-A for solar-terrestrial phys, Boulder, CO

From WR, Whitehead JD (1984) The calibration of an HF radar used for ionospheric research. Rad Sci 19: 423–428

Fuchs J (1938) Discussion. Report Nat Acad Sci Natl Res Coun Publ 581: 105

Fukao S, Sato T, Tsuda T, Kato S, Wagasuki K, Makihara T (1985) The MU radar with an active phased array system: 1. Antenna and power amplifiers, 2. In-house equipment. Rad Sci 20: 1155–1618, 1169–1176

Gallet R (1964) Whistlers, geophysics. In: DeWitt C, Hieblot J, Lebeau A (eds) The earth's environment. Gordon and Breach, New York

Gardiner GW, Lane JA, Rishbeth H (1982) Radio and space research at Slough 1920–1981. Radio Electr Engr 52: 111

Gardner FF, Pawsey JL (1953) Study of the ionospheric D-region using partial reflections. J Atmos Terr Phys 3: 321

Garriott OK, Little CG (1960) The use of geostationary satellites for the study of ionospheric electron content and ionospheric radio-wave propagation. Planet Space Sci 13: 829–838

Garriott OK, Smith FL, Yuen PC (1965) Observations of ionospheric electron content using a geostationary satellite. Planet Space Sci 13: 829–838

Garriott OK, da Rosa AV, Ross WJ (1970) Electron content obtained from Faraday rotation and phase path length variations. J Atmos Terr Phys 32: 705–728

Gates HP (1949) Observations of long distance pulse propagation. Interim Rep US Navy Electr Lab Rep 116

Geisweid KH, Little CG, Oberländer K, Schwentek H, Umlauft G (1968) Ionosphere-absorption measurements. Ann IQSY Vol 1 Geophys Measurements Tech, Observational Schedules and Treatment of Data MIT Press

Gekker IR (1982) The interaction of strong electromagnetic fields with plasmas. Clarendeon Press, Oxford

George PL (1971) The global morphology of the quantity $\int Nv \cdot dh$ an in the D- and E-regions of the ionosphere. J Atmos Terr Phys 33: 1893–1906

Georges TM (1967) Evidence for the influence of atmospheric waves on ionospheric motions. J Geophys Res 72: 422–425

Georges TM (1968) Ionospheric effects of atmospheric waves. ESSA Tech Rep IER 57-ITSA 54, US Gov Print Office, Washington DC

Georges TM, Stephenson JJ (1969) HF radar signatures of traveling ionospheric irregularities, 3-D ray-tracing simulation. Rad Sci 4: 679–696

Gilbert GN (1977) Growth and decline of a scientific specialty: the case of radar meteor research E $\oplus$ S Trans Am Geophys Univ 58: 273–277

Gilliland CR (1965) Sweep-frequency backscatter with calibrated amplitude. Rep SU-SE65-095, Tech Rep 111, Electr Labs, Stanford University

Gillmor CS (1976) The history of the term 'ionosphere'. Nature (Lond) 262 (29 July 1976): 347–48

Gillmor CS (1982) Wilhelm Altar, Edward Appleton, and the magnetoionic theory. Proc Am Philos Soc 126 (5): 395–440

Gillmor, CS (1984) The history of geophysics, Vol 1. Am Geophys Union, 149 pp

Ginzburg VL (1957) Nonlinear interaction of radio waves propagating in plasma. Zh Exsp Teor Fiz 35: 1573, also in Sov Phys JETP 8: 1100

Giraud A, Petit M (1978) Ionospheric techniques and phenomena. Reidel, Dardrechyt, Holland

Gladden SC (1959) A history of vertical-incidence ionosphere sounding at the National Bureau of Standards. NBS Tech Note No 28, US Dep Comm.

Gockel H (1948) Mesures d'absorption des ondes courtes par la couche de l'ionosphere. Ann Geophys 4: 232

Gonzales CA, Behnke RA, Woodman RF (1982) Doppler measurement with a digital ionosonde: technique and comparison of results with incoherent scatter data. Radio Sci 17: 1327–1333

Gordon WE (1958) Incoherent scattering of radio waves by free electrons with applications to space exploration by radar. Proc IRE 46: 1824

Gordon WE (1964) Arecibo ionospheric observatory. Science 146: 26–30

Gordon WE, Carlson HC (1974) Arecibo heating experiments. Rad Sci 9: 1041–1047

Gordon WE, Duncan L (1980) Review of radio-frequency, non-linear effects on the ionosphere. Proc Suzdal Conf

Gordon WE, Carlson HC, Showen RL (1971) Ionospheric heating at Arecibo: first tests. J Geophys Res 76: 7808–7813

Gray AM, Ecklund WL (1974) The Anchorage, Alaska, real-time auroral radar monitor: system description and some preliminary analyses. NOAA Tech Rep ERL 306–AL 9, Boulder, CO

Greenhow JS, Neufeld EL (1961) Winds in the upper atmosphere. Q J R Meteorol Soc 87: 472–489

Greenwald RA, Hunsucker RD (1983) Current and future programs for high-latitude radio wave research. Radio Sci 18: 1189–1194

Greenwald RA, Weiss W, Nielson E, Thomson NR (1978) Stare: a new radar auroral backscatter experiment in northern Scandinavia. Rad Sci 13: 1021–1039

Greenwald RA, Baker KB, Hutchins RA, Hanuise C (1985) An HF phased array radar for studying small-scale structure in the high-latitude ionosphere. Rad Sci 20: 63–74

Gregory JG, Manson AH, Ress DT (1970) Radiowave absorption at an auroral latitude. Can J Phys 48: 809

Grubb RN (1972) In: Leitinger R (ed) Future applications of satellite beacon measurements. Univ Graz, Austria pp 103, 115

Grubb RN, Fritz RB, Jones JE (1976) In: Mendillo M (ed) The geophysical use of satellite beacon observations. Boston Univ, p 717

Grubb RN (1979) The NOAA SEL HF radar system (ionospheric sounder). NOAA Tech Memo ERL SEL-55

Gurevich AV (1978) Nonlinear phenomena in the ionosphere, Vol 10. In: Physics and chemistry in space. Springer, Berlin Heidelberg New York

Gurevich AV, Migulin VV (1982) Investigations in the USSR of non-linear phenomena in the ionosphere. J Atmos Terr Phys 44: 1019–1024

Gurnett DA, Brice NM (1966) Ion temperature in the ionosphere obtained from cyclotron damping of proton whistlers. J Geophys Res 71: 3639–3652

Gurnett DA, Shawhan SD, Brice NM, Smith RL (1965) Ion cyclotron whistlers. J Geophys Res 70: 1665–1688

Hagen JB, Behnke RA (1976) Detection of the electron component of the spectrum in incoherent scatter of radio waves by the ionosphere. J Geophys Res 81: 3441–3443

Hagenbuch KM (1973) Theoretical models for electron energy relaxation in the lower ionosphere. J Geophys Res 78: 8345–8350

Hagn G (1978) Calibration of a 9-ft rod antenna for MF and HF EMC measurements. 1978 Int Symp EMC Record, IEEE Catalog No 78-CH-1304-5 EMC

Hagn G (1979) Interim DOD radio noise and electromagnetic interference (EMI) Survey Handbook. Final Rep Contract MDA 904:77-C-0555, SRI Proj 6512

Haines D, Kitrosser DF, Reinisch BW, Gorman FG (1989) A portable ionosonde in support of reliable communication. In: Albrecht HJ, Richter JH (eds) AGARD Conf Proc CP-453

Haldoupis C (1989) A review on radio studies of auroral E-region ionospheric irregularities. AGARD Conf Proc, München, FRG, CP 441: 31-1 to 31-20

Haldoupis C, Sofko GJ, Koehler JA, Prikryl P, McKibben MJ (1988) High resolution CW interferometry for auroral ionospheric studies: preliminary results. J Geophys Res 93: 5777–5786
Hanuise C, Villain JP, Crochet M (1981) Spectral studies of F-region irregularities in the auroral zone. Geophys Res Lett 8: 1083
Harang L, Stoffregen W (1938) Scattered reflections of radio waves from a height of more than 100 km. Nature (Lond) 142: 832
Harang L, Stoffregen W (1961) Investigation of auroral echoes. Planet Space Sci 5: 33–45
Hargreaves JK (1966) The coverage of satellite passes from a ground station: some results for use in planning combined satellite and ground-based studies. Planet Space Sci 14: 617
Hargreaves JK (1969) Auroral absorption of HF radio waves in the ionosphere: a review of results from the first decade of riometry. Proc IEEE 57: 1348–1373
Hargreaves JK (1970).... In: Dieminger W, Hartmann (eds) Future application of satellite beacon experiments. Max-Planck-Inst Aeronomie Lindae, pp 16–1
Hargreaves JK (1978) On the presentation of ATS-6 electron-content data. J Atmos Terr Phys 40: 493–496
Hargreaves JK (1979) The upper atmosphere and solar-terrestrial relations. Van Nostrand Reinhold, New York
Hargreaves JK, Jarvis MJ (1986) The multiple riometer system at Halley, Antarctica. Brit Ant Surv Bull 72: 13
Hargreaves JK, Poulter EM, Hunsucker RD (1980) ATS-6 electron-content measurements from Alaska: recovery from a storm in summer 1979. Proc Satallite Beacon Symp, Warsaw, Poland
Hargreaves JK, Feeney MT, Burns CJ (1985) Statistics of auroral radio absorption in relation to prediction models. AGARD Conf Proc CP-382, 7.3-1 to 7.3-10
Hartsfield WL, Ostrow SM, Silberstein R (1950) Backscatter observations by the CRPL, Aug 1947 to Mar 1948. J Res NBS 44: 199–214
Harvey CC, Etcheto J, DeJavel Y, Manning R, Petit M (1978) The ISEE electron density experiment. IEEE Trans Geosci Electro GE-16: 231–238
Hayden EC (1968) Ground scatter in review. In: Thrane E (ed) Scatter propagation of radio waves. Part 1 – AGARD CP-37
Heaviside O (1902) "Telegraphy", Encyclopedia Brittanica, 10th Edition, 33: 215
Helliwell RA (1956) Whistlers and related ionospheric phenomena. Stanford Univ Press, Stanford, CA
Helliwell RA (1958) Whistlers and ELF emissions. Geophys and the IGY (Geophys Monogr No 2), AGU, Washington DC
Helliwell RA (1959) Whistler paths and electron densities in the outer ionosphere. Proc Symp Phys processes in the Sun-Earth environment, DTRE Rep 1025: 165–175
Helliwell, RA (1965) Whistlers and related ionospheric phenomena. Stanford Univ Press, Stanford, CA
Helliwell RA (1983) Controlled stimulation of VLF emissions from Siple station, Antarctica. Rad Sci 18: 801–814
Helliwell RA (1987) VLF wave injection experiments from Siple station, Antarctica. Paper presented at the XXIInd General Assembly of URSI Tel Aviv, Israel
Helliwell RA (1988) VLF wave stimulation experiments in the magnetosphere for Siple Station, Antarctica. Rev Geophys 26: 551–578
Helliwell RA, Inan US (1982) VLF wave growth and discrete emission triggering in the magnetosphere: A feedback model. J Geophys Res 87: 3537–3550
Helliwell RA, Katsufrakis JP (1978) Controlled wave-particle interaction experiments. In: Lanzerotti LJ, Park CG (eds) Upper atmosphere research in Antarctica. Antarctica Res Ser 29, AGU, Washington DC
Hellrich FV, Ferraro AJ (1965) A mathematical technique for the conversion of radio wave interaction data to D-region electron density contours. J Geophys Res 70: 1195–1205
Hense U (1969) Tages-und Jahresgänge der ionosphärischen Absorption im Kurzwellenbereich. Tech Mitt RFZ (GDR) 13: 20
Herman JR (1979) Electromagnetic ambients and man-made noise Vol 3. Multivolumen EMS Encycl Ser, Don White Consultants, Gainsville, VA
Hertz H (1893) Electric waves. Translated into English by DE Jones, Macmillan, London
Hewish A (1952) The diffraction of galactic radio waves as a method of investigating the irregular structure of the ionosphere. Proc R Soc A214: 494–514

Hewish A (1989) A users' guide to scintillation, J Atmos Terr Phys 51: 743–750
Hey JS, Parsons SJ, Phillips JW (1946) Fluctuations in cosmic radiation at radio frequencies. Nature (Lond) 158: 234–234
Hibberd FH (1955) Ionospheric self-interaction of radio waves. J Atmos Terr Phys 6: 268–279
Hibberd FH (1957) Self-distortion of radio waves in the ionosphere, near the gyro-frequency. J Atmos Terr Phys 11: 102–110
Hines CO (1957) Heavy ion effects in audio-frequency radio propagation. J Atmos Terr Phys 11: 36–42
Hines CO (1960) Internal atmospheric gravity waves at ionospheric heights. Can J Phys 38: 1441–1481
Hines CO, Rao RR (1968) Validity of three-station methods of determining ionospheric motions. J Atmos Terr Phys 30: 979–993
Hoag JB, Andrew VJ (1928) A study of short time multiple echoes. Proc Inst Radio Engrs 16: 1386–1374
Hoegy WR, Benson RF (1988) DE/ISIS conjunction comparisons of high-latitude electron density features. J Geophys Res 93: 5947–5954
Hoffman RA (ed) (1981) Dynamics explorer. Space Sci Instr 5: 345, Reidel, Dordrecht
Hook JL (1968) Morphology of auroral zone radiowave absorption in the Alaska sector. J Atmos Terr Phys 30: 1341
Hook JL (1972) Wind patterns at meteor altitudes (75–105 km) above College, Alaska associated with midwinter stratospheric warmings. J Geophys Res 77: 3856–3868
Huang YA (1978) Lunar variation of total electron content at Lunping. J Atmos Terr Phys 40: 1219–1221
Hultqvist B (1966) Ionospheric absorption of cosmic radio noise. Space Sci Rev 5: 771 Reidel, Dordrecht, Holland
Hultqvist B (1987) The Viking project. Geophys Res Lett 14: 379–382
Hultqvist B, Lundin R (1987) Some Viking results related to dayside magnetosphere–ionosphere interactions. Ann Geophys 5: 503–509
Hunsucker RD (1964a) Auroral absorption effects on a transpolar synchronized step-frequency circuit. Proc IEEE 52: 369
Hunsucker RD (1964b) Auroral absorption effects on an HF arctic propagation path. Radio Sci 68D: 717–721
Hunsucker RD (1969) Remote sensing of the midlatitude ionosphere with a narrow-beam high-frequency radar. PhD Diss, Univ Colorado-Boulder, 154 pp
Hunsucker RD (1970) An atlas of oblique-incidence high-frequency backscatter ionogram of the midlatitude ionosphere. ESSA Tech Rep ERL 162-ITS 104, US Gov Printing Office, Washington DC, 155 pp
Hunsucker RD (1971a) Characteristics signatures of the midlatitude ionosphere observed with a narrow-beam HF backscatter sounder. Rad Sci 6: 535–548
Hunsucker RD (1971b) Analysis of a backscatter signature obtained with a high-resolution HF radar system. Rad Sci 6: 763–768
Hunsucker RD (1974) Simultaneous riometer and incoherent scatter radar observations of the auroral D region. Radio Sci 9: 335–340
Hunsucker RD (1975) Chatanika radar investigation of high latitude E-region ionization structure and dynamics. Radio Sci 10: 277–288
Hunsucker RD (1982) Atmospheric gravity waves generated in the high latitude ionosphere: a review. Rev Geophys Space Phys 20: 293–315
Hunsucker RD, Hargreaves JK (1988) A study of gravity waves in ionospheric electron content at L = 4. J Atmos Terr Phys 50: 167–172
Hunsucker RD, Tveten LH (1967) Large traveling-ionospheric disturbances observed at mid-latitudes utilizing the high-resolution HF backscatter technique. J Atmos Terr Phys 29: 909–916
Hunsucker RD, Bates HF, Belon AE (1972) Observations of simultaneous auroral D and E layers with incoherent-scatter radar. Nature-Phys Sci 239: 102–104
Hunsucker RD, Delana BS, Hargreaves JK (1984) ATS-6 Measurements of ionospheric storm-time behavior of TEC and other parameters at College, AK. Proc Beacon studies of the earth's environment, Radio Sci Div Nat Phys Lab, New Delhi, India
Huxley LGH (1950) Ionospheric cross-modulation at oblique incidence. Proc R. Soc A. Lond 200: 486–511

Huxley LGH (1953) Alternative development of the theory of radio wave interaction. Proc R Soc A, Lond 218: 520–536

Huxley LGH, Ratcliffe JA (1949) A survey of ionospheric cross-modulation. Proc Inst Elec Engrs 96-Pt 433–440

Huxley LGH, Ratcliffe JA (1950) Discussion on "A survey of ionospheric cross-modulation". Proc Inst Elec Engrs 97: 165

IEEE journal on oceanic engineering (1986) Special issue on HF Radar, OE-11, April 1986

IEEE Spectrum (1988) Special issue on acousto optics, August 1988

Igarshi K. Ogawa T, Ose M, Fujii R. Hitasawa T (1982) A new VHF Doppler radar experiment at Syowa station. Antarctica Mem Nat Inst Pol Res 22: 258–267, Science 6: 763–768

Isham B, Kofman W, Hagfors T, Nordling J, Thide B, Lattoz C, Stubbe P (1990) New Phenomena observed by EISCAT during an RF ionospheric modification experiments. Radio Sci 25: 251–262

Jackson JD (1975) Classical electrodynamics. Wiley, New York

Jackson JE (1986) Alouette-ISIS program summary. NSSDC/WDC-A-R&S 86-09, National Space Science Data Center/World Data Center A for rockets and satellites NASA Goddard Space Flight Center, Greenbelt, MD

Jackson JE (1988) Results from Alouette 1, Explorer 20, Alouette 2 and Explorer 31. NSSDC/WDA-A A-R&S 88-10

Jackson JE Kane JA (1959) Measurement of ionospheric electron densities using a RF probe technique. J Geophys Res 64: 8

Jackson JE, Schmerling ER, Whittaker JH (1980) Mini-review on topside sounding. IEEE Trans Ant Prop AP-28: 284–288

Jacobs JA, Watanabe T (1966) Doppler frequency changes in radio wave propagating through a moving ionosphere. Rad Sci 1: 257–264

James HG (1989) ISIS 1 Measurements of high-frequency backscatter inside the ionosphere. J Geophys Res 94: 2617–2629

Jarvis MJ, Dudeney JR (1986) Reduction of ambiguities in HF radar results through a revised receiving antenna array and sounding pattern. Radio Sci 21: 151–158

Jaye WE, Chesnut WG, Craig B (1969) Analysis of auroral data from the Prince Albert Radar Observatory. SRI Final Rep Contr 601932, SRI, Menlo Park, California

Johnson RC, Jasik H (eds) (1984) Antenna engineering handbooks, 2nd edn McGraw-Hill, New York

Johnson RC, Jasik H (1987) Antenna applications reference guide. McGraw-Hill, New York

Johnson K-O, Persson K, Schmidt W, Turunen T (1984) EISCAT experiment preparation manual EISCAT Tech Note 84/41, EISCAT Sci Assn, S-981 27 Kiruna, Sweden

Jones IL (1958) The height variation of drift in the E-region. J Atmos Terr Phys 12: 68–76

Jones TB, Daves K, Wieder B (1972) Observations of D-region modifications at low and very low frequencies, Nature 238: 33–34

Jones TB, Robinson T, Stubbe P, Kopka H (1984) Frequency dependence of anomalous absorption caused by high power radio waves. J Atmos Terr Phys 46: 147–153

Jordon EC (ed) (1985) Reference data for engineers: radio, electronics, computer and communications. 7th edn. Howard W Sams & Co. Inc, Indianapolis, Indiana

Jordan EC, Balmain KG (1968) Electromagnetic waves and radiating systems, 2nd edn. Prentice-Hall, Englewood Cliffs, NJ

Jursa, AS (ed) (1985) Handbook of geophysics and the space environment, Air Force Geophysics Laboratory Air Force Systems Command United States Air Force, Avail. from National Technical Information Service (NTIS) 5285 Port Royal Road Springfield, VA 22161

Kaiser TR (1956) Radio investigations of aurorae and related phenomena. In: Armstrong EB, Dalgarno A. (eds) The Airglow and the Aurorae. Pergamon Press, Lond

Kantor IJ (1974) High frequency induced enhancements of the incoherent scatter spectrum at Arecibo 2. J Geophys Res 79: 199–208

Kasuya I, Koseki T (1969) IONO-FAX – a device for ionospheric observation. JEE, August 1969

Kato S (1963) Theory of movement of irregularities in the upper atmosphere. Planet Space Sci 11: 823

Kato S (1980) Dynamics of the upper atmosphere. Reidel, Dordrecht

Kato S, Matsushita S (1968) Space charge waves and ionospheric irregularities. J Atoms Terr Phys 30: 857–869

Kato S, Ogawa T, Tsuda T, Sato T (1984) The middle and upper atmosphere radar: First results using a partial system. Rad Sci 19: 1475–1484

Kazantsev AN (1959) Absorption and electron distribution in the F_2 layer determined from measurements of transmitted radio signals from earth satellites. Planet Space Sci 1: 130–135

Kelly JD (1979) Radar measurements of temperatures and ionic composition in the high-latitude ionosphere. PhD Diss, Univ Alaska Fairbanks, AK

Kelly JD (1983) Sondrestrom radar-initial results. Geophys Res Lett 10: 1112–1115

Kelly JD, Vickrey JF (1984) F-region ionospheric structure associated with antisunward flow near the dayside polar cusp. Geophys Rev Lett 11: 907–910

Kelley MC (1989) The earth's ionosphere. Plasma physics and electrodynamics. Int-Geophys Ser 43, Academic Press, San Diego, New York 487 pp

Kelso JM (1964) Radio ray propagation in the ionosphere. McGraw-Hill, New York

Kennel CF, Petschek HE (1966) Limit on stably trapped particle fluxes. J Geophys Res 71: 1–28

Kennelly AE (1902) On the elevation of electrically conducting strata of the earth's atmosphere. Electrical world and Engineering 15: 473

Kent GS, Wright RWH (1968) Movements of ionospheric irregularities and atmospheric winds. J Atmos Terr Phys 30: 657–691

Kent SA (1959) High frequency fading observed on the 40 Mc/s wave radiated from artificial satellite 1957a. J Atmos Terr Phys 16: 10–20

Kersley L, Pryse SE, Wheadon NS (1988) Amplitude and phase scintillation at high latitudes over northern Europe. Radio Sci 23: 320–330

Kisabeth JL (1970) Ionospheric drift measurements in the auroral zone. Rep UAG R-209 Geophys Inst Univ Alaska Fairbanks

Kiwamoto Y, Benson RF (1979) Nonlinear Landau damping in the ionosphere. J Geophys Res 84: 4165–4174

Klobuchar JA, Allen RS (1970) A first-order prediction model of total-electron-content ground path delay for a midlatitude ionosphere. AFCRL-70-0403, AD711365

Klobuchar JA, Iyer KN, Vats HO, Rastogi RG (1977) A numerical model of equatorial and low latitude TEC for use by satellite tracking systems for ionospheric corrections. Indian J Radio Space Phys 6

Knowles SH, Andrews M (1987) Holographys of the ionosphere at HIPAS, NRL Memo Rep 6084, Naval Res Labs, Washington DC

Kokjer KJ, Roberts TD (1986) Networked meteor-burst data communications. IEEE Comm Mag 24: 23–29

Kolosov AA et al. (1987) Over the horizon radar, ARTECH House, Boston Lond, 286 pp

Kopka H, Stubbe P, Jones TB, Robinson T (1982) Nonlinear reflectivity of high-power radio waves in the ionosphere. Nature (Lond) 295: 680

Kossey PA, Turtle JP, Pagliarulo RP, Klementti WI, Rasmussen JE (1983) VLF reflection properties of the normal and disturbed polar ionosphere in northern Greenland. Rad Sci 18: 907–916

Kostinski AB, Boerner W-M (1986) On foundations of radar polarimetry. IEEE Trans Ant Prop AP-34: 1395

Krall NA, Trivelpiece AW (1973) Principles of plasma physics. McGraw-Hill, New York

Kramm R, Schlegel K Weiss W (1972) Partial reflection measurements with FM-CM – a preliminary investigation. Z Geophys 38:953

Kraus JD (1966) Radio astronomy. McGraw-Hill, New York

Kraus JD (1988) Antennas, 2nd edn. Cygnus-Quasar Books, PO Box 85, Powell, OH, USA

Krishnaswamy S, Detrick DL, Rosenberg TJ (1985) The inflection point method of determining riometer quiet day curves. Radio Sci 20: 123–130

La Belle J (1989) Radio noise of auroral origin. J Atmos Terr Phys 51: 197–211

Landmark B, Lied F (1961) Observations of the D-region from a study of ionospheric cross-modulation. J Atmos Terr Phys 23: 92–100

Larmor J (1924) Why wireless waves can bend round the earth. Philospherical Magazine 48: 1025–1036

Lauber WR, Pike CJ (1979) Radio noise survey procedures for a communication site 0.15–30 MHz. CRC Rep No 1325, Ottawa, Alberta, Canada

Lauter EA, Sprenger K (1966) Kleinheubacher Ber 11: 211

Lawrence RS, Little CG, Chivers HJA (1964) A survey of ionospheric effects upon earth-space propagation. Proc IEEE Jan: 4–27

Leadabrand RL, Baron MJ, Petricheks J, Bates HF (1972) Chatanika, Alaska, auroral zone incoherent-scatter facility. Rad Sci 7: 756

Lee HS, Ferraro AJ (1969) Winter D-region electron concentration and collision frequency features obtained with high power interaction measurements. J Geophys Res 74: 1184–1194

Lee HS, Ferraro AJ, Kissick WA, Dale SP (1977) Intercomparison of electron density synthesis methods for two ground-based techniques – wave interaction and partial reflection – 1: Theoretical evaluation of the methods. J Atmos Terr Phys 39: 195–209

Lee MC (1981) Ion line enhancement in ionospheric heating experiments. Phys D Appl Phys 14: 851–860

Lee MC, Kuo SP (1983) Excitation of upper-hybrid waves by a thermal parametric instability. J Plasma Phys 30: 463–478

Leinbach H (1962) Interpretations of the time variation of polar cap absorption associated with solar cosmic ray bombardments. Geophys Inst Univ Alaska, Rep No 2, Contract AF (604–5577)

Leinbach H, Basler RP (1963) Ionospheric absorption of cosmic radio noise at magnetically conjugate auroral zone stations. J Geophys Res 68: 3375–3382

Leitinger R, Klobuchar JA (1984) Special papers: ionospheric studies by means of beacon satellites. Rad Sci 19: 685

Leonard RS (1959) A low-power VHF radar for auroral research. Proc IRE 47: 320–322

Lerfald GM, Little CG, Parthasarathy R (1964) A study of D-region electron density profiles during auroras, using the multiple-frequency cosmic noise technique. J Geophys Res 69: 2857–2860

Lewis EA, Rasmussen JE, Kossey PA (1973) Measurements of ionospheric reflectivity from 6 to 35 kHz. J Geophys Res 78: 3903–3912

Little CG (1952) The origin of the fluctuations on galactic radio noise. PhD Thesis, Univ Manchester

Little CG (1954) High latitude ionospheric observations using extra-terrestrial radiowaves. Proc IRE (correspondence) 42: 1700–1701

Little CG, Leinbach H (1959) The riometer – a device for the continuous measurement of ionospheric absorption. Proc IRE 47: 315–320

Little CG, Rayton WM, Roof RB (1956) Review of ionospheric effects at VHF and UHF. Proc IRE 44: 992–1018

Little CG, Reid GC, Stiltner E, Merritt RP (1962) An experimental investigation of the scintillation of radio-stars observed at frequencies of 223 and 456 Mc/s from a location close to the auroral zone. J Geophys Res 67: 1763–1784

Little CG, Lerfald GM, Parthasarathy R (1964) Extension of cosmic noise absorption to lower frequencies using polarized antennas. Radio Sci 68D: 859–865

Livingston RC, Rino CL, Owen J, Tsunoda RT (1982) The anisotrophy of high-latitude nighttime F-region irregularities. J Geophys Res 87: 10,519–10,562

Lovell ACB, Clegg JA, Ellyett CD (1947) Radio echoes from the auroral borealis. Nature (Lond) 160: 372–373

Lund DS, Hunsucker RD, Bates HF, Murcray WB (1967) Electron number densities in auroral irregularities: comparison of backscatter and satellite data. J Geophys Res 72: 1053–1059, 1133

Machin KE, Ryle M, Vonberg DD (1952) The design of an equipment for measuring small radio-frequency noise powers. Proc IEE 99: 127–134

Mahajan KK, Kar J (1988) Planetary ionospheres. Spaces Sci Rev 47: 303–397

Manson AH (1966) Comments on the theory of the differential absorption experiment. J Geophys Res 71: 3783–3785

Manson AH (1967) Inherent limitation of the wave interaction experiment. J Geophys Res 72: 1654

Manson AH, Meek CE (1985) Middle atmosphere (60–110 km) tidal oscillations at Saskatoon, Canada (52°N, 107°W), during 1983–1984. Rad Sci 20: 1441–1451

Maresca JW, Barnum JR (1982a) Theoretical limitations of the sea of the detection of low Doppler targets by over-the-horizon radar. IEEE Trans Ant Prop AP-30: 837–845

Maresca JW, Barnum JR (1982b) Estimating wind speed from HF skywave radar sea backscatter. IEEE Trans Ant Prop AP-30: 846–852

Mass J (1963) Radio astronomical and satellite studies of the atmosphere. Aarons J (ed) North-Holland Publ Co, Amsterdam

Mathews JD, Breakall JK, Sulzer MP (1988) The moon as a calibration target of convenience for VHF-UHF radar systems. Rad Sci 23: 1–12

Matthews JD (1986) Incoherent scatter radar probing of the 60–100 km atmosphere and ionosphere. IEEE Trans Geosci Remote Sens GE-24: 765–776
Maxwell JC (1873) A treatise on electricity and magnetism, Vols I, II. Oxford Univ Press, London
McClure JP, Farely DT, Cohen R (1970) Ionospheric electron concentration measurements at the magnetic equator, 1964–1966. ESSA Tech Rep ERL 186-AL4
McKinley DWR (1961) Meteor science and engineering. McGraw-Hill, New York
McNamara AG, McDiarmid DR, Sofko GJ, Koehler JA, Forsyth PA, Moorcroft DM (1983) BARS – a dual bistatic auroral radar system for the study of electric field in the Canadian sector of the auroral zone. Adv Space Res 2: 145–148
McNichol RWE (1949) The fading of radio waves of medium and high frequencies. Proc IEE 96: 517–524
Meek CH, Manson AH (1987) Medium frequency interferometry of Saskatoon, Canada. Phys Scr A 35: 917–921
Meyer-Vernet N, Perche C (1989) Toolkit for antennae and thermal noise near the plasma frequency. J Geophys Res 94: 2405–2415
Migulin VV, Gurevich AV (1985) Investigations in the USSR of non-linear phenomena in the ionosphere. J Atmos Terr Phys 47: 1181–1187
Millman GH (1967) A survey of tropospheric, ionospheric and extra-terrestrial effects on radio propagation between the earth and space vehicles. In: Blackbrand WT (ed) Propagation factors in space communications. AGARD Conf Proc No 3, Technivision, Maidenhead, England
Millman GH, Moceyunas AJ, Sanders AE, Wyrick RF (1961) The effect of Faraday rotation on incoherent backscatter observations. J Geophys Res 66: 1564–1568
Mitra AP, Shain CA (1953) The measurement of ionospheric absorption using observations of 18.3 Mc/s cosmic radio noise. J Atmos Terr Phys 4: 204–218
Mitra SK (1952) The upper atmosphee, 2nd edn. Asiatic Soc, Calcutta
Mitra SN (1949) A radio method of measuring winds in the ionosphere. Proc IEE 96: 441
Mögel H (1930) Telefunken Zeitung 56: 14
Montbriand LE (1988) Auroral backscatter observed at HF from Ottawa. Radio Sci 23: 850–864
Montgomery DC, Tidman DA (1964) Plasma kinetic theory. McGraw-Hill, New York
Moore RK (1951) A VHF propagation phenomenon associated with aurora. J Geophys Res 56: 97–105
Morgan MG (1957) Whistlers and dawn chorus. Ann IGY 1957–1958, 3: 315–336
Morgan MG (1959) Whistlers. J Atmos Terr Phys 15: 54–57
Morgan MG, Pratt B (1975) A reliable ionosonde for continuous automatic operation. Rad Sci 10: 859
Mott H (1986) Polarization in antennas and radar. Wiley, New York
Muldrew DB (1978) The role of field-aligned ionization irregularities in the generation of the HF-induced plasma line at Arecibo. J Geophys Res 83: 2552–2560
Muldrew DB, Showen RL (1977) Height of the HF-enhanced plasma line at Arecibo. J Geophys Res 82: 4793–4804
NCAR CEDAR Data Base Catalog. National Center for Atmospheric Research, Boulder, Colorado
Nichols HW, Schelling JC (1925) The propagation of electric waves over the earth. Bell System Tech J 4: 215
Nielsen E (1980) The STARE system and some of its applications: the IMS source book. Russel CT, Southwood DJ (eds) AGU, Washington DC, pp 213–224
Nielson E, Guttler W, Thomas EC, Stewart CP, Jones TB, Hedberg A (1983) SABRE: a new radar auroral backscatter experiment. Nature (Lond) 304: 712–714; Observations and ionospheric electric fields. Adv Space Res 3: 131–144
Nishino M, Tanaka Y (1987) Observation of auroral LHR noise by the sounding rocket S-310JA-6. Plant Space Sci 35: 127–137
Obara T, Kitayama M, Mukai T, Kaya N, Murphree JS, Cogger LL (1988) Simultaneous observations of sun-aligned polar caps arcs in both hemispheres by Exos-C and Viking. Geophys Res Lett 15: 713–716
Ogawa T, Igarshi K, Ose M, Kuratani Y, Fujii R, Hirasawa T (1983) Preliminary results of new 50 MHz Doppler radar experiment at Syowa station. Mem Nat Inst Polar Res 26: 193–201
Oliver WL, Fukao, Takami T, Yamamoto M, Tsuda T, Navamura T, Kato S (1990) Thermospheric meridional winds measured by the middle and upper atmosphere. J Geophys Res 95: 7683–7692

Onwuke DN (1974) A study of the ionosphere electron content at Ibadan, 1965–68. J Atmos Terr Phys 36: 1259–1267

Owren L, Fremouw EJ, Hunsucker RD (1964) Radio-star scintillations and spread-F in the auroral zone. AGARDograph 95: 313–332, Technivision Press

Oya H (1970) Sequence of diffuse plasma resonances observed on Alouette 2 ionograms. J Geophys Res 75: 4379–4285

Oya H (1971) Verification of theory on weak turbulence relating to the sequence of diffuse plasma resonances in space. Phys Fluids 14: 2487–2499

Park D, Carpenter D (1978) Very low frequency radio waves in the magnetosphere. In: Lanzerotti LJ, Parrs CG (eds) Upper atmospheric research in Antarctica. Res Ser Vol 29 Monog 72, AGU

Parthasarathy R, Hessler VP (1964) Periodic covariance of radiowave absorption, earth current and other associated phenomena in the auroral zone. Geophys Res 69: 2867–2871

Parthasarathy R, Basler RP, De Witt RN (1959) A new method for studying the auroral ionosphere using earth satellites. Proc IRE 47: 1660

Parthasarathy R, Lerfald GM, Little CG (1983) Derivation of electron density profiles in the lower ionosphere using radio absorption measurements at multiple frequencies. J. Geophys Res 68: 3581–3588

Pawsey JL (1935) Further investigations of the amplitude variations of downcoming wireless waves. Proc Camb Phil Soc 31: 125

Perkins FW (1974) A theoretical model for short-scale field aligned plasma density striations. Radio Sci 9:1065

Peterson AM (1949) The interpretation of long scatter echo patterns. J Geophys Res 54: 284

Peterson AM (1951) The mechanism of F-layer propagated backscatter. J Geophys Res 56: 221–237

Peterson AM, Egan RD, Pratt DS (1959) The IGY three-frequency backscatter sounder. Proc IRE 47: 300–314

Phelps AV (1960) Propagation constants for electromagnetic waves in weakly ionized dry air. J Appl Phys 31: 1723–1729

Phelps AV (1969) Laboratory studies of electron attachment and detachment processes of aeronomic interest. Can J Chem 47: 1783

Phillips GJ, Spencer M (1955) The effect of ansiometric amplitude patterns in the measurement of ionospheric drifts. Proc Phys Soc London B68: 481–492

Phillips ML (1974) Ground-based vertical-incidence ionograms. IEEE Trans Antennas Prop AP-22:785

Pierce GW (1910) Principles of wireless telegraphy. McGraw-Hill, New York

Pierce JA, Mimno HR (1940) Reception of radio echoes from distant ionospheric irregularities. Phys Rev 57: 95–105

Piggott WR (1953) The reflection and absorption of radio waves in the ionosphere. Proc IEE 100:61

Piggott WR (1964) Studies of ionospheric absorption, Research in Geophysics Vol 1 Sun, upper atmosphere, and space. MIT Press, Cambridge, Massachusetts

Piggott WR (1975) High latitude supplement to the URSI handbook on ionogram interpretation and reduction. Rep UAG-50 World Data Center A for solar-terrestrial physics. NOAA. 325 Broadway Boulder, Colorado

Piggott WR, Rawer K (eds) (1972) URSI handbook of ionogram interpretation and reduction. Rep UAG-23, World Data Center A for Solar-Terrestrial Phys NOAA, US Dep Comm, Boulder, Co 80303

Pittott WR, Rawer K (eds) (1978) Revision of Chapters 1–4 of URSI Handbook. Rep UAG-23A, WDC-A, Boulder, Co

Piggot WR, Beynon WJG, Brown GM, Little CG (1957) The measurement of ionospheric absorption. Ann IGY III: 177, Pergamon Press, Lond

Pineo VC, Kraft LG, Briscoe HW (1960) Ionospheric backscatter observations at 400 Mcs. J Geophys Res 65: 1620–1621

Poole AWV (1985) Advanced sounding 1; the FMCW alternative. Rad Sci 20:1609

Poole AWV, Evans GP (1985) Advanced soundings; 2, First results from an advanced chirp ionosonde. Rad Sci 20: 1617

Potter RK (1951) Analysis of audio-frequency atmospherics. Proc IRE 39 (9): 1067–1069

Poulter EM, Hargreaves JK, Bailey GJ, Moffett RJ (1981) Electron content modeling: in the significance of protonospheric content. Planet Space Sci 29: 869–883

Preece WH (1994) Earth currents. Nature (Lond) 49 (1276): 554
Presnell RI, Fair BC, Kunnes HW, Larson AG, Stafford NS (1972) NDS project 609 Island radar phased array. SRI Fin Rep 2, Proj 7882, Stanford Res Ins Menlo Park, CA
Providakes JF (1985) Radar interferometer observations and theory of plasma irregularities in the auroral ionosphere. PhD Thesis, Cornell University, Ithaca, NY
Providakes JF, Farley DT, Schwartz WE, Riggin D (1985) Plasma irregularities associated with a morning discrete auroral arc: radar interferometer observations and theory. J Geophys Res 90: 7513–7523
Providakes J, Farley DT, Fejer BG et al. (1988) Observations of auroral E-region plasma waves and electron heating with EISCAT in a VHF radar interferometer. J Atmos Terr Phys 50: 339–356
Providakes JP, Swartz WE, Farley DT, Fejer BG (1983) First VHF auroral radar interferometer observations. Geophys Res Lett 10: 404–404
Pulinets SA, Selegey VV (1986) Ionospheric plasma modification in the vicinity of a spacecraft by powerful radio pulses in topside sounding. J Atmos Terr Phys 48: 149–157
Quäck E, Mögel H (1929) Short range echoes with short waves. Proc IRE 17: 824–829
Radio Research Laboratories, Ministry of Posts and Telecommunication, 2–1, Nukui-Kitamachi 4-Chomw, Koganei-shi Tokyo, Japan. Results from Ionospheric Sounding Satellites A and B (1978–1982)
Radio Science (1973) Special issue on ionospheric in-situ measurements (April)
Radio Science (1984) Special issue: Ionospheric studies by means of beacon satellites. 19: 685–804
Ramo S, Whinnery JR, Van Duzer T (1965) Fields and waves in communication electronics. Wiley, New York
Ranta H, Ranta A (1977) Study of latitudinal, diurnal and seasonal variation of ionospheric absorption according to observations of the riometer network in Finland. Geophysica 14: 182
Ranzi I, Dominici P (1963) Backscatter sounding during ionospheric storms. In: Gassmann GJ (ed) The effect of disturbances of solar origin on communications. AGARDograph 59, Macmillan, pp 143–154
Rasmussen JE, Lewis EA, Kossey PA (1975) Low-frequency wave-reflection properties of the equatorial ionosphere. AFGL Tech Rep 75-0615, Hanscom AFB, MA
Ratcliffe JA (1956) Some aspects of diffraction theory and their application to the ionosphere. Rep Progr Phys 19: 188–267
Ratcliffe JA (1959) The magnetoionic theory and its applications to the ionosphere. A monograph. Cambridge Univ Press, Lond
Ratcliffe JA (ed) (1960) Physics of the upper atmosphere. In: Booker HG (ed) Radar studies of the aurora. Academic Press, Lond New York, pp 355–377
Ratcliffe JA, Pawsey JL (1933) A study of the intensity variations of downcoming radio waves. Proc Camb Phil Soc 29: 301
Ratcliffe JA, Shaw IJ (1948) A study of the interaction of radio waves. Proc Soc A Lond 193: 311–343
Rawer K (1952) The ionosphere – Its significance for geophysics and radio communication. Ungar, New York
Rawer K (ed) (1976) Manual on ionospheric absorption measurements. Rep UAG-57 World Data Cent A Solar-Terrestrial Phys, Boulder Colorado, USA
Rees MH (1989) Physics and chemistry of the upper atmosphere. Cambridge Univ Press, Cambridge, UK
Reference data for radio engineers (1975) 6th edn. Sams, ITT
Reinisch B (1986) The digisonde 256 system and ionospheric research. In: Haggard R (ed) INAG Ionospheric Station Inf Bull 48
Reinisch BW (1986) New techniques in ground-based ionospheric sounding and studies. Rad Sci 21: 331–342
Revah I (1969) Vents de petite échelle observés au moyen des trainées météoriques. Ann Geophys 25: 1–45
Rice DD, Hunsucker RD, Lanzerotti LJ, Maelennon CJ, Craven JD, Frank L (1988) An observation of atmospheric gravity wave cause and effect during the October 1985 WAGS campaign. Radio Sci 23: 919–930
Rietveld MT, Stubbe P, Kopka H (1989) On the frequency dependence of ELF/VLF waves produced by modulated ionospheric heating. Radio Sci 24: 270–278

Riggin D, Swartz WE, Providakes JF, Farley DT (1986) Radar studies of long-wavelength waves associated with midlatitude sporadic-E layers. J Geophys Res 91: 8011–8024
Rino CL (1979a) A power law phase screen model for ionospheric scintillation, 1. Weak scatter. Rad Sci 14: 1135–1145
Rino CL (1979b) A power law phase screen model for ionospheric scintillation, 2. Strong scatter. Rad Sci 14: 1147–1157
Rino CL (1980) Numerical computations for a one-dimensional power law phase screen. Rad Sci 15: 41–48
Rino CL, Livingston RC (1982) On the analysis and interpretation of spaced-receiver measurements of transionospheric radio waves. Rad Sci 17: 845–854
Rino CL, Owen J (1980) The time structure of transionospheric radio wave scintillation. Rad Sci 15: 479–490
Rishbeth H, Garriott OK (1969) Introduction to ionospheric physics. Academic Press, Lond New York
Rishbeth H, Williams PJS (1985) The EISCAT ionospheric radar: the system and its early results. Q J Astron Soc 26: 478–512
Robinson RM, Vondrak RR, Craven JD, Frank LA, Miller K (1989) A comparison of ionospheric conductances and auroral luminosities with the Chatanika radar and DE1 auroral images. J Geophys Res 94: 5382–5396
Rogers GL (1956) A new method of analyzing ionospheric movement records. Nature (Lond) 177: 613–614
Rogers GL, Ireland W (1980a) Ionospheric holography I, the holographic interpretation of ionospheric data. J Atmos Terr Phys 42: 385–396
Rogers GL, Ireland W (1980b) Ionospheric holography II, the analysis of a set of ionospheric holograms. J Atmos Terr Phys 42: 397–406
Rogister A, D'Angelo N (1970) Type II irregularities in the equatorial electrojet. J Geophys Res 75: 3879–3887
Romick GJ, Killeen TL, Torr DG, Tinsley BA, Behnke RA (1987) CEDAR: An aeronomy initiative. EOS 68: 19–21, January 1987
Rose G (1965) Über die Bestimmung der D-Schicht-Absorption für eine 2000 km lange Meßstrecte and deren Endpunkte, Forschungs berichte des Landes Nordthein Westfalken, Nr 1511. (Pages unknown RDH)
Rosenberg, TJ (1990) Personal communication
Rowe JN, Ferraro AJ, Lee HS (1969) Changes in electron density and collision frequency at University Park, Pennsylvania during the stratospheric warming of 1967/68. J Atmos Terr Phys 31: 1077–1084
Rowe JN, Ferraro AJ, Lee HS (1970) Observations of electron density during a solar flare. J Atmos Terr Phys 32: 1609–1614
Rowe JN, Mitra AP, Ferraro AJ, Lee HS (1974) An experimental and theoretical study of the D-region-II: a semi-empirical model for mid-latitude D-region. J Atmos Terr Phys 36:755–785
Rumi GC (1962a) Experimental Luxemburg cross-modulation at high latitude, low height – Part I: Theoretical aspects. IRE Trans Ant Prop AP-10: 594–600
Rumi GC (1962b) Experimental Luxemburg: cross-modulation at high latitude, low height – Part II: Experimental aspects. IRE Trans Ant Prop AP-10: 601–607
Salpeter EE (1963) Density fluctuations in a non-equilibrium plasma. J Geophys Res 68: 4870–4872
Samson JC, Greenwald RA, Ruohoniemi A, Frey A, Baker KB (1990) Goose Bay radar observations of earth reflected, atmospheric gravity waves in the high latitude ionosphere. J Geophys Res 95: 7693–7709
Sander KF, Reed GAL (1986) Transmission and propagation of electromagnetic waves, 2nd edn. Cambridge Univ Press, Cambridge Lond
Sato (1988) Ionospheric incoherent scatter measurements with the middle and upper atmosphere radar: techniques and capability. Radio Sci 24: 85
Sayers J (1970) In-situ probes for ionospheric investigations. J Atmos Terr Phys 32: 663–691
Schey HM (1973) Div, Grad, Curl and all that – an informal text on vector calculus. Nortar, New York

Schiffmacher EA (1966) The riometer – a relative ionospheric opacity meter. NBS Rep 7605
Schlegel K (1984) HF and VHF coherent radars for investigation of the high latitude ionosphere. MPAE Rep W-100-84-33, Max-Planck-Inst Aeronomie, D-3411 Katlenlburg-Lindau, FRG
Schmerling ER, Langille RC (eds) (1969) Special issue on topside sounding and the ionosphere. Proc IEEE 57: 859–860
Schmidt G (1972) Determination of the height of ionospheric irregularities with the holographic method. Z Geophys 38: 891
Schmidt G, Tauriainen A (1975) The localization of ionospheric irregularities by the holographic method. J Geophys Res 80: 4313–4324
Schultz LD, Gallet R (1970) A survey and analysis of normal ionospheric absorption measurements-obtained for radio pulse reflections. ESSA Prof Pap 4, Inst Telecommun Sci Boulder Colorado, US Gov Printing Office, Washington DC
Schwentek H (1958) Bestimmung eines Kennwertes für die Absorption der Ionosphäre aus einer automatisch-statischen Analyse von Feldstarkregistrierungen. Archiv Elektr Übertragung 12: 301
Secan JA, Fremouw EJ, Robins RE (1987) A review of recent improvements to the WBMOD ionospheric scintillations mode. In IES87 Proceedings. US Library of Congress, Cat No 87-619868
Sen HK, Wyller AA (1960) On the generalization of the Appleton-Hartree magnetoionic formulas. J Geophys Res 65: 3931–3950
Shain CA (1951) Galactic radiation at 18.3 Mcs. Austr J Sci Res 4: 258
Shapley AH (ed) (1970) Atlas of ionograms. Rep UAG-10, World Data Cent A ESSA, US Dept Comm, Boulder, CO
Shaw IJ (1951) Some further investigations of ionospheric cross-modulation. Proc Phys Soc B Lond 65: 1–20
Shearman EDR (1956) The technique of ionospheric investigation using ground backscatter. Proc IEE 103b: 210–221
Shearman EDR, Harwood J (1958) Sporadic-E as observed by backscatter technique in the United Kingdom. In: Landmark B (ed) Sporadic-E ionization. AGARDograph 34: 111–128
Shoucri M, Morales GJ, Maggs JE (1987) Modification of the ionospheric electron velocity distribution function due to resonant absorption of hf waves. J Geophys Res 92: 246–254
Showen RL, Kim DM (1978) Time variations of HF-induced plasma waves. J Geophys Res 83: 623–628
Silberstein R (1949) Oblique incidence propagation work of the CRPL. J Geophys Res 54: 288
Silberstein R (1954) Sweep-frequency backscatter-some observations and deductions. Trans IRE AP-2: 56–63
Skolnik MF (1980) Introduction to radar system 2nd edn. McGraw-Hill, New York
Skomal (1978) Man-made noise. Van Nostrand Reinhold, New York
Smith EK (1974) A study of ionospheric scintillation as it affects satellite communication. OT Tech Memo 74–186, US Dep Comm
Smith FG, Little CG, Lovell ACB (1950) Origin of the fluctuations in the intensity of radio waves from galactic sources. Nature (Lond) 165: 422–423
Smith N (1939) The relation of radio sky-wave transmission to ionosphere measurements. Proc IRE 27: 332–347
Smith RA, Bourne IA, Loch RG et al. (1965) Radio wave interaction using gyro-waves. AFGL-65-460, Fin Rep Univ New England, NSW Australia
Snyder WF, Bragaw CL (1986) Achievement in radio SP555, Nat Bur Stan Boulder, CO
Sofko GJ, Koehler JA, Haldoupis C, McKibben MJ, McNamara AG (1987) Doppler radio observations of 3-meter irregularities in the polar cap E-region. J Geophys Res 92: 1271–1276
Spatz DE, Franke SJ, Yeh KC (1988) Analysis and interpretation of spaced receiver scintillation data recorded at an equatorial station. Radio Sci 23: 361–374
Spaulding AD, Washburn JS (1985) Atmospheric radio noise: worldwide levels and other characteristics. NTIA Rep 85–173, Natl Telecom Inf Admin Boulder, CO
Sprenger K, Grafe A (1984) Radio aurora observations at medium latitude over two solar cycles. JATP 46: 673–684
Stathacopoulos AD, Barry GH (1974) Geometrical considerations in the design of communications circuits using field aligned ionospheric scatter. Rad Sci 9: 1021–1024

Steel JG (1964) The effect of the ground backscatter coefficient on observations of sporadic-E over sea, land and mountains. J Atmos Terr Phys 26: 322–324
Stein S (1958) The role of ionospheric layer tilts in long-range high frequency radio propagation. J Geophys Res 63: 217–241
Stoker PH (1987) Riometer absorption and spectral index for precipitating electron with exponential spectra. J Geophys Res 92: 5961–5968
Stone WR (1976) A holographic radio camera technique for the three dimensional reconstruction of ionospheric inhomogeneities. J Atmos Terr Phys 38: 583–592
Stone WR (1978) The concept, design and operation of a demonstration holographic radio camera. PhD Diss, Appl Phys Info Sci Univ California San Diego
Storey LRO (1953) An investigation of whistling atmospherics. Phil Trans R Soc Lond A 246: 113–141
Stratton JA (1941) Electromagnetic theory. McGraw-Hill, New York
Strφm GB, Brekke A, Bratteng O, Klokkervoll FN (1986) Polar F-region plasma flow as observed by a spaced receiver system at Ny-Alesund, Svalbard. Ann Geophys 04: 107–112
Stubbe P, Kopka H (1979) Ionospheric modification experiments in northern Scandinavia – description of the heating project. Rep MPAE-W-02-79-04, Max Planck-Inst Aeronomie, 3411 Katlenburg-Lindau 3, FRG
Stubbe P, Kopka H (1980) Modification of the F-region by powerful radio waves. In Deehr CS, Holtet JA (eds) Exploration of the polar upper atmosphere. Reidel, Dordrecht, pp 83–98
Stubbe P, Kopka H, Lauche M et al. (1982) Ionospheric modification experiments in northern Scandinavia. J Atmos Terr Phys 44: 1025–1041
Stubbe P, Kopka H, Thide B, Derbloom H (1984) Stimulated electromagnetic emission: a new technique to study the parametric decay instability in the ionosphere. J Geophys Res 89: 7523–7536
Stubbe P, Kopka H, Rietveld MT et al. (1985) Ionospheric modification experiments with the Tromsφ heating facility. J Atmos Terr Phys 47: 1151–1163
Sugar GR (1964) Radio propagation by reflection from meteor trails Proc IEEE 52
Sulzer MP (1986) A phase modulation technique for a sevenfold statistical improvement in incoherent scatter data taking. Rad Sci 21: 737–744
Sulzer MP, Tomko A, Ferraro AJ, Lee HS (1976) Complementary cross modulation: a new technique for ionospheric modification studies. J Geophys Res 81: 4754–4756
Swift DW (1974) Effect of collisions on the plasma line of the electron density fluctuations spectrum. Phys Fluids 17: 1240–1248
Szuszewicz EP, Fejer B, Roelof E et al. (1988) SUNDIAL: a worldwide study of interactive ionospheric processes and their roles in the transfer of energy and mass in the Sun-Earth system. Ann Geophys 06: 1–18
Tanaka Y, Nishino M (1988) The propagation of auroral hiss observed on the ground as deduced from direction-finding measurements. Planet Space Sci 36: 259–269
Tauriainen A (1982) Application of wave field reconstruction of VHF radio waves in investigating single, isolated ionospheric irregularities. Rad Sci 17: 684–692
Taylor AH, Young LC (1928) Studies of high-frequency radio wave propagation. Proc IRE 16: 561–578
Taylor AH, Young LC (1929) Studies of echo signals. Proc IRE 17: 1491–1507
Tellegin BD (1933) Interaction between radio-waves? Nature (Lond) 131:840
Thomas JA, McNichol RWE (1960) A highly directive rotating array for 16 MHz. Nature (Lond) 187: 398
Thome GD, Blood DW, (1974) First observations of RF backscatter from field-aligned irregularities produced by ionospheric heating. Rad Sci 9: 917–922
Thomson JJ (1906) Conduction of electricity through gases. Cambridge Univ Press, Lond, p 321
Thrane EV (1969) Measurement of electron density in the undisturbed middle latitude D-region. In: Jones TB (ed) AGARD Conf Proc No 13 Technovision Services Slough pp 5–8
Timofeev E, Miroshnikov Y (1982) Altitude characteristics of radar aurora as seen by a 90 MHz radar system operated at Karmaselga, Karella. J Geophys Res 51: 44–54
Titheridge JE (1972) Determination of ionospheric electron content from the Faraday rotation of geostationary satellite signals. Planet Space Sci 20: 353–369

Titheridge JE (1988) The real height analysis of ionograms: a generalized formulation. Rad Sci 23: 831–849
Tomko AA, Ferraro AJ, Lee HS (1980) D-region absorption effects during high power radio wave heating. Radio Sci 15: 675–682
Triska P (1962) Measurements of ionospheric absorption using obliquely incident radio waves. Studia Geoph Geod 6: 235
Triska P (1967) Geofsikikalni Sbornik 287: 495
Trizna DB, McNeal GD (1985) Applications of spectral estimation techniques to radar doppler processing simulation and analysis of HF skywave data. Naval Res Lab Rep 8950
Tsunoda RT (1988) High latitude F-region irregularities: a review and synthesis. Rev Geophys 26: 719–760
Tsunoda RT, Baron MJ, Owen J, Towle DM (1979) Altair: an incoherent scatter radar for equatorial spread-F studies. Rad Sci 14: 1111–1119
Tsutsui M, Ogawa T, Kamide Y, Kroehl HW, Hausman BA (1988) A method of estimating horizontal vectors of ionospheric electric field deduced from HF Doppler data. Rad Sci 23: 119–128
Turunen T (1985) The GEN-system for the EISCAT incoherent scatter radars. EISCAT Tech Note 85/44, EISCATS Sci Assn, S-981 27 Kiruna, Sweden
Tveten LH (1961) Ionospheric motions observed with high frequency backscatter sounders. J Res NBS 65D: 115–127
Tveten LH, Hunsucker RD (1969) Remote sensing of the terrestrial environment with an HF radio high-resolution azimuth and elevation scan system. Proc IEEE 57: 487–492
Tveten LH (1967) Ionospherically propagated sea scatter. Science 157 (3794): 1302
URSI Handbook on ionogram interpretation and reduction (1972) 2nd edn. NOAA/WDC-A Rep UAG-23, Environ Data Serv
Utlaut WF (1961) Effect of antenna radiation angles upon HF radio signals is propagated over long distances. J Res NBS, D.-Radio Propagation 65D: 167–174
Utlaut WF (1970) An ionospheric modification experiment using very high power, high frequency transmission. J Geophys Res 73: 6402–6405
Utlaut WF (1972) Modifying the ionosphere with radio waves. New Sci 10 Aug 1972
Utlaut WF (guest ed) (1974) Special issue: ionospheric modification by high power transmitters. Radio Sci 9: 881–1089
Utlaut WF (1975) Ionosphere modification induced by high-power HF transmitters – a potential for extended range VHF-UHF communications and plasma physics research. Proc IEEE 63: 1022–1043
Utlaut WF, Cohen R (1971) Modifying the ionosphere with intense radio waves. Science 174: 245–254
Utlaut WF, Violette EJ (1974) A summary of vertical incidence radio observations of ionosphere modification. Rad Sci 9: 895–903
Vacchione JD, Franke SJ, Yeh KC (1987) A new analysis technique for estimating zonal irregularity drifts and variability in the equatorial F-region using spaced receiver scintillation data. Rad Sci 22: 745–756
Valverde JF (1958) Motions of large-scale traveling disturbances determined from high-resolution backscatter and vertical incidence records. Sci Rep No 1, Stanford Electr Res Lab, Stanford Univ Stanford, California
Vilenskii IM (1962) On the theory of interaction of radio waves in the ionosphere. Izv Vyssh Uchebn Zaved Radiofizikal 5: 468
Villain JP, Greenwald RA, Baker KB, Ruohoniemi JM (1987) HF radar observations of E-region plasma irregularities produced by oblique electron streaming. J Geophys Res 92: 12327–12342
Villain JP, Hanuise C, Greenwald RA, Baker KB, Ruohoniemi JM (1990) Obliquely propagating ion acoustic waves in the auroral E region: Further evidence of irregularly production by field-aligned electron streaming. J Geophys Res 95: 7833–7846
Villard OG (1976) The ionospheric sounder and its place in the history of radio science. Rad Sci 11: 847
Villard OG, Peterson AM (1952) Scatter sounding: a technique for studying the ionosphere at a distance. IRE Trans Antennas and Propagation, PGAP 3: 186–201
Villard OG, Peterson AM, Manning LA (1952) A method for studying sporadic-E clouds at a distance. Proc IRE 40: 992–994

Vogan EL, Campbell LL (1957) Meteor signal rates observed in forward scatter. Can J Phys 35: 1176–1189

Volland H (1982) (ed) CRC handbook of atmospherics. CRC Press, Boca Raton, FL, USA, Vol 1 and 2

Wagner LS (1990) Characteristics of mid-latitude sporadic-E observed with a wideband HF channel probe. Rad Sci 25 (in press)

Wagner LS, Goldstein JS, Chapman EA (1983) Wideband HF channel prober: System description. NRL Rep 8622, Naval Res Lab, Washington DC

Wagner LS, Gloldstein JA, Meyers WD (1988) Wideband probing of the trans-auroral HF channel: solar minimum. Radio Sci 23: 555–568

Wait JR (1970) Electromagnetic waves in stratified media, 2nd edn Pergamon, Elmsford, New York

Wait JR (1985) Electromagnetic wave theory. Harper & Row, New York

Wakai N, Ishizawa K (1962) A method for the simultaneous recording of the echo amplitude with h'-f curves in the ionospheric sounding and some applications to the measurement of ionospheric parameters. J Rad Res lab 9: 421

Walden DC, Winkelman JR, Matheson LD, Schultz LD, Merrill RG (1984) High-frequency radar software reference manual for product one. NOAA Tech Memo ERL SEL-67

Walker ADM, Greenwald RA, Baker KB (1986) HF radar observations of pulsations near the magnetospheric cusp. J Geophys Res 91: 8919–8928

Walker ADM, Greenwald RA, Baker KD (1987) Determination of the fluctuation level of ionospheric irregularities from radar backscatter measurements. Rad Sci 22: 689–705

Wang ST, Tetenbaum D, Balsey BB, Obert RL, Avery SK, Avery JP (1988) A meteor echo detection and collection system for use on VHF radars. Radio Sci 23: 46–54

Watson-Watt RA, Herd JF, Bainbridge-Bell LH (1933) Applications of the cathode ray oscillograph in radio research. His Majesty's Stationary Office, Lond

Watts JM, Davies K (1960) Rapid frequency analysis of fading radio signals. J Geophys Res 65: 2295–2302

Watt TM, Oetzel GN, Stanley GM (1968) Short pulse investigation of ionospheric structure at high frequencies. Fin Rep Contr E22-128-68(N), SRI-I, Menlo Park, CA

Weisbrod S, Lee HS, Ferraro AJ (1964a) Preliminary measurements of radio wave phase interaction. J Geophys Res 69: 1448–1450

Weisbrod S, Ferraro AJ, Lee HS (1964b) Investigation of phase interaction as a means of studying the lower ionosphere. J Geophys Res 69: 2337–2348

Weitzen JA (1986) Predicting the arrival of meteors useful for meteor burst communication. Radio Sci 21: 1009–1020

Weitzen JA (1989) Effects of polarization coupling loss mechanism on design of meteor scatter antennas for short- and long-range communication. Radio Sic 24: 549–558

Wernik AW, Liu C-H, Yeh KC (1983) Modeling of spaced-receiver scintillation measurements. Rad Sci 18: 743–764

Whitney HE, Makik C, Aarons J (1969) A proposed index for measuring ionospheric scintillation. Planet Space Sci 17: 1069

Whitney HE, Aarons J, Allen RS, Seeman DR (1972) Estimation of the cumulative amplitude probability distribution function of ionospheric scintillations. Rad Sci 7: 1095–1104

Wickersham AF (1964) Identification of ionospheric motions, detected by the high-frequency backscattering technique. J Geophys Res 69: 457–463

Wickwar VB, Kelly JD, de la Beaujardiere O, Leeger CA, Steenstrup F, Dawson CH (1984) Sondrestrom overview. Geophys Res Lett 11: 883–886

Wickwar VB, Smith RA (1988) Global ionosphere-thermosphere coupling and dynamics. WITS Handbook Vol 1: 169–178 SCOSTEP Secretariat, Univ Illinois, 1406 W Green St, Urbana, IL 61801

Widdel HB (1957) Beobachtungen an rückgestreuten Echoes bei Kurzwellen-Fernübertragung. Arch Elektr Übertragung 11: 429–439

Wilkins AF, Shearman EDR (1957) Backscatter sounding: an aid to radio propagation studies. J IRE 17: 601–616

Williams PJS, Taylor GN, (1974) The UK incoherent scatter radar. Rad Sci 9: 85–88

WITS Handbook, Vol 1 (1988) Liu CH, Eduards B (eds) World ionosphere/thermosphere study. SCOSTEP Secretariat, University of Illinois Urbana, Ill

WITS Handbook, Vol 2 (1989) Liu CH (ed) World ionosphere/thermosphere study. SCOSTEP Secretariat, University of Illinois, Urbana, Ill

Wong AY (1977) Electromagnetic wave interactions with inhomogeneous plasmas. In: Schwart HJ, Hora H (eds) Laser interactions and related plasma phenomena. Plenum, New York, pp 783–840

Wong AY (1990) Foreword: Ionospheric modification in the polar region (IMPR) Rad Sci 25: 1249

Wong AY, Taylor RL (1971) Parametric excitation in the ionosphere. Phys Rev Lett 27: 644–647

Wong AY, Santoru J, Darrow C, Wang L, Roederer JG (1983) Ionospheric cavitons and related nonlinear phenomena. Rad Sci 18: 815–830

Woodman RF (1971) Inclination of the magnetic field measured by incoherent scatter. J Geophys Res 76: 178

Wright JW (1968) The interpretation of ionospheric radio drift measurements, I. Some results of experimental comparisons with neutral wind profile. J Atmos Terr Phys 30: 919–930

Wright JW (1969) Some current developments in radio systems for sounding ionospheric structure and motions. Proc IEEE, 57 4: 481–486

Wright JW (1975) Development of systems for remote sensing of ionospheric structure and dynamics: functional characteristics and applications of the "Dynasonde". NOAA/SEL Preprint 206, NOAA/SEL, Boulder, CO

Wright JW (1977) Development of systems for remote sensing of ionospheric structure and dynamics: the Dynasonde data acquisition and dynamic display system. NOAA Spec Rep, NOAA/SEL, Boulder, CO

Wright JW, Hunsucker RD (1983) Estimation of turbulent energy dissipation, winds, and ionospheric structures from Dynasonde measurements. Rad Sci 18: 988–994

Wright JW, Knecht RW (1957) Atlas of ionograms. Cen Radio Prop Lab, Nat Bur Stan, Boulder, CO

Wright JW, Pitteway MLV (1979) Real-time data acquisition and interpretation capabilities of the Dynasonde. 1. Data acquisition and real-time display. Rad Sci 14 (5) September-October: 815–825

Wright JW, Plywaski W, Fedor LS (1971) Development of a system for remote sensing of ionospheric motions and microstructure. NOAA Tech Memo ERLTM-OD6, NOAA/ERL, Boulder, Co

Wright JW, Paul AK, Pitteway MLV (1980) On the accuracy and interpretation of Dynasonde virtual height measurements. Rad Sci 15 (3) May-June: 617–626

Wright JW, Kressman RI, Virdi TS, Collis PN (1988) Comparisons of EISCAT and dynasonde ionospheric measurements: simple to moderately structured plasma densities. J Atmos Terr Phys 50: 405–421

Yeh KC, Flaherty BJ (1966) Ionospheric electron content and its variations at temperate latitudes during the declining phase of the sunspot cycle

Yeh KC, Liu CH (1972) Theory of ionospheric waves. Academic Press, Lond New York

Yeh KC, Liu CH (1982) Radio wave scintillation in the ionosphere. Proc IEEE 70: 324–360

Yeh KC, Raymund TD (1991) On limitations of the tomographic technique applied to Ionospheric imaging. Radio Sci 26, in Press

Yerukhimov LM, Kovalev V, Ya, Myashnikov, Ye N, Rakhlin AV, Rubtsov LN (1988) The form of the spectrum of artificial large-scale inhomogeneities excited with the "Gissar" heating station. Geomagnetism Aeronomy 28 (5)

Subject Index

Absorption, of radio waves in the ionosphere 50–53
deviative 51
index for deviative 52
index for non-deviative 52
MF/HF approximation 51
total absorption (attenuation) 53
Absorption techniques 165–183
A1a vertical incidence pulse reflection 166–168
A1b oblique incidence pulse reflection 168–169
A2 cosmic noise absorption (Riometers) 169–177
A3 oblique field strength 177–180
A3a oblique field strength at frequencies above 2 MHz 179
A3b oblique field strength at frequencies in and below the MF broadcasting band 170–180
capabilities and limitations 183
partial reflection experiments (PRE) 180–182
URSI nomenclature on absorption 165
Acronyms (list) 241–245
Advanced ionospheric sounders (AIS) 77–90
capabilities 90–92
characteristics of some sounders 75–77
chirps ounder 89–90
Annotated bibliography 247–253
Appleton-Hartree equations 46–50
Application of specific techniques 230–233
Arecibo HF heating facility (see Ionospheric modification, heating facilities)
Arecibo ISR (see ISR facilities)
Atmospheric structure 3–15
terrestrial ionosphere 3–15
Attenuation, integrated 53

Breit and Tuve's theorem (see Radio propagation)

Coherent backscatter facilities
BARS 121 (Table 4.4)
CUPRI 94, 112, 121
DAR 121
Goose Bay 94, 114–117, 121
PACE 94, 115, 118–119
SABRE 94, 112, 121
SHERPA 121
Slope Point 121
SOUSY 121
STARE 94, 112, 121
Coherent scatter, from ionospheric irregularities 54–58
Complementarity 233–234
Concluding remarks 235
Controlled stimulation of VLF emissions 218–219
Conversion factors 238–240
CW (continuous wave) 242

Deviative absorption 51–52
Dielectrics and conductors (see Electromagnetic waves)
Doppler backscatter systems 211–213

Electromagnetic waves 20–36
attenuation 25
evanescent 30
exponential 30–31
group velocity 24
Maxwell's equations 20–23
ordinary plane wave 30
phase-shift 25
phase velocity 24
plane wave propagation 23–25
polarization in free space 31–36
Poynting vector 22
propagation characteristics of dielectrics and conductors 26–27
propagation constant, complex 25
propagation vector 27–31
wave equation 22–23
wave impedance 21, 24
wave number 24
Evanescent waves 30
Excelsior 235
Exponential waves 30–31

Fresnel zone
Cornu spiral 260
geometry of diffraction 258–259
integral 259
3D zone 265
Future trends 235

Goose Bay HF radar 94, 114–117
Ground constants 257
Group velocity (see Electromagnetic waves)

HAARP facility (see Ionospheric modification, heating facilities)
HF (high-frequency) 242
HF channel probes 222–225
High-latitude ionosphere 9–12, 14–15
electron density profiles 7–11, 15
HIPAS heating facility (see Ionospheric modification, heating facilities)
History of development of radio techniques 229–230
Holography 225–227

Incoherent scatter
Bragg formula 62
Debye length 59–60
Doppler shift 59–60
Faraday rotation 64
ion line 62
plasma line 62
radar cross-section for ISR 58
spectrum 59–63
Incoherent scatter radar (ISR) facilities
Arecibo 126–127
basic system consideration 122–123
capabilities 139
clear BMEWS 138
EISCAT 132–135
Jicamarca 125–126
limitations 140–141
Millstone Hill 131–132
MU radar 127–130
presently operating systems 123–124
Sondrestrom 135–137
In-situ ionospheric measurements
Langmuir probe 205
RF impedance probe 205–206
topside sounder, plasma resonance 206–207
Interference 16
Ionograms 67–68
Ionosondes 67–93
digisonde 82–87
history of development 67–77
ionogram 67–68
IPS-42 87–88
Limitations 92–93
NOAA sounder 79–82
single-pulse ionosonde 89–91
Ionosphere
high latitude 9–12, 15
low latitude 12–13
structure and properties 3–15
Ionospheric modification
capabilities 162–163
cross-modulation 142
cross-modulation experimental results 150–152
heating 142–144
heating facilities 155
Arecibo 158–159
HAARP 162
high latitude 159–162
HIPAS 161–162
low latitude 158–159
midlatitude 155–156
Monchegorsk 162
Tromso 159–161
USSR 156–158, 162
heating theory 152–155
limitations 163–164
Ionospheric imaging by radio 225–227
IRIS riometer 243

LHI (linear-homogeneous-isotropic) medium 23, 32
LOF 243
LORAN 243

Martyn's theorem (see Radio propagation)
Maxwell's equations (see Electromagnetic waves)
Meteoric ionization, backscatter 227–228
MST radar 243

Noise, external electric
atmospheric 254–256
galactic 254–256
man-made 254–256
Noise, internal electrical
noise factor of a radio receiver 17
noise figure 17
noise temperature 18–19
thermal noise in a resistor 16

Oblique backscatter sounders (OBS) (see Coherent backscatter)
direct OBS 107–115
aspect angle 110
azimuth and elevation scan system 106
Chapman layer 102, 103

characteristics of a typical system 107–108
electrojet 110
electrostatic ion waves 110
from E-region 110–112
fixed frequency, rotating antenna 104
gradient drift 110
history of development 109–110
plasma instabilities 110
plasma waves 109
radio aurora 109–111
range time at fixed frequency 106
two-stream instabilities 110
Groundscatter
basic groundscatter theory 98–103
Chapman layer 102–103
characteristics of a typical system 107–108
different groundscatter techniques 104–107
historical development of OBS technique 96–98
least-time focusing (LTF) 101
skip distance 100–103
transmission curve 101
OTH radar 244

Partial reflection technique (see Absorption techniques)
Phase velocity (see Electromagnetic waves)
Physical constants 238
Polarization
ellipse 33–34
in magnetoionic medium 48–49
left-hand-circular (LHC) 33
Poincare sphere 35–36
right-hand-circular (RHC) 33
Stokes parameters 35
unpolarized wave 31
vertical and horizontal 31
Poynting vector 22, 31
Properties of ionospheric regions 6
terrestrial ionosphere 3–15

Radar equation
derivation 36–38
distributed targets 39
maximum range 37–39
Radio propagation, in magnetized plasma
Appleton-Hartree equations 46–50
characteristic waves 49
dispersion equation 47
gyrofrequency 47
longitudinal and transverse propagation 49
longitudinal angular gyrofrequency 48
polarization equation 48
QL and QT approximations 49–50
transverse angular gyrofrequency 48
RBE (radio beacon experiment) 244
RFI 244
Riometer 245

Tromso heating facility (see Ionospheric modification, heating facilities)

URSI nomenclature 165

Wave equation (see Electromagnetic waves)

MIX
Papier aus verantwortungsvollen Quellen
Paper from responsible sources
FSC® C105338

If you have any concerns about our products,
you can contact us on
ProductSafety@springernature.com

In case Publisher is established outside the EU,
the EU authorized representative is:
Springer Nature Customer Service Center GmbH
Europaplatz 3, 69115 Heidelberg, Germany

Printed by Libri Plureos GmbH
in Hamburg, Germany